CLIMATE ADAPTATION ENGINEERING

Risks and Economics for Infrastructure Decision-Making

CLIMATE ADAPTATION ENGINEERING

Risks and Economics for Infrastructure Decision-Making

Edited by

EMILIO BASTIDAS-ARTEAGA

MARK G. STEWART

ELSEVIER

Butterworth-Heinemann
An imprint of Elsevier

Butterworth-Heinemann is an imprint of Elsevier
The Boulevard, Langford Lane, Kidlington, Oxford OX5 1GB, United Kingdom
50 Hampshire Street, 5th Floor, Cambridge, MA 02139, United States

Library of Congress Cataloging-in-Publication Data
A catalog record for this book is available from the Library of Congress

British Library Cataloguing-in-Publication Data
A catalogue record for this book is available from the British Library

ISBN: 978-0-12-816782-3

Publisher: Matthew Deans
Acquisition Editor: Ken Mccombs
Editorial Project Manager: Gabriela D. Capille
Production Project Manager: Kamesh Ramajogi
Cover Designer: Greg Harris

Typeset by SPi Global, India

Contents

Contributors

Yekenalem Abebe
University of British Columbia, School of Engineering, Kelowna, BC, Canada

Morshed Alam
Department of Civil and Construction Engineering, Centre for Sustainable Infrastructure, Swinburne University of Technology, Hawthorn, VIC, Australia

Emilio Bastidas-Arteaga
Research Institute in Civil and Mechanical Engineering, UMR CNRS 6183, Université de Nantes, Nantes Cedex, France

Sigridur Bjarnadottir
VSÓ Consulting, Reykjavik, Iceland

Alessandro Contento
Department of Civil and Environmental Engineering, MAE Center, University of Illinois at Urbana-Champaign, Urbana, IL, United States

Axel Creach
Department of Geography and Planning, Faculty of Arts and Humanities, Sorbonne Universite, Paris, France

Paolo Gardoni
Department of Civil and Environmental Engineering, MAE Center, University of Illinois at Urbana-Champaign, Urbana, IL, United States

Boulent Imam
Department of Civil and Environmental Engineering, University of Surrey, Guildford, United Kingdom

Yue Li
Case Western Reserve University, Cleveland, OH, United States

Denis Mercier
Department of Geography and Planning, Faculty of Arts and Humanities, Sorbonne Universite, Paris, France

Kazuyoshi Nishijima
DPRI, Kyoto University, Kyoto, Japan

Maria Nogal
School of Engineering, Trinity College Dublin, Dublin, Ireland

Alan O'Connor
School of Engineering, Trinity College Dublin, Dublin, Ireland

Sophie Pardo
Laboratoire d'Economie et de Management de Nantes-Atlantique, Universite de Nantes, Nantes, France

Paraic C. Ryan
Civil, Structural and Environmental Engineering; Environmental Research Institute, University College Cork, Cork, Ireland

Jay Sanjayan
Department of Civil and Construction Engineering, Centre for Sustainable Infrastructure, Swinburne University of Technology, Hawthorn, VIC, Australia

Mark G. Stewart
Centre for Infrastructure Performance and Reliability, School of Engineering, The University of Newcastle, Callaghan, NSW, Australia

Solomon Tesfamariam
University of British Columbia, School of Engineering, Kelowna, BC, Canada

Dimitri V. Val
Institute for Infrastructure & Environment, EGIS, Heriot-Watt University, Edinburgh, United Kingdom

Hao Xu
Department of Civil and Environmental Engineering, MAE Center, University of Illinois at Urbana-Champaign, Urbana, IL, United States

Daniil Yurchenko
Institute of Mechanical, Process & Energy Engineering, EPS, Heriot-Watt University, Edinburgh, United Kingdom

Patrick X.W. Zou
Department of Civil and Construction Engineering, Centre for Sustainable Infrastructure, Swinburne University of Technology, Hawthorn, VIC, Australia

Preface

Climate change and actions taken to mitigate CO_2 emissions or reduce the vulnerability of infrastructure are often based on worst-case thinking, probability and cost neglect, opportunity costs, and risk aversion. This can result in a frightened public, costly policy outcomes, and wasteful expenditures.

Our background in probabilistic risk assessment for built infrastructure lends itself to risk-based decision-making for extreme events. And climate change ticks the box as an extreme event in need of deeper thinking. About 10 years ago we began discussing how policy outcomes may be improved by a rational and scientific approach to the risks posed by climate change, and how these risks may be better understood, and if possible, ameliorated with respect to infrastructure damage and disruption to communities.

Mark Stewart was fortunate enough to lead the Climate Adaptation Engineering for Extreme Events (CAEx) Cluster which received $3 million of National Research Flagship Collaboration Funding from the Australian Commonwealth Scientific and Industrial Research Organisation (CSIRO) from 2012 to 2016. Emilio Bastidas-Arteaga was leading and involved in the following research programs: probabilistic evaluation of the cost effectiveness of climate adaptation strategies for existing reinforced concrete structures (funded by Campus France in 2013); and international network on risk assessment and climate adaptation of civil engineering structures and buildings (funded by the Pays de la Loire Regional Council from 2014 to 2017). These projects aimed to provide climate-conscious engineering knowledge and solutions to reduce the impact of potential climate change on the performance of buildings and infrastructure and provided the catalyst for this book.

The edited volume will explain how risk and decision-making analytics can be applied to the wicked problem of climate change adaptation. There is increasing research that takes into account the changing climate risks and life-cycle costs in engineering to reduce the vulnerability or increase the resiliency of infrastructure–we refer to this as '*climate adaptation engineering.*' Climate-adaptation engineering is defined as measures taken to reduce the vulnerability or increase the resiliency of built infrastructure to a changing climate, this may include, for example, enhancement of design standards, retrofitting or strengthening of existing structures, utilisation of new materials, changes to land use, and changes to inspection and maintenance regimes. A recent IPCC report states that "vulnerability is a key factor in disaster losses, yet it is not well accounted for." To fill this knowledge gap is the key focus of the book.

The book will introduce the key concepts needed to assess the risks, costs, and benefits of future proofing infrastructure to a changing climate. This will include: climate scenarios and their likelihoods, hazard modeling, infrastructure vulnerability, resilience or exposure, social and economic loss models, risk reduction from adaptation measures, and decision theory. This will be followed by case studies authored by experts from France, Australia, United States, Canada, United Kingdom, Ireland, Iceland, and Japan. These case studies will describe succinctly the practical aspects of risk assessment when deciding on the most cost-efficient measures to reduce infrastructure vulnerability to wind, flood, deterioration, and heat-induced climate hazards for houses, buildings, bridges, harbors, roads, and electricity infrastructure in the developed and developing worlds. We believe that the insights provided by the case studies have worldwide applicability, and the book will challenge preconceived notions while providing critical guidance on evidence-based policy. The book will finish by providing a closing discussion of political, economic, and social imperatives that affect the policy-making decision process under a changing climate.

We are most grateful to the authors for their time and effort in preparing their chapters and in their support for our endeavor. Their case studies reveal the depth and breadth of their research expertise. And more importantly, the ability of the authors to translate sophisticated mathematical and probabilistic models into outcomes readily understandable to a general readership.

Finally, such an endeavor was not possible without the patience and support of our families, so a special thanks to Elodie and Xiaoli for their constant encouragement and to Julian, Elisa and Gabriel that are meaningful sources of motivation and inspiration.

Emilio Bastidas-Arteaga
Mark G. Stewart

Acknowledgements

This book has benefited from interaction and conversations with John Mueller, C.K. Shum, Craig Jenkins, David Rosowsky, Lizhengli Peng, Michael Netherton, Mauricio Sánchez-Silva, Alaa Chateauneuf, Franck Schoefs, Thomas de Larrard and Frédéric Duprat. And of course, from the contributors of the chapters.

We are also grateful for the support by Commonwealth Scientific and Industrial Research Organisation (CSIRO), The University of Nantes, the Pays de la Loire Regional Council, Campus France, the Mershon Center for International Security Studies at Ohio State University, the Cato Institute, the Research Institute in Civil and Mechanical Engineering, and the Centre for Infrastructure Performance and Reliability at the University of Newcastle.

PART I

Introduction

CHAPTER ONE

Introduction to Climate Adaptation Engineering

Mark G. Stewart*, Emilio Bastidas-Arteaga†
*Centre for Infrastructure Performance and Reliability, School of Engineering, The University of Newcastle, Callaghan, NSW, Australia
†Research Institute in Civil and Mechanical Engineering, UMR CNRS 6183, Université de Nantes, Nantes Cedex, France

1.1 Introduction

Climate change arouses much fear and anxiety in society; and for good reasons, if climate projections are correct, a changing climate will cause sea-level rise, more flooding, and more intense storms and hurricanes, droughts, and other climate extremes. This will affect every nation, and populations in developing countries will be hit hardest. This can, in the worst case, lead to energy and food scarcity, increase the spread of disease, mass migration of 'climate refugees', and weaken fragile governments. Urban communities are particularly vulnerable to a changing climate, and "Rapid urbanization and the growth of megacities, especially in developing countries, have led to the emergence of highly vulnerable urban communities, particularly through informal settlements and inadequate land management" (IPCC, 2012).

The focus of this chapter (and this book) is on technological innovation and adaptive behaviours—that is, the future proofing of infrastructure to climate change for future generations, economies, and environments. The impact of climate change on infrastructure performance is a temporal and spatial process, but most existing models of infrastructure performance are based on a stationary climate. For example, the World Bank states that "Despite the ability to quantify future risk (albeit with uncertainty), risk assessments typically fail to account for changing climate, population, urbanization, and environmental conditions" (World Bank, 2016). Hence, there is a need to quantify the costs and benefits of adaptation strategies. Climate adaptation engineering involves estimating the risks, costs and benefits of climate adaptation strategies, and assessing at what point in time climate adaptation becomes economically viable. Climate adaptation measures aim to reduce the vulnerability or increase the resiliency of built infrastructure to

Climate Adaptation Engineering
https://doi.org/10.1016/B978-0-12-816782-3.00001-2

a changing climate, this may include, for example, enhancement of design standards, retrofitting or strengthening of existing structures, utilisation of new materials, and changes to inspection and maintenance regimes. Engineers have a unique capability to model infrastructure vulnerability, and these skills will be essential to modelling future climate impacts, and measures to ameliorate these losses.

The climate change literature places more emphasis on impact modelling than on climate adaptation engineering modelling. This is to be expected when the current political and social environment is focused on mitigating (reducing) CO_2 emissions. The impacts on people and infrastructure may be considerable if there is no climate adaptation engineering to the existing and new infrastructure. Some posit that climate change may even be a threat to national security, but Stewart (2014) suggests that climate change threats to US national security are modest and manageable. On the other hand, higher temperatures in higher latitude regions such as Russia and Canada can be beneficial through higher agricultural yields, lower winter mortality, lower heating requirements, and a potential boost to tourism (Stern, 2007).

There is seldom mention of probabilities, quantitative measures of vulnerability, or the likelihood or extent of losses in 'risk' and 'risk management' reports on climate change and infrastructure. While useful for initial risk screening, intuitive and judgement-based risk assessments are of limited utility to complex decision-making since there are often a number of climate scenarios, adaptation options, limited funds, and doubts about the cost-effectiveness of adaptation options. In this case, the decision maker may still be uncertain about the best course of action, and so a detailed risk analysis is required (e.g., AS 5334-2013). For this reason, there is a need for sound system and probabilistic modelling that integrates the engineering performance of infrastructure with the latest developments in stochastic modelling, structural reliability, and decision theory. Such an approach is a logical extension of disaster risk management. The emphasis of the book is built infrastructure. This accords with the World Bank (2016) for the need for the "construction of buildings, infrastructure, and urban developments should consider how design, construction practices, and construction materials will affect disaster risk in both current and future climates".

The cost to mitigate CO_2 emissions is considerable. Stern (2007) estimated that to stabilise CO_2 levels at 550 ppm (by reducing total emissions to three quarters of today's levels by 2050), it would cost 1.0%–3.5% of gross domestic product (GDP), with a central estimate of approximately 1%. The mean estimate would result in an annual mitigation cost of approximately

$720 billion. This is a stupendous sum, and so a pivotal question becomes: is this the best option, or are there others? This is the question that Bjorn Lomborg posed to a group of experts—they found that climate change action ranked very low when compared with other hazard and risk-reducing measures, in this case the benefit-to-cost ratio (BCR) for CO_2 mitigation was only 0.9 (not cost-effective), but increased to 2.9 for a mix of mitigation and adaptation strategies (Lomborg, 2009). Yohe et al. (2009) found that a global investment of $18 billion per year in 'R&D (research and development) and mitigation' can halve 'business as usual' CO_2 emissions by 2100. Such actions would reduce the impact of climate change by at least 60%. The key here is R&D where innovation can be an important driver to reducing CO_2 emissions.

Some of the more dire predictions of food and energy insecurity, and mass migration can be ameliorated by funding climate adaptation measures in the developing world. Adaptation measures to reduce vulnerability of infrastructure, coastal zones, agriculture, forestry, fisheries, and human health to climate change hazards would include: flood control dikes and levees, dams, cyclone shelters, storm and flood-resistant housing, improved communication infrastructure, resettlement of populations to lower risk zones, and improved health care. The World Bank (2010) estimated that the cost to the developing world of adapting to an approximately 2°C warmer world by 2050 is approximately $75 billion per year. This represents about one-tenth of 1% of world GDP. Clearly, investing in targeted adaptation measures has the potential to dramatically reduce the impact of climate change.

As we have seen, CO_2 mitigation costs can be high, and the benefits of reduced CO_2 levels will take decades to accrue. Modest and sustained investments in R&D, CO_2 mitigation, and adaptation will lessen the worst impacts of climate change. Hence, a mix of mitigation and adaptation is desirable to cope with a changing climate. There are uncertainties, risks, upsides, and downsides that need to be factored into any decision. And we are talking about decisions that will involve many hundreds of billions of dollars of expenditures, so there is a need to explore the full range of options to effectively compare costs and benefits. There is no certainty about the future which makes decision-making for climate change incredibly challenging. There is clearly a need for action, the question is what should we be doing now? What decisions can be deferred? and to when? And perhaps most importantly—what information do we need to be able to make better decisions?

There are strongly held beliefs about climate change, vested interests, wasteful expenditures, and poor thinking at times. There is a need to bring a scientific, analytic, and objective risk-based approach to this serious issue. Poor infrastructure decisions made today can lead to long-term adverse impacts over many decades to come.

The chapter will describe how risk-based decision support is well suited to optimising climate adaptation strategies related to the design, construction, operation, and maintenance of built infrastructure. Fig. 1.1 describes the overall concept outlined in this book. An important aspect is assessing when climate adaptation becomes economically viable, if adaptation can be deferred, and decision preferences for future costs and benefits (many of them intergenerational). Stochastic methods are used to model infrastructure vulnerability, effectiveness of adaptation strategies, exposure, and costs. Case studies to follow in other chapters will detail how state-of-the-art risk-based approaches will help 'future proof' built infrastructure to a changing climate.

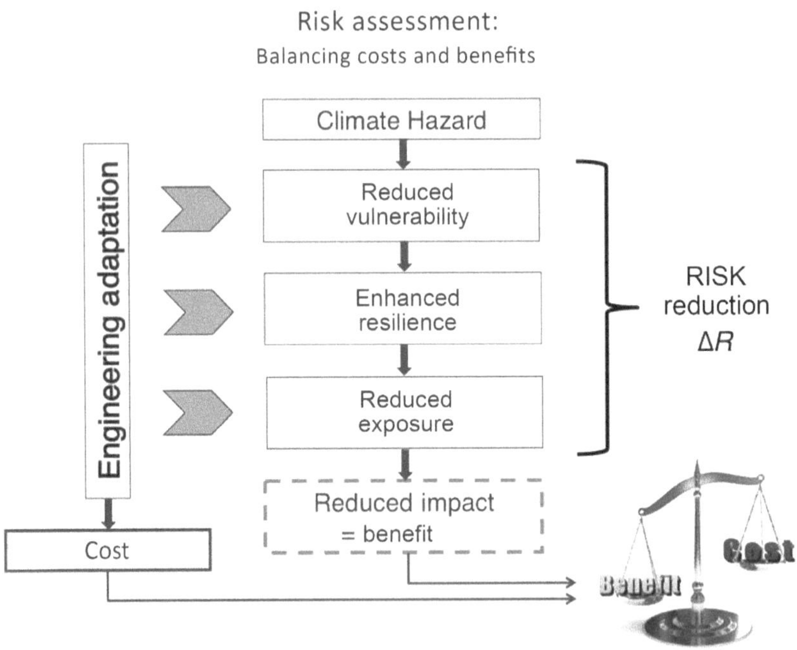

Fig. 1.1 Flowchart of risk-based decision support.

1.2 Climate Change Impact

The 2014 Intergovernmental Panel for Climate Change (IPCC) Fifth Assessment Report (AR5) concluded that the "Warming of the climate system is unequivocal, and since the 1950s, many of the observed changes are unprecedented over decades to millennia. The atmosphere and ocean have warmed, the amounts of snow and ice have diminished, sea level has risen, and the concentrations of greenhouse gases have increased" (IPCC, 2014). What is less certain is the impact that rising temperatures will have on rainfall, wind patterns, sea-level rise, and other phenomena.

There is considerable literature on the impact of climate change. We start with the latest IPCC AR5 report released in 2013. For the sake of brevity, the main changes to climate by 2100 relevant to this book are (IPCC, 2013) as follows:

- Temperatures to increase from 1995 levels, anywhere from 0.3°C to 4.8°C.
- Sea-level rise of 26–82 cm.
- More intense tropical cyclones and other severe wind events.
- Precipitation may increase in high latitudes, but will decrease in Central America, Southern Africa, and Southern Europe.
- Enhanced monsoon precipitation.
- More frequent hot and fewer cold temperature extremes over most land areas.

The IPCC (2014) then suggests with a high or very high confidence level that these changes to climate will increase drought affected areas, hundreds of millions of people will be affected by coastal flooding, increases in risks of fire, pests, and disease outbreak, will have significant consequences on food and forestry production, food insecurity, and so on. When these impacts are projected to monetary units the damages are staggering. In the United States, one study estimates that $238 billion to $507 billion worth of property will be below sea level by 2100, and that average annual losses from hurricanes and other coastal storms along the Eastern Seaboard and the Gulf of Mexico will grow by more than $42 billion due to sea-level rise alone (Risky Business, 2014). The 2006 review by economist Nicholas Stern (2007) predicts that if no action is taken against climate change, the mean loss of GDP would be 2.9% and 13.8% each year ('now and forever') by 2100 and 2200, respectively. This is equivalent to worldwide losses of up to $10 trillion each year by 2200.

Not surprisingly, some consider it to be highly pessimistic in its assumptions (Lomborg, 2006; Mendelsohn, 2006). However, the Australian Garnaut Review predicted that unmitigated climate change would reduce Australian GDP by approximately 8% by 2100 (Garnaut, 2008), and a more recent study projects that unmitigated warming is 'expected to reshape the global economy' and reduce global incomes by 23% by 2100 (Burke et al., 2015). A 2014 White House report states that a delay of implementing mitigation policies "that results in warming of 3°C above preindustrial levels, instead of 2°, could increase economic damages by approximately 0.9% of global output. To put this percentage in perspective, 0.9% of estimated 2014 US GDP is approximately $150 billion. Moreover, these costs are not one-time, but are rather incurred year after year because of the permanent damage caused by increased climate change resulting from the delay" (White House, 2014).

A reality check is often needed, as these are often worst-case losses, and the losses described above are all predicated on 'business as usual'. According to the Risky Business Report, this "assumes no new national policy or global action to mitigate climate change and an absence of investments aimed at improving our resilience to future climate impacts. Taking these policy and adaptive actions could significantly reduce the risks we face". It is difficult to fathom that the world will continue on its current path to the next millennium with no changes in the way we emit CO_2 emissions, how we live, how we work, how we travel, how we value the environment, and so on. The history of human kind is one of constant change and innovation—it has never been static not even for a generation.

These staggering losses also do not reflect wealth creation, human capital, and new improved technologies. Goklany (2008) states that these "often reduce the extent of the human health and environmental 'bads' associated with climate change more than temperature increases exacerbate them". Weather and climate-related fatality rates and economic losses are also 3–10 times higher in developing countries (IPCC, 2012). Clearly then, if people are wealthier in the future, their well-being will be higher, and "the argument that we should shift resources from dealing with the real and urgent problems confronting present generations to solving potential problems of tomorrow's wealthier and better positioned generations is unpersuasive at best and verging on immoral at worst" (Goklany, 2008). This is not to say that we should place our trust in market forces alone, but to recognise that economic development, CO_2 mitigation and adaptation are interlinked, and that "Vulnerability too changes with urban and socioeconomic development. Some people become less vulnerable because of improved construction and a more prosperous economic situation. But in many areas, structural vulnerability continues to increase because of unregulated building practices and unplanned development" (World Bank, 2016).

The observed increase in weather-related losses in the United States, Australia, and elsewhere is more a function of increased exposure with more people moving to vulnerable coastal locations than climate-change increases in wind speed or flood levels (Crompton and McAneney, 2008; IPCC, 2012). The apparent lack of a climate change signal in disaster losses is backed by the IPCC: "Economic losses due to extreme weather events have increased globally, mostly due to increase in wealth and exposure, with a possible influence of climate change (*low confidence* in attribution to climate change)" (IPCC, 2014).

Moreover, climate impacts will not be sudden, but gradual in their appearance. For instance, hurricane wind speeds are predicted to increase at worst by 10% in 50 years due to climate change, or a miniscule 0.2% per year (Bjarnadottir et al., 2011). This is in contrast with the population growth rate in Florida and other southern states which is double that of the United States at large, and is running at above 2% per year. This suggests that there will be time to adapt to a changing climate.

There is significant regional variability in climate change-induced hazards. A CSIRO study based on AR5 climate projections found that wind speeds are projected to fall by on average 1.8% in the southern city of Adelaide by 2090, but expected to increase by 2.2% in the Far North Queensland city of Cairns. Similarly, the number of days over 40°C is expected to increase by only 0.7 days per year for Cairns, but 83 days per year for the Central Australian city of Alice Springs (Webb and Hennessy, 2015). Large fluctuations in climate hazard in time and space necessitate tailoring engineering adaptation measures for local conditions. A 'one-size' fits all approach will not be optimal.

1.3 Climate Adaptation Engineering

The design and construction of infrastructure has evolved over many millennia so today we are able to predict with relative ease the likelihood and size of natural hazards, and take steps to design houses, buildings, bridges, power stations, dams, and other infrastructure to withstand these anticipated hazards. Earthquakes, tropical cyclones or hurricanes, storm surge, floods, and blizzards are often the low-probability high-consequence hazards of interest. Over the past century building standards have been developed and continually improved—with the prevention of building collapse and catastrophic loss (ultimate limit state) being the main driver for change. Moreover, while uncertainties and knowledge gaps still exist, disaster risks in the developed world are, in general, at an acceptable level. This is particularly the case for life-safety risks where, for example, the annual fatality rate from earthquakes in New Zealand is close to the generally acceptable risk of 1×10^{-6} or one in a million (e.g., Taig, 2012, Stewart and Melchers, 1997). However, the seismic fatality rate in China in the decade 2001–10 is 50 times higher at 5×10^{-5} (data sourced from Li et al., 2015).

Often the huge loss of life in the developing world is due to the poor quality of construction. In Bangladesh, the quality of concrete is poor (Koehn and Ahmmed, 2001). In Turkey, higher than expected earthquake damage is attributed to project errors, poor quality of construction, unlicensed modifications to buildings, and so on (Irtem et al., 2007). A magnitude 7.0 earthquake in Haiti in 2010 killed more than230,000 people, mainly because of poor building construction, whereas a larger earthquake in densely populated Kobe, Japan, in 1995 killed around 6000, and a magnitude 6.9 earthquake in 1989 in the San Francisco Bay area killed some 63 people. Surveying the damage caused by an earthquake in China in 2008, in which many schools collapsed, killing hundreds of children, a field team of Australian and Hong Kong earthquake experts observed that "many buildings had inadequate construction quality including insufficient reinforcement, poor detailing and poor quality concrete" (Wibowo et al., 2008). In addition, building codes have been bypassed with the complicity of corrupted officials and construction site staff. As Penny Green noted for Turkey, 'Violations were part of a well-entrenched political process,' and she quotes an adviser to the mayor in one of the worst hit earthquake areas of Turkey, who admits, "The project managers, they take bribes, we do it ourselves. There is no project inspection" (Green, 2005).

On the other hand, large economic losses often arise from natural disasters in the developed world. For example, in 2012 Hurricane Sandy (also known as 'Superstorm Sandy') damaged over 750,000 residences in New Jersey and New York, and caused more than $50 billion in losses (Huffington Post, 2013). Loss of life numbered around 100, mostly from drownings. The 2010–11 earthquakes in Christchurch killed 185 people, most of them were victims of the collapse of two multistorey buildings, and caused over $30 billion in damages—or 20% of the New Zealand GDP. The widespread damage across the central business district and suburbs led to over 750,000 insurance claims being lodged, 64% of businesses were forced to close temporarily, and 11% were forced to close permanently (Potter et al., 2015). With the exception of two buildings that collapsed, other buildings performed as expected and did not collapse—so life safety was ensured. However, the widespread damage to building reduced their functionality and this loss was the main contributor to the huge economic losses suffered by the New Zealand Economy, and to the massive social dislocation of residents.

A key emphasis of this book, thus, is on how climate adaptation engineering can reduce damage to infrastructure that in turn can ameliorate the social and economic disruption of climate change hazards to the built environment. It aims to provide practical and climate-conscious engineering knowledge and solutions to reduce the impact of potential climate change on the performance of buildings and infrastructure, including safety, serviceability, and durability. Examples of climate adaptation engineering to be presented in this book are as follows:

- using probabilistic models for the assessment of climate change effects for infrastructure and buildings subjected to hurricanes under a changing climate (Chapter 2);
- installing underground storage tanks to enhance urban drainage and reduce flooding (Chapter 3);
- specifying higher durability recommendations to reduce climate change increases in corrosion (Chapter 4);
- changing inspection and replacement practices for timber power poles to reduce losses from climate-induced increases in wind speeds and rates of timber decay (Chapter 5);
- increasing bridge foundation depths to reduce the risk of scouring from floods (Chapter 6);
- integrating the interdependency between several infrastructure systems to increase its resilience in the case of extreme events (Chapter 7);
- combining housing and protection structure adaptations to reduce flooding vulnerability, (Chapter 8);
- enhancing design standards for new houses and retrofitting existing houses to reduce damage from extreme wind events (Chapter 9);
- upgrading construction quality and practices to increase housing resilience in the developing world (Chapter 10);
- upgrading building energy efficiency ratings for houses using insulation, sealing, and phase change materials to reduce heat stress during heatwaves (Chapter 11); and so on.

The above case studies are cognizant of the overriding principles for effective adaptation that include (IPCC, 2014) the following:

1. Adaptation is place- and context-specific, with no single approach for reducing risks appropriate across all settings.
2. Adaptation planning and implementation can be enhanced through complementary actions across levels, from individuals to governments.
3. A first step towards adaptation to future climate change is reducing vulnerability and exposure to present climate variability. Strategies include actions with cobenefits for other objectives.
4. Adaptation planning and implementation at all levels of governance are contingent on societal values, objectives, and risk perceptions. Recognition of diverse interests, circumstances, social-cultural contexts, and expectations can benefit decision-making processes.
5. Indigenous, local, and traditional knowledge systems and practices, including indigenous peoples' holistic view of community and environment, are a major resource for adapting to climate change.
6. Decision support is most effective when it is sensitive to context and the diversity of decision types, decision processes, and constituencies.
7. Integration of adaptation into planning and decision-making can promote synergies with development and disaster risk reduction.
8. Poor planning, overemphasising short-term outcomes, or failing to sufficiently anticipate consequences can result in maladaptation.

Climate adaptation engineering integrates the previously mentioned items on a general risk-based decision support framework that could be applied to many infrastructure applications. Fig. 1.2 summarises the major steps in developing risk-based decision support for assessing the risks, costs, and benefits of climate adaptation measures. The following sections will describe the different components of this framework.

1.4 Climate Change Emission Scenarios

Future climate is projected by defining carbon emission scenarios in relation to changes in population, economy, technology, energy, land use, and agriculture—a total of four scenario families, that is, A1, A2, B1, and B2 were used in the IPCC's Third and Fourth Assessment Reports in 2001 and 2007, respectively. The A1 scenarios indicate very rapid economic growth, a global population that peaks in mid-century and declines thereafter, and the rapid introduction of new and more efficient technologies, as well as substantial reduction in regional differences in per capita

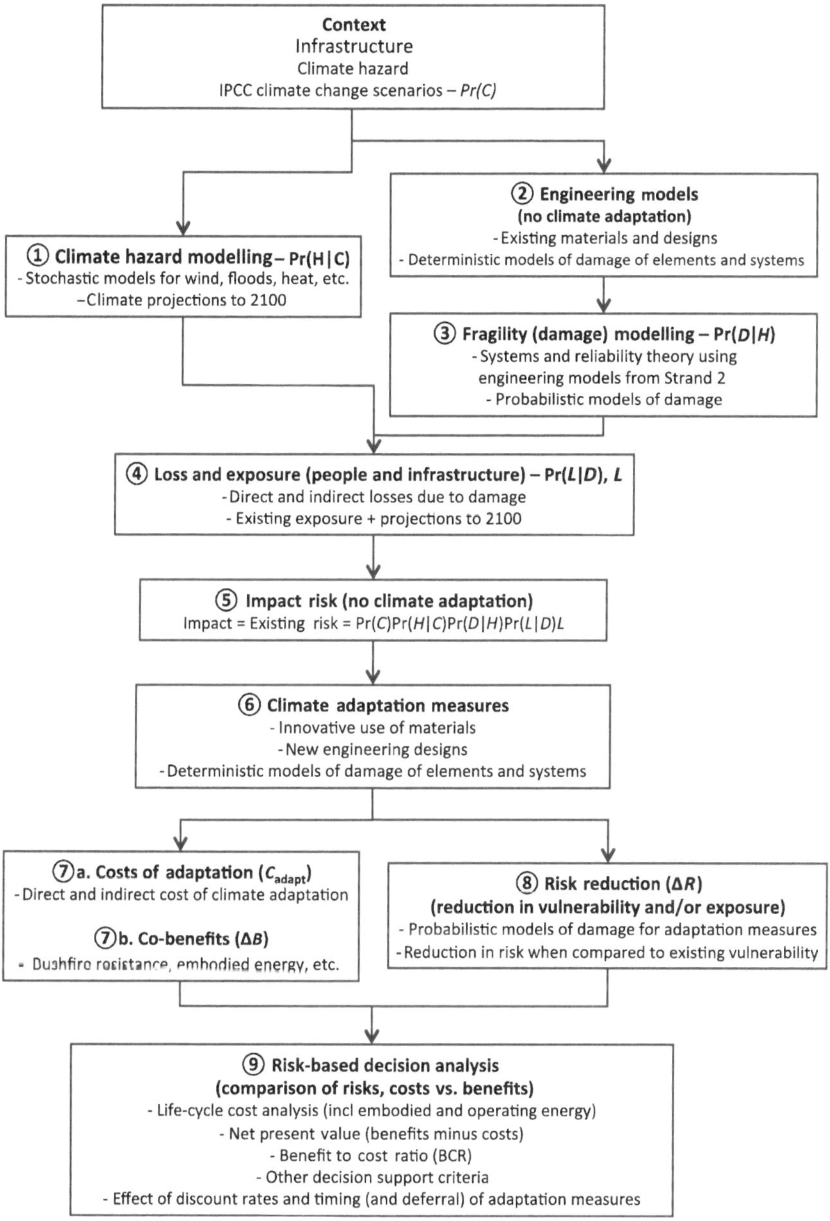

Fig. 1.2 Flowchart of decision support framework for assessing cost-effectiveness of adaptation measures.

income. Subcategories of A1 scenarios include A1FI and A1B, which represent the energy in terms of fossil intensive and a balance across all sources, respectively. The A2 scenarios are based on a very heterogeneous world. The underlying theme is that of strengthening regional cultural identities, high population growth, and less concern for rapid economic development. The B1 scenarios are more integrated and more ecologically friendly than A1 scenarios with rapid changes towards a service and information economies, and reductions in material intensity and the introduction of clean and resource efficient technologies.

The IPCC Fifth Assessment Report (IPCC, 2013) uses Representative Concentration Pathways (RCPs). The four RCPs, RCP 2.6, RCP 4.5, RCP 6, and RCP 8.5, are named after a possible range of radiative forcing values in 2100 (2.6, 4.5, 6.0, and 8.5 W/m^2, respectively), where RCP 8.5, RCP 6.0, and RCP 4.5 are roughly equivalent to A1FI, A1B, and A1B to B1 emission scenarios, respectively. The selected RCPs were considered to be representative of the literature, and included a strict mitigation scenario leading to a low forcing level (RCP 2.6) with CO_2 concentrations reaching 421 ppm by the end of the century, two (medium) stabilisation scenarios (RCP 4.5—CO_2 concentration of 538 ppm, RCP 6.0—CO_2 concentration of 670 ppm) and one scenario with very high greenhouse gas emission (RCP 8.5—CO_2 concentration of 936 ppm). The RCPs can thus represent a range of 21st century climate policies, and are shown in Fig. 1.3.

Peters et al. (2013) showed that current emissions are tracking slightly above RCP 8.5. However, the 12 December 2015 COP21 Paris Agreement

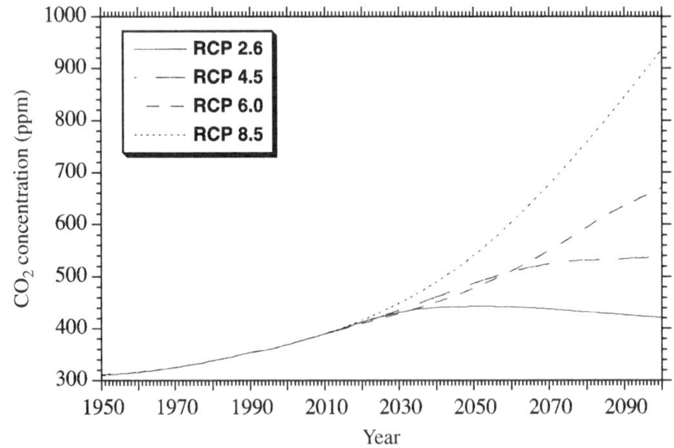

Fig. 1.3 Projected CO_2 concentrations.

reaffirmed the goal of limiting the global temperature increase to well below 2°C (UN, 2015a). Only the RCP 2.6 scenario is 'unlikely' to cause temperatures to exceed 2°C by the end of the century, hence, if CO_2 emissions are mitigated to the 2015 Paris Agreement (and beyond), the most likely emission scenario for this century is RCP 2.6.

1.5 Climate and Hazard Modelling

The performance of infrastructure facilities such as bridges, buildings, dams, offshore structures, etc. is affected by environmental conditions, which are characterised by climate/weather variables (e.g., temperature, humidity, precipitation, wind speed). It may be affected by mean values of these variables as well as their variability, in particular extreme weather events (e.g., floods, storms, heat and cold waves). In the following, current approaches to modelling weather variables and extreme weather events in conditions of climate change are briefly described. The case studies in this book will each describe in more detail climate and hazard modelling, their uncertainties, and their effect on impact and adaptation modelling.

Atmosphere–Ocean General Circulation Models (AOGCMs) are currently the main tool for climate change studies (IPCC, 2012). However, they are computationally demanding which limits their spatial resolution. Downscaling is possible to grids of 25–50 km, but are still too large to capture extreme events such as tornadoes or extreme rainfall. Uncertainties associated with future emission scenarios are usually not quantified and future climate projections are produced separately for individual scenarios (Stewart et al., 2014). Selecting an AOGCM to be used in an impact assessment is not a trivial task, given the variety of models. Weather systems cannot be perfectly modelled due to Chaos components, and climate models are even harder due to the scale of the data, and so no single model can be considered the best. Not surprisingly, uncertainties in climate projections are considerable (Bastidas-Arteaga and Stewart, 2015, 2016). This makes impact and adaptation modelling even more challenging. For more details see Stewart et al. (2014).

A major climate hazard is coastal flooding induced by extreme water level events along low lying, highly populated coastlines due to presently and continuously rising sea levels. Statistical analysis based on extreme value theory has been employed to estimate probabilities of extreme water levels and assess suitable design levels. Estimation of the average recurrence interval (ARI) and the annual exceedance probability (AEP) needs sea-level measurements over a long period of time, greater than 30 years, traditionally observed at tide gauges. Over the last five decades, several statistical analysis methods for estimating the ARI and AEP have been developed (cf. Haigh et al. 2010). There is, however, no universally accepted method available at transnational or even national scales. Therefore, the most applicable method has to be chosen based on different stretches of coastline and length of sea-level records.

Over regions where long period of tide-gauge data are not available or the spatial distribution of tide-gauge sites is sparse along coastlines, modelling of extreme sea levels can be made from sea-level fields produced by ocean circulation models. A two-dimensional (2D) depth-averaged barometric ocean circulation model is usually configured over coastal oceans. The mode is driven by tidal forcing and atmospheric forcing. The tidal component including the Earth geocentric tide and other nonlinear tidal constitutes are simulated using the existing global tidal modes, and the harmonic analysis algorithm. The storm surge component is simulated through the model driven by wind stress and the atmospheric pressure at the sea level, for which the wind stress calculated from the wind velocity at 10 m above the sea surface and the sea-level pressure can be taken from the climate forecast system reanalysis fields. To achieve the fine structure of the wind and pressure associated with tropical storms, the reanalysis field is enhanced by adding the idealised wind and pressure profiles during cyclones (e.g., Zhang and Sheng, 2013). The total sea levels due to the combination of tides and storm surges use Monte Carlo simulation methods. Other natural hazards, such as El Niño Southern Oscillation (ENSO), follow periodic patterns that are very difficult to forecast. To deal with this challenge, Cárdenas-Gallo et al. (2016) proposed to forecast future ENSO events using Markov switching autoregressive processes that could constantly be updated with environmental measures.

In recent years, satellite data with high resolution and homogeneous global coverage have already played an important role in monitoring extreme sea level, especially storm surge. However, advanced satellite products such as wind speed, wind direction from high-resolution scatterometry, and sea state information from coastal altimetry have not yet been widely used in the modelling of extreme sea levels. Since these satellite data have been collected for more than 20 years, assimilating them into numerical models will have the potential to improve the estimation of extreme sea level (e.g., Cheng et al., 2012; Deng et al., 2011, 2013; Idris et al., 2014). Nonetheless, uncertainties are considerable, and remote sensing technologies are increasingly relied upon for measuring and predicting sea-level rise and other phenomena. Wind speed projections also face distinct measuring and prediction challenges (e.g., Harper, 2013).

As discussed above, climate projections are subject to considerable uncertainty that depend on CO_2 emission scenarios and accuracy of general circulation models (GCM). These uncertainties can be classified into three types (Bastidas-Arteaga and Stewart, 2015; Hawkins and Sutton, 2009; Madsen, 2013):

- *Internal uncertainty* is related to the natural variability of the climate system without considering any anthropogenic climate change effect. There are weather disturbances of different duration, size and location that turn climate into a chaotic system. Consequently, it is currently impossible to predict future climate at different scales (daily, monthly, yearly, etc.) even for the more complete climate models and short-time windows.
- *Model uncertainty* (also known as response uncertainty) is associated to the fact that GCMs simulate different changes in climate in response to a given radiative forcing. This kind of uncertainty depends mainly on the simplifications and assumptions that are implemented for each GCM to simulate natural systems.
- *Scenario uncertainty* is related to the assumptions made to define each climate change scenario that determine the future radiative forcing used in climate projections (e.g., future emissions of greenhouse gases, population growth, introduction of clean technologies, changes in land use, etc.).

Fig. 1.4 illustrates how these uncertainties interact over time for surface temperature projections and two different scales: global (Earth) and regional (British Isles) (Hawkins and Sutton, 2009). At a global scale, it is observed that model and internal uncertainties are initially predominant (Fig. 1.4A). However, scenario uncertainties grow considerably and become the most important source of uncertainties after 50 years. A regional scale changes the relative importance of uncertainties (Fig. 1.4B). Internal uncertainty has initially the largest importance because regional weather is largely affected by random weather and climate fluctuations. Model uncertainties have the largest importance from 20 to 70 years. The importance of scenario uncertainties grows significantly during the latter part of the century (after 70 years).

The effect of these uncertainties on climate projections can be considerable. Fig. 1.5 shows that climate projections for temperature vary by up to 4°C when six climate models were compared for Sydney in Australia

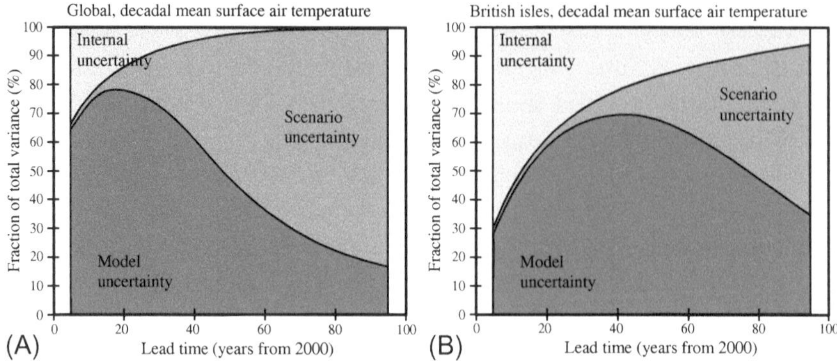

Fig. 1.4 Relative importance of uncertainty sources over time for decadal mean surface air temperature projections: (A) global (Earth) and (B) British Isles. *(Adapted from Hawkins, E., Sutton, R., 2009. The potential to narrow uncertainty in regional climate predictions. Bull. Am. Meteorol. Soc. 90(8), 1095–1107.)*

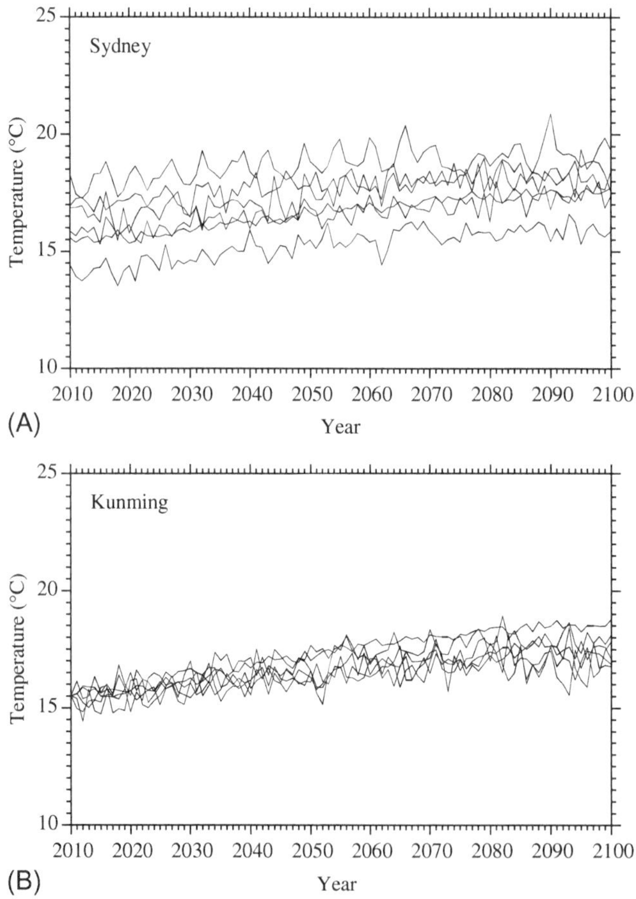

Fig. 1.5 Projected median temperatures for six GCM predictions for RCP4.5, for (A) Sydney (Australia), and (B) Kunming (China).

and Kunming in China: CSIRO-Mk3.6.0, ACCESS model (Australia), IPSL CM5A-LR (France), MICRO5 (Japan), bcc-csm1-1 (China), and CNRM-CM5 (European Centre).

The complexity of these uncertainties implies several considerations for the assessment of climate change effects on civil infrastructure:

- Use of several climate trajectories from a same GCM to account for internal uncertainty.
- Use of several climate trajectories for various GCMs to account for model uncertainty.
- Consider several climate change scenarios to account for scenario uncertainty. A scenario of no change in climate may make economic sense as a 'no regrets' policy even if climate predictions are wrong.
- Verify that climate change projection be representative of climate at the scale of the study (local, regional, global). For instance, downscaling is required to represent climate at a local scale.

1.6 Risk-Based Decision Support

1.6.1 Key Issues

Cyclones, earthquakes, tsunami, and floods are natural hazards that cause significant human, economic, and social losses. Added to this are 'man-made' hazards such as climate change and terrorism. These hazards are low-probability high-consequence events which in recent times are more commonly referred to as 'extreme events'. Extreme events illicit extreme reactions—risk aversion, probability neglect, cost neglect, worst-case thinking—that may distort the decision-making process in an effort by policy makers to be seen to be 'doing something' irrespective of the actual risks involved. Policy making in these circumstances becomes a 'risky business' (Hardaker et al., 2009). If rational approaches to public policy making are not utilised, then politically driven processes "may lead to raising unnecessary fears, wasting scarce resources, or ignoring important problems" (Paté-Cornell, 2002).

There are a number of issues and questions related to controversial and emotive questions often encountered with climate change, and are discussed further below.

1.6.1.1 Worst-Case Thinking

Worst-case thinking, or hyperbole, tends to dominate the thinking of many climate change experts. In 2014, Mayor Bill de Blasio of New York at a UN summit proclaimed that "We know humanity is facing an existential threat" from climate change (Grynbaum, 2014). The notion that a threat short of all-out nuclear war could be existential to humanity is hard to fathom. In the same year, the United Nations Secretary-General Ban Ki-Moon warned that "The human environmental and financial cost of climate change is fast becoming unbearable" (UN, 2015b). And according to John Knox, UN Special Rapporteur on Human Rights and the Environment, climate change is "the greatest threat to human rights in the 21st century" (UCL, 2016). If business as usual predictions are biased towards impending doom, then this justifies any response no matter the cost in loss of civil liberties, quality of life, and treasure.

1.6.1.2 Cost Neglect

While it is not difficult to list threats and vulnerabilities, what is more challenging is to ascertain the cost to reduce these threats and vulnerabilities, and to decide who pays, and when. There is a notion that safety is infinitely good, and no cost is too high. There is no attempt to compare costs against benefits.

1.6.1.3 Probability Neglect

Many analysts base their findings on threats or scenarios that they assume will occur. There is no consideration of the likelihood that a specific CO_2 emission scenario will occur, or that mitigation or adaptation will be effective. For example, a US 2014 climate risk assessment report predicts trillions in dollars of damage due to climate change for the business as usual scenario, that is, the United States continues in its current path (Risky Business, 2014). There is no attempt to quantify the likelihood that CO_2 emissions will continue unabated for the next 85 years, CO_2 mitigation measures will be implemented, adaptation measures are implemented, or the impact of improved or game-changing technologies. Sunstein (2003) terms this as 'probability neglect' and that "people's attention is focused on the bad outcome itself, and they are inattentive to the fact that it is unlikely to occur". There is no certainty with predictions, nicely summed up by physicist Niels Bohr: "Prediction is very difficult, especially if it's about the future".

1.6.1.4 Opportunity Costs

Policy makers who act before they carefully consider the implications of their actions can result in undesirable outcomes which are often referred to as 'opportunity costs'. A CO_2 mitigation strategy that reduces economic growth, particularly in developing countries, may reduce their ability to adapt.

1.6.1.5 Acceptable Risk

The notion of acceptable risk is rarely raised in public discussions. The world is not risk free. The generally accepted level of annual fatality risk is 1 in a million (e.g., Stewart and Melchers, 1997), see, for example, Gardoni and Murphy (2014) for a fuller discussion on risk acceptability. The probability that an American will be killed by a hurricane stands at about 1 in 7 million per year, and 1 in 2.8 million per year for a heat-related death (NOAA, 2016).[1] By comparison, an American's chance of being killed in an automobile crash is about 1 in 8000 a year, the chances of becoming a victim of homicide is about 1 in 22,000, and the chances of being killed by lightning is 1 in 7 million. How much should we be willing to reduce (or even maintain) a risk that is already very low, and is the risk reduction worth the cost?

These are issues without easy answers, but the questions need to be asked. They are also other similar controversial and emotive issues such as terrorism, nuclear power plant accidents, and other extreme events (Mueller and Stewart, 2011a,b, 2016). All too often climate change studies assume there is certainty about the future, and so suffer from probability neglect, as well as cost neglect by ignoring the large costs involved to mitigate CO_2 emissions.

1.6.2 Definition of Risk

Risk is a measure of expected loss, and quantifies the effect of uncertainty on factors that influence this loss. The standard definition of risk is:

$$(\text{Risk}) = (\text{Hazard}) \times (\text{Vulnerability}) \times (\text{Consequences}) \qquad (1.1)$$

where

1. Hazard—probability there will be a climate hazard.
2. Vulnerability—probability of damage or loss given the hazard.
3. Consequences—loss or consequence if the hazard is successful in causing damage.

Eq. (1.1) can be re-expressed as:

$$E(L) = \sum \underbrace{\Pr(C)\Pr(H|C)}_{\text{Hazard}} \underbrace{\Pr(D|H)\Pr(L|D)}_{\text{Vulnerability}} \underbrace{L}_{\text{Loss}} \qquad (1.2)$$

where $\Pr(C)$ is the annual probability that a specific climate scenario will occur, $\Pr(H|C)$ is the annual probability of a climate hazard (wind, heat,

[1] Ten year average 2006–15. United States population of 310 million.

etc.) conditional on the climate, $\Pr(D|H)$ is the probability of infrastructure damage or other undesired effect conditional on the hazard (also known as fragility) for the baseline case of no extra protection (i.e., 'business as usual'), $\Pr(L|D)$ is the conditional probability of a loss (economic loss, loss of life, etc.) given the occurrence of the damage, and L is the loss or consequence if full damage occurs. The product $\Pr(D|H)\Pr(L|D)L$ refers to the expected loss given the occurrence of the hazard. In some cases, 'damage' may equate to 'loss' and so a vulnerability function may be expressed as $\Pr(L|H)$ which is equal to the product $\Pr(D|H)\Pr(L|D)$. The summation sign in Eq. (1.2) refers to the number of possible climate scenarios, hazards, damage levels, and losses. If the loss refers to a monetary loss, then $E(L)$ represents an economic risk.

The expected loss after climate adaptation is derived from Eq. (1.2) as:

$$E_{adapt}(L) = \sum (1 - \Delta R)E(L) - \Delta B \tag{1.3}$$

where ΔR is the reduction in risk caused by climate adaptation (or other protective) measures, $E(L)$ is the 'business as usual' risk given by Eq. (1.2), and ΔB is the cobenefit of adaptation such as reduced losses to other hazards, increased energy efficiency of new materials, etc. Costs of adaptation, timing of adaptation, discount rates, future growth in infrastructure, and spatial and time-dependent increase in climate hazards need to be included in any risk analysis (Fig. 1.2).

1.6.2.1 Climate and Hazard Projections

There are significant challenges in characterising (in probabilistic terms) climate impact and adaptation in time and space. Quite rightly, there has been substantial research on climate variability as this will be the driver to climate impact. To project spatially dependent future climates under different emission scenarios, various climate models have been developed (see Section 1.5). The IPCC suggests that it is necessary to use multiple AOGCMs to take into account the uncertainties of models in any impact assessment. The IPCC also states that probabilities or likelihood are not assigned to individual RCP scenarios (IPCC, 2013). However, the estimation of $\Pr(C)$ may be based on expert opinion about the likelihood of each emission scenario, and multiple AOGCMs may be used to infer the probabilistic characterisation of $\Pr(H|C)$ for future climate projections.

1.6.2.2 Fragility and Vulnerability

Infrastructure fragility or vulnerability can be expressed in terms of structural damage or other losses, and are derived from fitting curves to damage data from historical damage records (i.e., empirical models and insurance data) or from engineering models (e.g., Stewart and Melchers, 1997).

Insurance or building performance data are often used to derive vulnerability models which are often expressed in terms of $\Pr(L|H)$. For example, Fig. 1.6 shows a vulnerability model for Australian houses subject to floods derived from insurance loss records. In this case, the hazard H is the water depth above the floor. Empirical models have drawbacks such as, lack of damage data (Ham et al., 2009), lack of capability to examine the effect of changes in building design and construction methods on damages, lack of ability to examine the effectiveness of building adaptation measures for climate change (Zhang et al., 2014), and they tend to focus on losses (vulnerability) and not on damage (fragility). There are also a number of issues associated with utilising claim data such as access to the insurance claim data, insurance valuation cost and the actual damage cost, and insurance claim databases that do not disaggregate losses between building exterior and interior (Pita et al., 2013). Most importantly, empirical vulnerability curves are based on what has happened in the past. They cannot assess changes in fragility or vulnerability due to future changes in design standards, materials, or construction practices. This highlights the need for developing fragility models based on engineering and structural reliability methods. It is noted

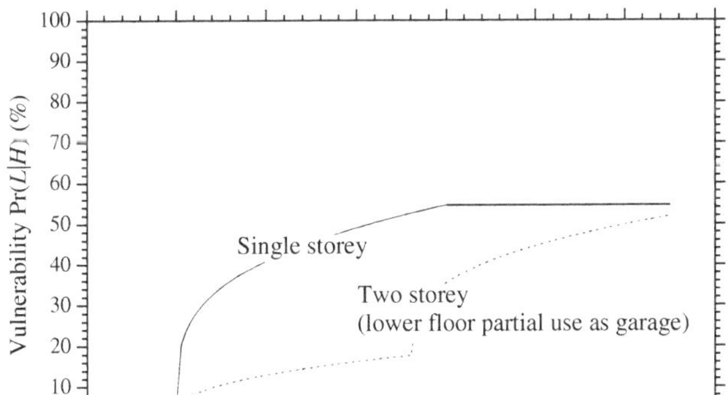

Fig. 1.6 Flood vulnerability curves for residential construction in Brisbane. *(Data from Mason, M., Phillips, E., Okada, T., O'Brien, J., 2012. In: Analysis of Damage to Buildings Following the 2010/2011 East Australian Floods, NCCARF, Griffith University, Australia.)*

however that, as with all models, engineering fragility models should be validated or benchmarked with empirical models based on past events where possible to give more confidence in modelling assumptions and realism.

The stochastic modelling of infrastructure fragility is $\Pr(D|H)$ and is the probability of damage conditional on a specific wind speed, flood level, temperature, or other hazard:

$$\Pr(D|H) = \Pr(R(\mathbf{X}) - H < 0) \tag{1.4}$$

where $R(\mathbf{X})$ is the function for resistance or capacity, $\mathbf{X}$ is the vector of all relevant variables that affect resistance, and H is the known hazard level. Fragility modelling will require probabilistic information on materials, dimensions, model errors, deterioration, and other input variables ($\mathbf{X}$) into engineering models which define the resistance function $R(\mathbf{X})$—these variables vary with time and space.

A key challenge, at least for engineers, is the development of fragility models for damage prediction. Most damage and loss from floods and storms are not due to major structural failure or collapse, but due to water ingress from damaged roofs or walls, or rising water levels. There is much work on predicting reliabilities for the ultimate limit state (collapse) where life safety is the major criterion. However, modelling of damage and serviceability limit states is a less tractable problem as this requires advanced simulation modelling to accurately track component and member performance and failure, load sharing, failure of other components/members due to load redistribution, and damage progression leading to economic and other losses.

Another challenge is that infrastructure, particularly houses, are very complex systems comprising hundreds to thousands of components and members of differing materials. Issues such as poor detailing and workmanship contribute to most damage—so the engineering and stochastic models need to consider these variables—such as screw fasteners being spaced too far apart, or some not connected to purlins and battens, etc. These are more challenging to model stochastically than more conventional 'engineered' constructions such as bridges, towers, etc., where materials are more uniform, and workmanship subject to more quality control measures. Stewart et al. (2018) have conducted structural reliability analyses to assess the roof envelope fragility $\Pr(D|H)$ of contemporary timber-framed houses built in the Australian city of Brisbane, see Fig. 1.7. In this case, Monte Carlo simulation and structural reliability methods were used to stochastically model spatially varying pressure coefficients, roof component failure for

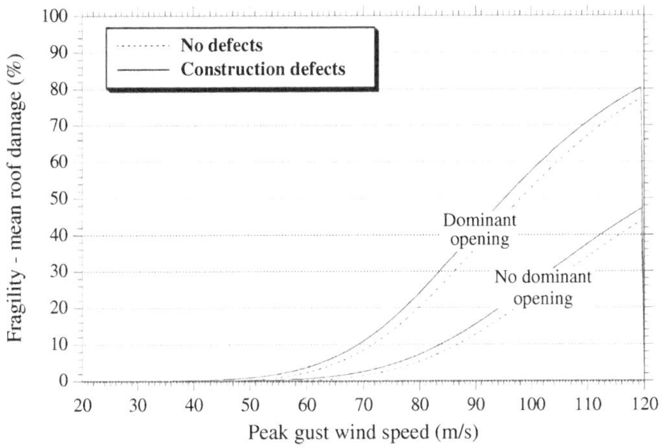

Fig. 1.7 Fragility curves for Australian timber-framed housing (Stewart et al., 2018).

1600 roof fasteners and 500 battens, load redistribution and spatial variability across the roof as connections progressively fail, loss of roof sheeting as a critical number of connections fail, and changes in internal pressure coefficient with increasing roof sheeting loss. The fragility of the roof envelope is vulnerable to dominant openings on the windward wall, and also to construction defects.

1.6.2.3 Losses

Exposure and loss data relates to direct and indirect loss or consequence due to location and extent of infrastructure damage, for existing exposure and future projections. A probability of loss $\Pr(L|D)$ and loss L needs to consider direct and indirect losses, however, most existing studies consider direct losses related to infrastructure damage and content losses. For example, Fig. 1.8 shows a typical direct loss function for wind vulnerability (Hazus, 2014), for roofing and building interior losses. It is observed that the trend between extent of damage (D) and loss is nonlinear, and that losses reach close to their maximum value when damage is only 20%. Clearly, losses accumulate rapidly for low levels of damage because "once the envelope is breached, most of the damage to the interior of the building is a function of the amount of water that enters the building" (Hazus, 2014).

Indirect losses caused by business interruption, clean-up, loss during reconstruction, extra demands on social services, and changes to demand and supply of intermediate consumption goods, postdisaster inflation,

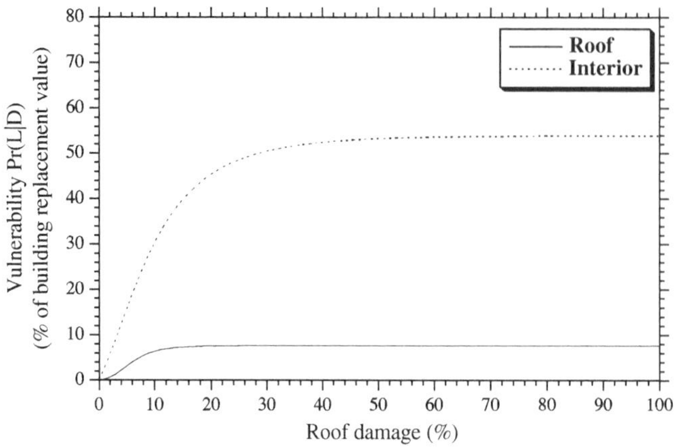

Fig. 1.8 Components of building loss due to wind.

etc., can also be significant (e.g., NAS, 1999; Hallegatte, 2008; Walker, 2011). The data is very limited to accurately quantify how indirect losses increase with vulnerability. Indirect losses were estimated for Hurricane Katrina using an adaptive regional input–output model where damage to houses was \$20 billion, contents \$7 billion, \$17 billion damage to government, and \$63.5 billion to the private sector—total damage to fixed capital was \$107 billion (Hallegatte, 2008). The total indirect loss was \$42 billion or 39% of direct losses. Hallegatte (2008) estimated that indirect losses could exceed 100% of direct losses for a damaging event twice as bad as Hurricane Katrina. An Australian assessment of direct and indirect costs shows indirect costs of 9%–40% of direct losses for bushfire, cyclones, and floods (BTE, 2001).

There is often a high level of postdisaster inflation (or demand surge) of building costs in Australia (e.g., Walker, 2011) which can lead to higher insurance and home owner losses. Walker (2011) estimated that the post-disaster inflation was close to 100% for Cyclone Tracy.

Finally, resiliency is a term that is increasingly being applied to disaster risk reduction. It may be defined as the ability of the system to restore functionality after a damaging event, and the time needed to achieve full restoration of the system is affected by social, economic, and political aspects (e.g., Faber et al., 2014), or it may be defined more broadly to capture vulnerability, exposure, or loss. Resiliency in one way or another has been included in most risk assessments, particularly for low-probability high-consequence events, when assessing loss likelihoods and magnitudes. For example, an urban community that has ready access to emergency services that can temporarily place tarpaulins over damaged roofs will reduce water ingress losses and allow inhabitants to remain in their homes—the community will be able to recover more quickly from such a disaster, and so direct and indirect losses could be minimised. This 'bouncing back' implies a return to the status quo, whereas, 'bouncing forward' leads to continually improving conditions which is a more desired outcome (IPCC, 2012). While the term 'resiliency' may not appear explicitly in risk modelling, it is modelled implicitly in nearly all cases.

1.6.2.4 Risk Reduction

Climate adaptation measures should result in risk reduction (ΔR) that may arise from a combination of reduced fragility, vulnerability ($\Pr(D|H)$ or $\Pr(L|D)$), and exposure (L). For instance, changes to planning may reduce the number of new properties built in a flood plain which will reduce L, or more stringent design codes may reduce the fragility of new infrastructure. For example, installing wind-rated doors to industrial buildings was found to reduce damage from extreme wind events by $\Delta R = 17\%$–33% (Stewart, 2016). Systems and reliability modelling are essential tools to quantify the level of risk reduction, and the extent of risk reduction due to adaptation measures will depend on the hazard, location, and timing of adaptation. For any climate adaptation measure the risk reduction ΔR can vary from 0% to 100% (or even a negative number for an ill-suited adaptation measure).

1.6.3 Decision Preference and Selection of Adaptation Strategies

Three criteria may be used to assess the decision preferences of adaptation strategies:

1. net present value (NPV)
2. benefit-to-cost ratio (BCR)
3. probability of cost-effectiveness or $\Pr(\text{NPV} > 0)$ or $\Pr(\text{BCR} > 1)$

These are not mutually exclusive, but complementary. The 'benefit' of an adaptation measure is the reduction in damages or losses associated with the adaptation strategy, and the 'cost' is the cost of the adaptation strategy. The net benefit or NPV is equal to benefit minus the cost which is also equivalent to the present value or life-cycle cost of an adaptation strategy (sum of damage and adaptation costs) minus the 'business as usual' or 'do nothing' present value. The decision problem is to maximise NPV:

$$\text{NPV} = \sum E(L)\Delta R + \Delta B - C_{\text{adapt}} \tag{1.5}$$

where C_{adapt} is the cost of adaptation measures including opportunity costs that reduces risk by ΔR, ΔB is the cobenefit, and $E(L)$ is the 'business as usual' risk given by Eq. (1.2). The cobenefits of adaptation (ΔB) may include reduced embodied energy and reduced carbon footprint over the life cycle of the facility. This might consider the initial embodied energy associated with the dwelling including footings, structure, and fit-out together with the recurrent embodied energy associated with refurbishment over the life cycle and the operational energy needed to operate a building.

The benefit-to-cost ratio is:

$$\mathrm{BCR} = \frac{\sum E(L)\Delta R + \Delta B}{C_{adapt}} \tag{1.6}$$

If NPV > 0 or BCR > 1 then there is a net benefit and so the adaptation measure is cost-effective. Fig. 1.9 shows how adaptation costs may increase with risk reduction, while benefits increase linearly with risk reduction according to Eqs. (1.5), (1.6). The optimal adaptation occurs when NPV is a maximum, leading to an optimal risk reduction. It is important to note here that such an analysis is aiming to find the 'sweet spot,' that is, the decision where risk reduction or level of protection is optimised, higher protection may be achieved but at a higher marginal cost (law of diminishing returns), and vice versa for a lower level of protection. Other notations and formulae can be used to provide optimal adaptation (e.g., Hall et al., 2012), but ultimately these also mostly rely on maximising NPV. At what

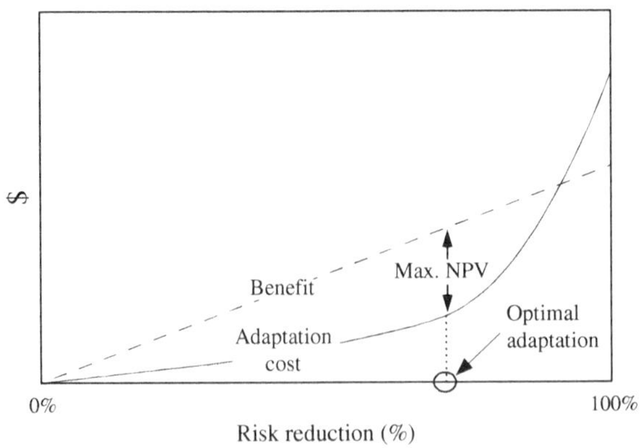

Fig. 1.9 Schematic of net present value (NPV) showing optimal adaptation.

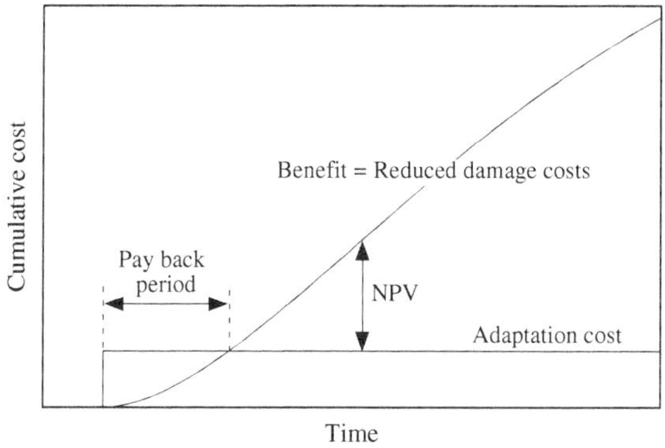

Fig. 1.10 Schematic of payback period.

point in time the benefits exceed the cost, that is, the payback period, is also an important decision metric for policy makers (see Fig. 1.10).

Confidence bounds of NPV or BCR can then be calculated if input parameters are random variables. The probability that an adaptation measure is cost-effective denoted herein as Pr(NPV > 0) or Pr(BCR > 1) may also be inferred.

If the probability that a specific climate scenario will occur Pr(C) is too unreliable, then a decision analysis based on scenario analysis where climate scenario probability is decoupled from Eq. (1.2) provides alternative decision-making criteria based on expected costs. If the loss refers to the fatality of an individual, then $E(L)$ represents an individual annual fatality risk which can be compared with appropriate societal risk acceptance criteria (Stewart and Melchers, 1997).

Hazard, vulnerability, loss, and adaptation costs are subject to considerable uncertainty. For this reason, calculations of risks, costs, and benefits will be imprecise. Hence, a 'break-even' analysis may be useful where minimum risk reduction or maximum cost necessary for adaptation measures to be cost-effective is selected such that there is 50% probability that benefits equal cost, that is, mean(NPV) = 0. For example, if the actual cost of adaptation exceeds the predicted break-even value, then adaptation is not cost-effective. Decision makers can then judge whether a protective strategy meets these break-even values.

1.6.3.1 Risk Preferences

Governments and their regulatory agencies normally exhibit risk-neutral attitudes in their decision-making as described by Eqs. (1.5) and (1.6).

This is confirmed by the US Office of Management and Budget (OMB) which specifically states that "the standard criterion for deciding whether a government program can be justified on economic principles is net present value—the discounted monetized value of expected net benefits (i.e., benefits minus costs)" and that "expected values (an unbiased estimate) is the appropriate estimate for use" (OMB, 1992), and also by many practitioners and researchers (e.g., Sunstein, 2002; Faber and Stewart, 2003; Ellingwood, 2006). This entails using mean or average estimates for risk and cost–benefit calculations, and not worst-case or pessimistic estimates. Paté-Cornell (2002) elaborates on this point by stating "if risk ranking is recognized as a practical necessity and if resource limitations are acknowledged, the maximum overall safety is obtained by ranking the risks using the means of the risk results (i.e., expected value of losses)."

This type of 'rational' approach to risky decision-making is challenging to governments and their agencies which might have other priorities and political concerns. Hardaker et al. (2009) note that 'policy making is a risky business,' and that "Regardless of the varied desires and political pressures, we believe that it is the responsibility of analysts forcefully to advocate rational decision methods in public policy making, especially for those with high risk. We believe that more systematic analysis of risky policy decisions is obviously desirable." Probability neglect is a form of risk aversion as decision makers are clearly averse to events of large magnitude irrespective of the probability of it actually occurring.

Utility theory can be used if the decision maker wishes to explicitly consider factor risk aversion or proneness into the decision process (e.g., Jordaan, 2005; Stewart et al., 2011). In this case, utility theory provides a means of evaluating the risk preferences of the interested parties under choice uncertainty. The objective of the decision-making process is to maximise the expected utility, and so an option is preferable if it has a higher utility. Multiattribute utility theory is capable of handling more than one interested party, nonmonetised parameters (such as risk preferences for quality of life), and risk-neutral, risk-averse, or risk-prone decision preferences.

While the appetite for risk will no doubt vary from person from person, governments and regulators need to act in the best interests of the public, and this means ensuring that public policy is geared towards achieving the highest outcomes. This is best achieved by adopting risk-neutral attitudes to decision-making, and not allowing emotion or the political imperative interfere with the process.

It is important to note that the issue of risk aversion is not a new one, but has been well researched and documented for politically sensitive and controversial decisions associated with terrorism, nuclear power safety, aviation safety, pharmaceutical benefits scheme, environmental pollution, etc. In these cases, risk acceptance criteria have been developed based on annual fatality risks and cost–benefit analysis using expected (mean) values. In principle, decisions related to climate adaptation measures should be made with similar risk-based methodologies.

As noted in Section 1.6.1, risk aversion is often a feature of public policy making (Stewart et al., 2011). This can lead to very different outcomes depending on the belief system of the decision maker. For instance, where risk is measured by government expenditures (and less concerned about climate impacts), a risk averse decision maker may wish the likelihood of cost-effectiveness to be high before investing in an adaptation measure, for example, that there is 90% likelihood that benefit exceeds the cost ($\Pr(\text{NPV} > 0) = 90\%$) so there is more certainty about a net benefit and small likelihood of a net loss. On the other hand, if the decision maker believes that climate change is the 'greatest challenge of the present century' the precautionary principle comes into play, so to avoid the potential risk of catastrophic climate change impacts the decision maker will support adaptation measures almost irrespective of their cost. Anticipating these types of contradictory risk preferences is not the matter for this book. What is needed, however, is a transparent decision-making framework that outlays in a systematic and rigorous manner risks, costs, and benefits. The effect of policy decisions will then be more apparent, as will the trade-offs involved in suboptimal decisions.

1.6.3.2 Discount Rates

Costs of adaptation, timing of adaptation, discount rates, future growth in infrastructure, and spatial and time-dependent increase in climate hazards need to be included in any risk analysis. Of particular interest is uncertainty about the level of discount rates. Infrastructure Australia recommends discount rates of 4%, 7%, and 10% for infrastructure projects (IA 2008). The Australian Government Office of Best Practice Regulation (OBPR) recommends that the discount rate for regulatory interventions is 7%, and that sensitivity analyses consider discount rates of 3% and 10% (OBPR, 2010). Projects with significant effects beyond 30–50 years are considered intergenerational, and so a time-declining discount rate may be appropriate. However, the Australian OBPR states that "there is no consensus about how to value impacts on future generations" and "Rather than use an arbitrarily lower discount rate, the OBPR suggests that the effects on future generations be considered explicitly" (OBPR, 2010). Nonetheless, the Australian Garnaut Review adopted discount rates of 1.35% and 2.65% (Garnaut, 2008). These relatively low discount rates were selected so as to not underestimate climate impacts on future generations. Countries and institutions worldwide use other discount rates. France recommend 2.5% and 1.5% discount rates for short-term (lifetime lower than 70 years) and long-term investments, respectively (Quinet, 2013), and the European Commission recommends a 5% discount rate (Harrison, 2010). Other discount rates vary from 3% (Germany) to over 10% (World Bank) (Harrison, 2010).

Discount rates are generally assumed constant with time. However, this may not be appropriate when considering intergenerational effects often associated with climate change policy decisions. For example, the UK Treasury recommends time-declining discount rates (e.g., Boardman et al., 2011) which places more emphasis on future benefits by reducing the discount rate. While there is some uncertainty about discount rates (e.g., Dasgupta, 2008), the selection of discount rates for public policy 'regulatory intervention' often associated with climate change policy is a matter best determined by government regulations specific to each country.

1.7 Summary

This chapter introduced the background, basic concepts, and principles that constitute the basis of climate adaptation engineering. It reviewed the main potential climate change effects and consequences in terms of vulnerability and costs. The literature review consensually concluded that climate change will increase human vulnerability and that worldwide income and GDP would be significantly reduced if mitigation or adaptation measures will be not implemented. Consequently, climate adaptation engineering aims to improve the response of built infrastructure to future potential climate change damage based on a risk-based decision support. This decision support starts evaluating risks by combining climate hazard modelling, to evaluate the demands under different climate change scenarios, as well as engineering and fragility models to quantify the structural response with and without adaptation measures. The outputs of this risk analysis are introduced into a decision framework that allows determining the cost-effectiveness of climate adaptation measures in terms of NPV, BCR, or probability of cost-effectiveness. The following chapters of this book will present various study cases where this risk-based decision support is applied.

Acknowledgements

The authors gratefully acknowledge the support of the Université de Nantes, and the Pays de la Loire Regional Council for supporting the project RI-ADAPTCLIM.

References

Bastidas-Arteaga, E., Stewart, M.G., 2015. Damage risks and economic assessment of climate adaptation strategies for design of new concrete structures subject to chloride-induced corrosion. Struct. Saf. 52, 40–53.

Bastidas-Arteaga, E., Stewart, M.G., 2016. Economic assessment of climate adaptation strategies for existing RC structures subjected to chloride-induced corrosion. Struct. Infrastruct. Eng. 12 (4), 432–449.

Bjarnadottir, S., Li, Y., Stewart, M.G., 2011. A probabilistic-based framework for impact and adaptation assessment of climate change on hurricane damage risks and costs. Struct. Saf. 33 (3), 173–185.

Boardman, A.E., Greenberg, D.H., Vining, A.R., Weimer, D.L., 2011. Cost-Benefit Analysis: Concepts and Practice. Pearson, Boston.

BTE, 2001. Economic Costs of Natural Disasters in Australia. Bureau of Transport Economics Report 103, Canberra.

Burke, M., Hsiang, S.M., Miguel, E., 2015. Global non-linear effect of temperature on economic production. Nature 527, 235–239.

Cárdenas-Gallo, I., Akhavan-Tabatabaei, R., Sánchez-Silva, M., Bastidas-Arteaga, E., 2016. A Markov regime-switching framework to forecast El Niño southern oscillation patterns. Nat. Hazards 81 (2), 829–843.

Cheng, Y., Andersen, O.B., Knudsen, P., 2012. Integrating non-tidal sea level data from altimetry and tide gauges for coastal sea level prediction. Adv. Space Res. 50 (6), 1099–1106.

Crompton, R.P., McAneney, J.K., 2008. Normalised Australian insured losses from meteorological hazards: 1967–2006. Environ. Sci. Pol. 11 (5), 371–378.

Dasgupta, P., 2008. Discounting climate change. J. Risk Uncertain. 37 (2–3), 141–169.

Deng, X., Griffin, D.A., Ridgway, K., Church, J.A., Featherstone, W.E., White, N.J., Cahill, M., 2011. Satellite Altimetry for Geodetic, Oceanographic, and Climate Studies in the Australian Region. Coastal Altimetry, Springer, Heidelberg, pp. 473–508.

Deng, X., Andersen, O.B., Gharineiat, Z., Stewart, M.G., 2013. Observing and modelling the high sea level from satellite radar altimetry during tropical cyclones. In: International Association of Geodesy Symposium, Potsdam.

Ellingwood, B.R., 2006. Mitigating risk from abnormal loads and progressive collapse. J. Perform. Constr. Facil. 20 (4), 315–323.

Faber, M.H., Stewart, M.G., 2003. Risk assessment for civil engineering facilities: critical overview and discussion. Reliab. Eng. Syst. Saf. 80 (2), 173–184.

Faber, M.H., Giuliani, L., Revez, A., Jayasena, S., Sparf, J., Mendez, J.M., 2014. Interdisciplinary approach to disaster resilience education and research. Procedia Econ. Finan. 18, 601–609.

Gardoni, P., Murphy, C., 2014. A scale of risk. Risk Anal. 34 (7), 1208–1227.

Garnaut, R., 2008. The Garnaut Climate Change Review: Final Report. Commonwealth of Australia, Cambridge University Press, Cambridge.

Goklany, I.M., 2008. What to do About Climate Change. Policy Analysis, No. 609. Cato Institute, Washington, D.C.

Green, P., 2005. Disaster by Design. Br. J. Criminol. 45 (4), 528–546.
Grynbaum, M.M., 2014. At U.N., de Blasio Warns of 'Existential Threat' From Climate Change. New York Times, September 23, 2014.
Haigh, I.D., Nicholls, R., Wells, N., 2010. A comparison of the main methods for estimating probabilities of extreme still water levels. Coast. Eng. 57 (9), 838–849.
Hall, J.W., Brown, S., Nicholls, R.J., Pidgeon, N.F., Watson, R.T., 2012. Proportionate adaptation. Nat. Clim. Chang. 2 (12), 833–834.
Hallegatte, S., 2008. An adaptive regional input-output model and its application to the assessment of the economic cost of Katrina. Risk Anal. 28 (3), 779–799.
Ham, H.J., Lee, S., Kim, H.S., 2009. Development of Typhoon Fragility for Industrial Buildings. In: The Seventh Asia-Pacific Conference on Wind Engineering, Taipei, Taiwan.
Hardaker, J.B., Fleming, E., Lien, G., 2009. How should governments make risky policy decisions?. Aust. J. Public Adm. 68 (3), 256–271.
Harper, B.A., 2013. Best practice in tropical cyclone wind Hazard modelling: in search of data and emptying the skeleton cupboard. In: 16th Australasian wind engineering society workshop, Brisbanepp. 1–10.
Harrison, M., 2010. Valuing the future: the social discount rate in cost-benefit analysis. In: Visiting Researcher Paper, Productivity Commission, Canberra.
Hawkins, E., Sutton, R., 2009. The potential to narrow uncertainty in regional climate predictions. Bull. Am. Meteorol. Soc. 90 (8), 1095–1107.
Hazus, 2014. Multi-hazard Loss Estimation Methodology—Hurricane Model. Hazus–MH 2.1 Technical Manual. Federal Emergency Management Agency, Mitigation Division, Washington, D.C.
Huffington Post, 2013. Hurricane Sandy's Impact, by the Numbers (INFOGRAPHIC), 10/29/2013.
IA, 2008. Outline of Infrastructure Australia's Prioritisation Methodology. Australian Government, Infrastructure Australia, Canberra. 24 September 2008.
Idris, N.H., Deng, X., Andersen, O.B., 2014. The importance of coastal altimetry retracking and detiding: a case study around the great barrier reef, Australia. Int. J. Remote Sens. 35 (5), 1729–1740.
IPCC, 2012. Managing the risks of extreme events and disasters to advance climate change adaptation. In: Field, C.B. et al., (Eds), A Special Report of Working Groups I and II of the Intergovernmental Panel on Climate Change. Cambridge University Press, Cambridge.
IPCC, 2013. Climate change 2013: the physical science basis. In: Stocker, T.F., Qin, D., Plattner, G.-K., Tignor, M., Allen, S.K., Boschung, J., Nauels, A., Xia, Y., Bex, V., Midgley, P.M. (Eds.), Contribution of Working Group I to the Fifth Assessment Report of the Intergovernmental Panel on Climate Change. Cambridge University Press, Cambridge and New York, NY 1535 pp.
IPCC, 2014. Climate change 2014: impacts, adaptation, and vulnerability. Part A: global and sectoral aspects. In: Field, C.B., Barros, V.R., Dokken, D.J., Mach, K.J., Mastrandrea, M.D., Bilir, T.E., Chatterjee, M., Ebi, K.L., Estrada, Y.O., Genova, R.C., Girma, B., Kissel, E.S., Levy, A.N., MacCracken, S., Mastrandrea, P.R., White, L.L. (Eds.), Contribution of Working Group II to the Fifth Assessment Report of the Intergovernmental Panel on Climate Change. Cambridge University Press, Cambridge and New York, NY. 1132 pp.
Irtem, E., Turker, K., Hasgul, U., 2007. Causes of collapse and damage to low-rise RC buildings in recent Turkish earthquakes. J. Perform. Constr. Facil. 21 (5), 351–360.
Jordaan, I., 2005. Decisions Under Uncertainty: Probabilistic Analysis for Engineering Decisions. Cambridge University Press, Cambridge.
Koehn, E., Ahmmed, M., 2001. Quality of building construction materials (cement) in developing countries. J. Archit. Eng. 7 (2), 44–50.

Li, Y.Y., Su, G.F., Yuan, H.Y., 2015. Statistical analysis on earthquake fatality vulnerability in China. In: Proceedings of the World Congress on Engineering 2015 Vol II WCE 2015, July 1–3, 2015, London.
Lomborg, B., November 2, 2006. Stern review: the dodgy numbers behind the latest warming scare. Wall Street J.
Lomborg, B., 2009. Global Crises, Global Solutions. Cambridge University Press, Cambridge.
Madsen, H., 2013. Managing structural safety and reliability in adaptation to climate change. In: Deodatis, G., Ellingwood, B.R., Frangopol, D. (Eds.), 11th International Conference on Structural Safety & Reliability. CRC Press, New York, pp. 81–88.
Mendelsohn, R.O., 2006. A critique of the stern report. Regulation 29 (4), 42–46.
Mueller, J., Stewart, M.G., 2011a. Terror, Security, and Money: Balancing the Risks, Benefits, and Costs of Homeland Security. Oxford University Press, Oxford and New York.
Mueller, J., Stewart, M.G., 2011b. The price is not right: The U.S. spends too much money to fight terrorism. Playboy 58 (10), 149–150.
Mueller, J., Stewart, M.G., 2016. Chasing Ghosts: The Policing of Terrorism. Oxford University Press, Oxford and New York.
NAS, 1999. The Impact of Natural Disasters: A Framework for Loss Estimation. National Academy of Sciences, Washington, D.C.
NOAA, 2016. Weather Fatalities. National Weather Service, Silver Springs, MD. http://www.nws.noaa.gov/om/hazstats.shtml. accessed 30.08.16.
OBPR, 2010. Best Practice Regulation Handbook. Office of Best Practice Regulation, Australian Government, Canberra.
OMB, 1992. Guidelines and Discount Rates for Benefit-Cost Analysis of Federal Programs (Revised). Circular No. A-94, October 29, 1992. Office of Management and Budget, Washington, DC.
Paté-Cornell, E., 2002. Risk and uncertainty analysis in government safety decisions. Risk Anal. 22 (3), 633–646.
Peters, G.P., Andrew, R.M., Boden, T., Canadell, J.G., Ciais, P., Quéré, C.L., Marland, G., Raupach, M.R., Wilson, C., 2013. The challenge to keep global warming below 2°C. Nat. Clim. Chang. 3 (1), 4–6.
Pita, G.L., Pinelli, J.P., Gurley, K.R., Hamid, S., 2013. Hurricane vulnerability modeling: development and future trends. J. Wind Eng. Ind. Aerodyn. 114, 96–105.
Potter, S.H., Becker, J.S., Johnston, D.M., Rossiter, K.P., 2015. An overview of the impacts of the 2010-2011 Canterbury earthquakes. Int. J. Disaster Risk Reduct. 14, 6–14.
Quinet, E., 2013. L'évaluation socioéconomique des investissements publics. Commissariat Général à la Strategie et à la Prospective (In French), Paris.
Risky Business, 2014. The Economic Risks of Climate Change in the United States. Riskybusiness.org.
Stern, N., 2007. The Economics of Climate Change: The Stern Review. Cambridge University Press, Cambridge.
Stewart, M.G., 2014. Climate change and national security: balancing the costs and benefits, dangerous world?. In: Preble, C., Mueller, J. (Eds.), Threat Perception and U.S. National Security. Cato Institute, Washington, D.C.
Stewart, M.G., 2016. Climate change impact assessment of metal-clad buildings subject to extreme wind loading in non-cyclonic regions. Sustain. Resilient Infrastruct. 1 (1–2), 32–45.
Stewart, M.G., Melchers, R.E., 1997. Probabilistic Risk Assessment of Engineering Systems. Chapman and Hall, London.
Stewart, M.G., Ellingwood, B.R., Mueller, J., 2011. Homeland security: A case study in risk aversion for public decision-making. Int. J. Risk Assess. Manage. 15 (5/6), 367–386.
Stewart, M.G., Val, D., Bastidas-Arteaga, E., O'Connor, A., Wang, X., 2014. Climate adaptation engineering and risk-based design and management of infrastructure.

In: Frangopol, D.M., Tsompanakis, Y. (Eds.), Maintenance and Safety of Aging Infrastructure. Taylor and Francis Publishers, pp. 641–684.
Stewart, M.G., Ginger, J.D., Henderson, D.J., Ryan, P.C., 2018. Fragility and climate impact assessment of contemporary housing roof sheeting failure due to extreme wind. Eng. Struct. 171, 464–475.
Sunstein, C.R., 2002. The Cost-Benefit State: The Future of Regulatory Protection. ABA Publishing, American Bar Association, Chicago.
Sunstein, C.R., 2003. Terrorism and probability neglect. J. Risk Uncertain. 26 (2–3), 121–136.
Taig, T., 2012. A Risk Framework for Earthquake Prone Building Policy. TTAC Limited and GNS Science, Wellington.
UCL, 2016. Climate change—The greatest threat to human rights in the 21st century—John Knox, UN Special Rapporteur on Human Rights and the Environment, https://www.ucl.ac.uk/global-governance/ggi-events/john-knox-climatechange (accessed 30.08.16).
UN, 2015a. Framework Convention on climate change. In: Conference of the Parties. Twenty-first session, Paris, 30 November to 11 December 2015, United Nations.
UN, 2015b. Climate summit: 'All hands on deck' declares Ban, calling for leadership, concrete action. UN News Centre, 23 September 2014.
Walker, G.R., 2011. Comparison of the impacts of cyclone Tracy and the Newcastle earthquake on the Australian building and insurance industries. Aust. J. Struct. Eng. 11 (3), 283–293.
Webb, L.B., Hennessy, K., 2015. Climate Change in Australia: Projections for Selected Australian Cities. CSIRO and Bureau of Meteorology, Australia.
White House, 2014. The cost of delaying action to stem climate change. In: The Council of Economic Advisers, Executive Office of the President of the United States, Washington, DC, July 2014.
Wibowo, A., Kafle, B., Kermani, A., Lam, N., Wilson, J., Gad, E., 2008. Damage in the 2008 China earthquake. In: Proceedings of 2008 Australasian Earthquake Engineering Society Conference, 2008, Ballarat, Victoria.
World Bank, 2010. Economics of Adaptation to Climate Change: Synthesis Report. The International Bank for Reconstruction and Development/World Bank, Washington, D.C.
World Bank, 2016. The Making of a Risker Future: How our Decisions Are Shaping Future Disaster Risk, Global Facility for Disaster Reduction and Recovery. World Bank, Washington, D.C.
Yohe, G.W., Tol, R.S.J., Richels, R.G., Blanford, G.F., 2009. Climate change. In: Lomborg, B. (Ed.), Global Crises, Global Solutions. Cambridge University Press, Cambridge.
Zhang, H., Sheng, J., 2013. Estimation of extreme sea levels over the eastern continental shelf of North America. J. Geophys. Res. Oceans 118 (11), 6253–6273.
Zhang, S., Nishijima, K., Maruyama, T., 2014. Reliability-based modelling of typhoon induced wind vulnerability for residential buildings in Japan. J. Wind Eng. Ind. Aerodyn. 124, 68–81.

Further Reading

AS5334, 2013. Climate Change Adaptation for Settlements and Infrastructure—A Risk Based Approach. Standards Australia, Sydney.
Mason, M., Phillips, E., Okada, T., O'Brien, J., 2012. Analysis of Damage to Buildings Following the 2010/2011 East Australian Floods. NCCARF, Griffith University.

PART II

Case Studies for Infrastructure

CHAPTER TWO

Risk Analysis for Hurricanes Accounting for the Effects of Climate Change

Alessandro Contento, Hao Xu, Paolo Gardoni
Department of Civil and Environmental Engineering, MAE Center, University of Illinois at Urbana-Champaign, Urbana, IL, United States

2.1 Introduction

In the last decades, the devastating impact of hurricanes revealed the vulnerability of large areas of the US East and Gulf coasts to this natural hazard. The high toll paid in terms of human lives and the structural damage due to different storm occurrences raised the awareness about hurricanes. The importance of developing mitigation and adaptation strategies to reduce the loss of lives and properties due to hurricane occurrences was formally recognised at the Federal level with the National Windstorm Impact Reduction Act Reauthorisation of 2015 (PL 114-52). The resulting National Windstorm Impact Reduction Program (NWIRP), a joint Program of several Federal agencies led by the National Institute of Standards and Technology (NIST), has as its main objectives the improvement of the understanding of windstorms and the development of cost-effective mitigation strategies to reduce their impact.

Additionally, the impact of hurricanes may be exacerbated by climate change and the population growth of coastal communities (Murphy et al., 2018). Understanding how climate change affects the behaviour of extreme storms, and in particular of hurricanes, is a challenging task. Large-scale average climate is typically simulated using Atmospheric General Circulation Models (AGCMs). The main issues when looking at climate change using AGCMs are: (1) the identification of events is challenging when events have limited temporal and spatial extension as in the case of hurricanes, and (2) the AGCMs typically have difficulties in capturing the underlying local physics. Because of these issues, identifying trends (e.g., possibly due to climate change) in hurricanes is more complex than identifying trends in large environment characteristics such as changes in

https://doi.org/10.1016/B978-0-12-816782-3.00002-4

temperature. The Intergovernmental Panel on Climate Change Fourth Assessment Report (IPCC AR4) (IPCC, 2007) was able to show that high-resolution AGCMs can reproduce approximately the frequency and distribution of hurricanes. More recently, the Intergovernmental Panel on Climate Change Fifth Assessment Report (IPCC AR5) (IPCC, 2013) showed that high-resolution AGCMs (LaRow et al., 2008; Zhao et al., 2009; Strachan et al., 2013) can simulate the year-to-year count variability of hurricanes for given observed sea-surface temperatures (SSTs). However, in both cases, such models underestimated the storm intensity. New models with improved parameterisations and higher resolution, such as the one presented by Mizuta et al. (2012), seem to be capable of providing an accurate distribution of the annual storm occurrences as well as simulating events with intensities comparable to those observed.

Although a general consensus has not been reached, the majority of the numerical analyses (e.g., Camargo, 2013; IPCC, 2013; Knutson et al., 2015) suggests an increase in the global mean intensity of the hurricanes, the number of the most intense occurrences, and the precipitation rate. Conversely, the global frequency of hurricanes is expected to slightly decrease or remain unaltered. A probable trend detected in the results of the numerical analyses is the poleward migration of the locations where the hurricanes land (Kossin et al., 2016). Another interaction between hurricanes and climate change is the potential effect of sea-level rise that sets a higher starting level for storm surge and may increase the extent of the vulnerable areas.

Foreseeing an increase in the already consistent damage due to hurricanes clearly motivates the need for long-term mitigation and adaptation strategies. For natural hazards, the terms 'mitigation' and 'adaptation' may assume different meanings than those in the literature on climate change. Consequently, there is the need to clarify how these terms are used and what is intended for mitigation and adaptation strategies when talking about the natural hazards affected by climate change. Independently from the choice of the adopted terminology, the choice of the mitigation and adaptation strategies should be based on risk assessment analyses. Such analyses require the use of a specific risk analysis framework that, starting from the modelling of the hazard, allows the propagation of uncertainties up to the loss estimation (Contento et al., 2017).

A framework for the risk analysis for hurricanes needs to consider three fundamental aspects of hurricanes (wind, rainfall, and storm surge), should be able to account for the effects of climate change, and should adopt wind, rainfall, and storm-surge models that are computationally efficient. Several wind and rainfall models in the literature can account for the effects of climate change and are computationally efficient (Vickery et al., 2000b; Emanuel et al., 2006; Mudd et al., 2017). However, currently models for storm surge that can account for the effects of climate change are generally computationally inefficient (Jelesnianski et al., 1992; Westerink et al., 1994). A few models that are computationally efficient (e.g., Irish et al., 2008; Jia and Taflanidis, 2013) are empirical, and their functional forms lack physical meaning and are not based on the understanding of the underlying physical phenomena. Moreover, because of the specific formulation used to construct such models, it is generally not possible to incorporate both results from simulations and historical observations in the model calibration. Consequently, there is the need to define efficient models that can account for the effects of climate change while capturing the physics of the phenomena and can incorporate data from both simulations and historical observations.

After clarifying the meaning of mitigation and adaptation strategies in the present context, the chapter summarises a general framework for hurricane risk analysis that can be used to compare mitigation and adaptation strategies. The chapter describes each step of the framework and the specifications of the models that are needed. The chapter also presents a novel probabilistic model proposed by Contento et al. (2018) to model the storm-surge height. Such a model accounts for the underlying physics of the phenomena, is computationally efficient, and overcomes the limitation of the available empirical models on the use of different data. Several examples of storm-surge height predictions show the ability and accuracy of the model. Lastly, the chapter suggests possible uses of the presented model for the evaluation of different mitigation and adaptation strategies.

Section 2.2 proposes a consistent use of the terms mitigation and adaptation when talking about natural hazards affected by climate change. Section 2.3 presents the general framework for hurricane risk analysis. Section 2.4 reviews the existing formulations for storm surge and presents the probabilistic model proposed by Contento et al. (2018). Finally, Section 2.5 presents how the information provided by the hurricane risk analysis can be used.

2.2 Mitigation Versus Adaptation

The terms 'mitigation' and 'adaptation' are widely used in the literature related to natural hazards and climate change. However, the meaning of mitigation and adaptation is inconsistent between the literature related to natural hazard and the literature on climate change. Looking at the dictionary definition, mitigation is "the action of reducing the severity, seriousness, or painfulness of something" (Stevenson, 2010). When referred to natural hazard and climate change, the word mitigation assumes different meanings. For natural hazards, mitigation refers to the reduction of the impacts (or consequences) of physical hazards through the reduction of exposure or vulnerability. Since natural hazards traditionally could not be influenced by human interventions, the focus has been only on reducing of the effects of the physical phenomenon and not on changing the intensity or the likelihood of occurrence of the phenomenon itself. Since climate change is at least in part due to human activities, mitigation typically focuses on the source of climate change and refers to the reduction of greenhouse emissions and the enhancement of the sinks of such gasses (IPCC, 2013). The dictionary definition of adaptation refers to "adjust[ing] to new conditions" (Stevenson, 2010). The term adaptation is typically not used in the context of natural hazards. In the context of climate change, adaptation typically refers to the efforts towards ameliorating the consequences of climate change by adjusting to new climate conditions and their effects (IPCC, 2013), see also Chapter 1.

For both natural hazards and climate change, mitigation and adaptation strategies can be defined at two different levels: (i) the level of the built or modified natural environment, and (ii) the community level. When looking at hurricanes as an example of natural hazards, mitigation strategies for the built or modified natural environment might consist in reducing the vulnerability of structures and infrastructure to high wind, torrential rain, and storm surge. At the community level, mitigation might include education actions meant to increase awareness and preparedness, and the adoption of building codes and regulations for land use. As an extreme measure, mitigation strategies may culminate in relocation (Olshansky, 2018). When looking at climate change, mitigation strategies are only defined at the community level. Examples are activities that increase carbon stocks, reduction of direct agricultural emissions, and prevention of deforestation and degradation of high-carbon ecosystems. The adaptation strategies for climate change are the set of actions meant to reduce the consequences of climate change and, as such, they correspond to the mitigation strategies for natural hazards.

As a result, the same activity (like erecting a levee) is called a mitigation strategy when considering hurricanes as natural hazards, while it is called an adaptation strategy in the context of climate change. Consequently, we need to clarify how these terms are used when talking about the natural hazards affected by climate change. We propose to use the term mitigation to refer to the activities targeted at influencing climate change insofar it affects the natural hazards. We also propose to use the term adaptation to refer to the activities targeted at reducing the impact of natural hazards (influenced or not by climate change). In particular, we propose the use of the terms 'environmental adaptation' and 'human adaptation' to differentiate between adaptation measures that modify the environment (built or natural) and those that involve changes in human activities (Cooper and Pile, 2013).

Given the inevitability of climate change, all coastal communities will be forced to adopt adaptation strategies. While farsighted legislators may anticipate adaptation strategies, adaptation and will most likely follow hurricane-induced disaster occurrences in the form of long-term postdisaster recovery (Olshansky, 2018). Be it before or after a disaster, the choice of adaptation strategies should be based on the results of risk analyses. Such risk analyses necessitate the use of a comprehensive framework, which combines the different aspects of the hazard (e.g., wind, rainfall, and storm surge in the case hurricanes), accounts for climate change, and adopts wind, rainfall, and storm-surge models that are computationally efficient.

2.3 Framework for Risk Analysis

This section presents a general framework for hurricane risk analysis that is based on the literature and current practice. The presented framework involves four steps. Step 1 models the hurricane activity and simulates synthetic storms needed to derive the landfall statistics. Step 2 models the three aspects of the hazard and provides possible hazard scenarios. Step 3 models damage, predicts losses, and provides the corresponding distributions and probabilities of exceedance. Step 4 uses the information coming from the previous steps for the comparison and development of mitigation and adaptation strategies. The framework accounts for the underlying uncertainties from the hazard up to the loss estimation and the development of mitigation and adaptation strategies (Gardoni, 2017; Murphy et al., 2018). To allow for both short- and long-term analyses, each step directly or indirectly can account for the effects of climate change. Additionally, the framework adopts models that are mostly physics based (which is important to improve the applicability of the models to cases other than those used in the model calibration) and computationally efficient (which is important given the necessity of large numbers of simulations).

The remaining of this section briefly describes each step. Section 2.4 reviews the existing formulations for storm surge and presents the probabilistic model proposed by Contento et al. (2018) that can be used in Step 2. Section 2.5 presents the details of Step 4 describing how the information coming from the first three steps can be used in the development of mitigation and adaptation strategies.

2.3.1 Hurricane Activity Models (Step 1)

Developing a hurricane model and simulating hurricane occurrences is the first step of a hurricane risk analysis (Lin et al., 2012, 2014). Hurricane models are needed to simulate synthetic storms to be used in a Monte Carlo simulation for deriving the landfall statistics (i.e., the hurricane characteristics at landing). Since hurricane models are function of variables depending on climate change, such as SSTs, the effects of possible mitigation strategies will directly affect the landfall statistics.

There are several models available in the literature to describe the genesis, track, and intensity of hurricanes (e.g., Russell, 1971; Vickery et al., 2000b; Emanuel et al., 2006; Lee and Rosowsky, 2007). These models can be empirical (Neumann, 1991, Vickery and Twisdale, 1995), mixed empirical and physics based (Vickery et al., 2000b, Powell et al., 2005, Lee and Rosowsky, 2007), or physics based (Emanuel et al., 2006). In this chapter, we distinguish between empirical and physics-based models. We call empirical models those constructed by fitting an arbitrary model form to simulated or field data. In contrast, physics-based models are those constructed based on the understanding of the underlying physics of the phenomena. We also distinguish between deterministic and probabilistic models. Deterministic models do not account for the underlying uncertainties, while probabilistic models do. Empirical model could be either deterministic (e.g., when the parameters in the model calibrated from the data are fixed at their mean values), or probabilistic. Similarly, physics-based models can be either deterministic (e.g., when the models are constructed purely based on first principles) or probabilistic (e.g., when data are used in addition to the first principles to develop the model and the uncertainties are captured in the final model form). Each of the three kinds of model available in the literature for hurricane activities has strengths and limitations; extensive reviews on hurricane models can be found in Vickery et al. (2009a) and Lin et al. (2014).

Among the available models, the one presented by Vickery et al. (2000b), and improved by Vickery et al. (2009b), is well-suited for analyses that require a high number of simulations. Although this model proposes an empirical tracking model, it is still capable of retaining information about the physics of the problem through a proper choice of the regressors. This model was used by Mudd et al. (2014) for analyses that include the effects of climate change using projections of SSTs.

2.3.2 Hazard Scenario Models (Step 2)

After the simulation of the hurricane characteristics, hazard models are needed to simulate the resulting wind, rainfall, and storm surge. For the wind aspect, different wind profiles, either parametric (Holland, 1980; Holland et al., 2010; Chavas et al., 2015) or nonparametric (Thompson and Cardone, 1996; Kepert, 2010) are available in the literature. Generally, the surface wind field is obtained as the combination of an axisymmetric surface wind field and a background wind field.

The axisymmetric surface wind field is derived from a gradient wind field model (e.g., Batts et al., 1980; Holland, 1980; Jelesnianski et al., 1992; Emanuel, 2004; Emanuel and Rotunno, 2011). Such derivation can be done either by accounting for the surface friction with empirical correction terms, such as an empirical surface wind reduction factor (Schwerdt et al., 1979; Batts et al., 1980; Georgiou, 1985; Vickery et al., 2000a; Powell et al., 2003) and an inflow angle (Bretschneider, 1972), or using boundary layer models (e.g., Thompson and Cardone, 1996; Vickery et al., 2009b; Kepert, 2010). Boundary layer models provide the mean wind speed profile in the hurricane boundary layer. Such models tend to be more accurate and computationally demanding than those that use empirical correction terms. They are usually nonparametric with only a few exceptions. One example of a parametric model was developed by Vickery et al. (2009b) that empirically modelled the variation of the mean wind speed with height.

The background wind field is typically modelled as a function of the hurricane's translational velocity having the same direction and a proportional intensity. Some formulations use the full value of the hurricane's translational velocity (e.g., Powell et al., 2005; Mattocks and Forbes, 2008; Vickery et al., 2009b), while others use reducing factors that vary in each formulation (e.g., Jelesnianski et al., 1992; Phadke et al., 2003; Emanuel et al., 2006; Lin et al., 2012).

The rainfall models available in the literature are mostly empirical models (Lonfat et al., 2007; Tuleya et al., 2007; Mudd et al., 2017) based on the rainfall climatology and persistence model (R-CLIPER). These models are usually developed using recorded rainfall data associated with hurricane events. Langousis and Veneziano (2009) developed a model of tropical cyclone rainfall based on the vertical outflow of water vapour from the tropical cyclone boundary layer. Only a few studies in the literature proposed coupled studies for wind and rainfall (e.g., Mudd et al., 2017).

For the storm surge, the common practice is the use of models such as the sea, lake, and overland surges from hurricanes (SLOSH) model (Jelesnianski et al., 1992) or the advanced circulation (ADCIRC) model (Westerink et al., 1994). Usually these models have either low accuracy or high input requirements, are complex and time consuming to develop for a specific site, and are computationally expensive. Those that have low accuracy (but are easier to develop and more computationally efficient) are most suitable to forecast the storm surge promptly in the case of the occurrence of a real hurricane. Those that have higher accuracy are preferred for detailed probabilistic analyses. However, their computational inefficiency makes them unsuitable for performing the high number of simulations typically required by probabilistic analyses. Especially for the comparisons between possible climate change scenarios where multiple analyses are needed, simplified models fitted to available data are a viable alternative. However, such models available in the literature and sometime also referred to as 'metamodels' or 'surrogate models' (Jia and Taflanidis, 2013; Kim et al., 2015; Jia et al., 2016) have some limitations. First, the models can only be trained using data coming from either simulations or historical records. Second, they are only able to predict the storm surge at the same locations of the data used to calibrate the model. Section 2.4 focuses explicitly on this particular piece of the hurricane risk analysis framework and presents a novel probabilistic model for storm-surge prediction that is computationally efficient and overcomes such limitations (Contento et al., 2018).

2.3.3 Damage and Loss Models (Step 3)

The hazard scenarios obtained with the models described in Step 2 can be used to obtain predictions of annual damage and loss distributions. In general, the damage estimations are obtained using fragility curves that provide the likely damage states of the structures or infrastructure given the values of the intensity measures describing the wind, rainfall, and storm surge.

Similarly, the loss estimations are obtained using loss-ratio curves. Such curves estimate the losses in the form of a percentage of the initial values of the structures and infrastructure as a function of the intensity measures. These estimations can be obtained for each hurricane scenario. The resulting damage and losses can then be used to derive the corresponding distributions. An example of such procedure can be found in Contento et al. (2017).

The common practice has been to conduct damage and loss assessments (Scawthorn et al., 2006) separately for wind and storm surge with different models and tools (e.g., HAZUS see FEMA, 2009a,b, and CLARA see Johnson et al., 2013). However, this might lead to double counting the damage and losses (since a structure damaged by wind might be also counted as damaged by storm surge). There are only a few recent studies that estimate the damage and losses for the combined effects of wind and storm surge (one example of such studies is Li et al., 2012). Considering the combined effects of wind and storm surge avoids double counting damage and losses. In terms of available software, HAZUS-MH (FEMA, 2012) provides the possibility of performing joint wind-surge damage and loss analyses. To take advantage of the storm-surge probabilistic model developed by Contento et al. (2018) in computing the contribution from the storm surge to the damage and loss estimates, user defined hazard scenarios from Contento et al. (2018) can be used instead of the ones automatically generated by HAZUS using SLOSH and SWAN (Booij et al., 1996).

2.3.4 Development of Mitigation and Adaptation Strategies (Step 4)

The hazard scenarios obtained from Step 2 and the distributions of damage and losses obtained from Step 3 can be used for the development and comparison of mitigation and adaptation strategies (Stewart et al., 2014, see also Chapter 1). Specifically, mitigation strategies would affect the inputs to Step 1, and environmental adaptation strategies would affect the inputs to Steps 2 and 3. The results from Step 3 can be used in decision models for life-cycle risk assessment of structures (e.g., Lee and Ellingwood, 2017; Gardoni et al., 2016). Such decision models guide long-term decisions by incorporating the costs associated to the entire service life of an asset. Incorporating the results from Step 3 in the decision models allows us to consider the effects of mitigation and adaptation strategies in the cost estimates. Finally, the results from Step 3 can inform insurance and financial models, for example, to improve the loss- and uncertainty-dependent components of the insurance premium and catastrophe (CAT) bond pricing (Hofer et al., 2018).

2.4 Hazard Scenario Models: Storm-Surge Modelling

While there are a number of studies on wind field and debris movement during hurricanes and their associated damage, there are only a few studies on storm surge and its associated damage. State-of-the-art storm-surge models can use outputs from wind and pressure field models to generate storm-surge simulations, or observed wind and pressure fields for hindcasts of storm surge (e.g., Houston et al., 1999; Westerink et al., 2008; Bunya et al., 2010). Physics-based models use shallow-water equations (SWEs) to model the hydrodynamics of the storm surge (e.g., Jelesnianski et al., 1992, Westerink et al., 1994, Hubbert and McInnes, 1999, Roland et al., 2009). The most commonly adopted models for storm surge are the ADCIRC model (Westerink et al., 1994) and the SLOSH model (Jelesnianski et al., 1992). The ADCIRC model accurately simulates hurricane storm surge, but also astronomical tides. The model numerically integrates the depth-integrated barotropic SWEs in spherical coordinates over an unstructured grid. The grid has a wide range of element sizes (Westerink et al., 2008) that are chosen to obtain an extremely fine resolution near the coast that becomes coarser and coarser moving towards the deeper ocean, according to the required accuracy. The high-resolution numerical grid used by ADCIRC captures the complex spatial variability of the phenomenon (Dietrich et al., 2011); however, the model is highly computationally demanding. SLOSH was originally developed for real-time forecasting of hurricane storm surges on continental shelves. It solves the SWEs using the finite difference method over a grid (polar, elliptic or hyperbolic) centred on the region of interest (Jarvinen and Lawrence, 1985). The model requires the characteristics of the hurricane to derive a model of the wind field, which drives the storm surge. The main limit of SLOSH is that the accuracy of the storm-surge predictions is highly dependent on the accuracy of the meteorological input and the error in the estimate is about 20% using accurate hurricane predictions (Jelesnianski et al., 1992). Moreover, the model does not account for astronomical tides and wind waves. Lin et al. (2012) used ADCIRC and SLOSH together with grids having different extensions and resolutions, restricting the use of ADCIRC to the area for which the risk assessment requires higher accuracy in the predictions.

Although some of the models available in the literature are accurate, a trade-off between efficiency and accuracy is required when a high number of simulated scenarios are needed like in the case of probabilistic analysis. In this case, simplified models fitted to available data are a more efficient option. They are developed to approximate the dominant features of a complex model in a computationally efficient way (Asher et al., 2015). Approximate or empirical models can be divided into three main categories (Asher et al., 2015): (i) projection-based methods that project on a reduced subspace the governing equation of the original model; (ii) multifidelity-based methods that either reduce the resolution of the original model or simplify the underlying equations; and (iii) data-driven methods that approximate the original model output with a relation calibrated with inputs and outputs of the same original model. For storm surge, only empirical models (i.e., models that belong to the third category) can be found in the literature. They are calibrated using inputs and outputs obtained by a limited number of physics-based simulations of storm-surge scenarios. As outputs, they provide estimates of the storm surge at predefined locations (points) in the domain of the model, which are the locations of the observations used for the model calibration (Irish et al., 2008; Jia and Taflanidis, 2013; Kim et al., 2015; Jia et al., 2016). While computationally efficient, such empirical models generally lack physical meaning in their functional form.

Irish et al. (2008) developed an empirical model for storm-surge height as a function of the central pressure, radius of maximum wind speed, and forward velocity of the hurricane. The model was calibrated using ADCIRC simulations. Such simulations used synthetic hurricanes to drive the storm surge. The sustained near-surface winds used in the simulations were estimated using a coupled hurricane vortex-planetary boundary layer model (Thompson and Cardone, 1996). This empirical model presented a simplified formulation whose aim was to show the dependence of the storm surge on the size of the hurricane but not to obtain accurate estimates of storm surge. Jia et al. (2016) developed a Kriging model to predict storm surge over an extended coastal region. To improve the computational efficiency of the Kriging model, they integrated the principal component analysis (Jolliffe, 2002) in their formulation. The results presented in Jia et al. (2016) showed that their empirical model is capable of providing accurate predictions of both storm surge and wave height. However, the model was subject to a computationally demanding calibration that required 400 combined ADCIRC and SWAN simulations (Dietrich et al., 2011). Also, artificial neural networks (ANNs) have been used for storm-surge prediction. Kim et al. (2015) developed a time-dependent empirical model for storm surge using an ANN trained with high-fidelity simulations of two historical hurricanes. While ANNs are a good option for problems that are highly nonlinear or have a large amount of data, they are usually best suited for deterministic applications and may not significantly reduce the computational expenses. In general, common limitations of the empirical models on storm surge available in the literature are that: (i) they cannot include in the calibration data recorded or computed at different locations between storms (as a result, it is typically difficult to include both historical data and data from simulations because historical data are typically sparser than simulation data); and (ii) they can provide estimates of storm surge only in the points used for model calibration (as a result, a separate interpolation among the predictions is needed to estimate the storm-surge height over the region of interest).

The probabilistic model presented in Contento et al. (2018), and discussed in the next sections, can be a valid alternative for probabilistic analyses of storm surge. In addition, the model in Contento et al. (2018) is also able to capture the effects of climate change. The model combines a logistic regression model and a nonstationary random field. The logistic regression model estimates the probability of one location being wet (i.e., flooded), as a function of physical regressors affecting the storm surge. The random field estimates the distribution of the surge height over the wet locations. The random field is developed using the improved latent space approach (ILSA) proposed in Xu and Gardoni (2018). Specifically, the ILSA-based random field considers physical regressors as latent dimensions in addition to the spatial dimensions, and models the mean, variance, and spatial correlation of the storm surge as functions of such regressors. Therefore, such a model is particularly suited to model the nonstationarity in the storm-surge distribution on the basis of physical principles.

Differently from the empirical models available in the literature, the probabilistic model presented in Contento et al. (2018) can be trained using both results from historical records and high-fidelity simulations, and can provide predictions at locations different from those of the observations used for the model calibration. In addition, the model calibration is computationally efficient because it requires a limited amount of data for calibration to provide accurate predictions. The consequent advantage is that comparative analyses on the effects of climate change for different climate change scenarios can be performed training the model with results from a limited amount of simulations for each scenario.

2.4.1 Physics of the Storm Surge and Storm-Surge Data

The intense circular motion of the wind around the eye of a hurricane that occurs on the surface of the ocean generates a vertical circular motion in the oceanic waters (Resio and Westerink, 2008). As the storm approaches the coast, and the depth of the ocean decreases, the bottom of the ocean disturbs the vertical motion generating a storm surge, an anomalous rise of water above the expected astronomical tide. In the low-lying coastal areas, where the motion of the water is not prevented by natural or artificial barriers the water penetrates inland. Two different groups of factors, the geomorphological features of the landing site and the characteristics of the hurricane, influence the storm-surge intensity. For a given location, among the characteristics of the landing site, the one that is mentioned in the literature as having the highest impact on the storm-surge height is the mean altitude of the area affected by the storm surge. The hurricane characteristics that affect the storm surge are those related to the intensity of the hurricane (i.e., the wind speed, the central pressure, and the hurricane forward speed), to its track (i.e., the landfall location and the angle of approach to the coast), as well as the size of the storm.

To develop a probabilistic model for storm-surge predictions, it is fundamental to take into account the characteristics mentioned above. In the proposed model, this is done by introducing explanatory functions described in Section 2.4.2. To define the explanatory functions, the vector of storm-surge heights, **y**, at different locations in the region of interest has to be provided along with the vector of corresponding characteristics of the locations, **x**, and the vector of hurricane characteristics, **s**. The location characteristics considered in the model are the longitude, x_{1j_i}, latitude, x_{2j_i}, and altitude, x_{3j_i}, of each measurement location $j_i = (1, \ldots, m_i)$, for each hurricane $i = (1, \ldots, n)$. The hurricane characteristics considered are the longitude and latitude of the hurricane landing point, s_{1i} and s_{2i}; the heading angle of the hurricane at landing, s_{3i}; the central pressure (pressure in the eye of the hurricane), s_{4i}; the forward velocity, s_{5i}; and the radius of maximum wind speed, s_{6i}. Fig. 2.1 shows some of the characteristics included in **s** and **x**. Table 2.1 shows schematically what data are needed for the model calibration and how they are organised. In the table, the subscript o stands for 'observation' to clarify that the quantities are data needed to calibrate the model.

The storm-surge data used to show the capabilities of the model discussed in this chapter come from simulations performed by Kendra et al. (2017). The simulations are performed coupling hurricane wind/

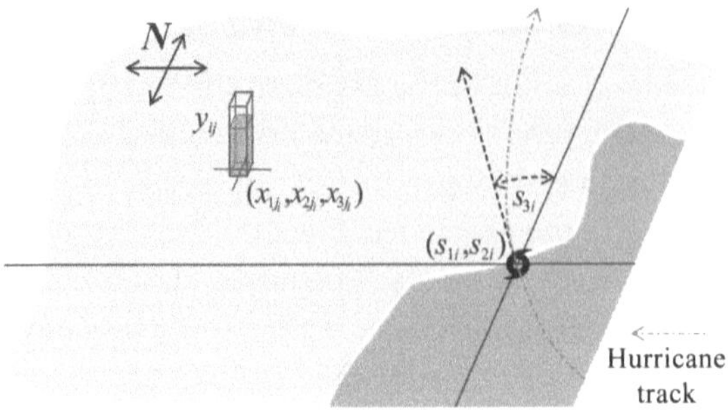

Fig. 2.1 Graphic representation of location and hurricane characteristics.

Table 2.1 Format of the Storm-Surge Observations

Hurricane event	Storm-surge observations y_o	Location characteristics x_o			Hurricane characteristics s_o					
		x_{o1}	x_{o2}	x_{o3}	s_{o1}	s_{o2}	s_{o3}	s_{o4}	s_{o5}	s_{o6}
1	y_{o11}	x_{o11_1}	x_{o21_1}	x_{o31_1}	s_{o11}	s_{o21}	s_{o31}	s_{o41}	s_{o51}	s_{o61}
	y_{o1j}	x_{o1j_1}	x_{o2j_1}	x_{o3j_1}						
	y_{o1m_1}	x_{o1m_1}	x_{o2m_1}	x_{o3m_1}						
i	y_{oi1}	x_{o11_i}	x_{o21_i}	x_{o31_i}	s_{o1i}	s_{o2i}	s_{o3i}	s_{o4i}	s_{o5i}	s_{o6i}
	y_{oij}	x_{o1j_i}	x_{o2j_i}	x_{o3j_i}						
	y_{oim_i}	x_{o1m_i}	x_{o2m_i}	x_{o3m_i}						
n	y_{on1}	x_{o11_n}	x_{o21_i}	x_{o31_n}	s_{o1n}	s_{o2n}	s_{o3n}	s_{o4n}	s_{o5n}	s_{o6n}
	y_{onj}	x_{o1j_n}	x_{o2j_n}	x_{o3j_n}						
	y_{onm_n}	x_{o1m_n}	x_{o2m_n}	x_{o3m_n}						

precipitation, hydrologic, hydrodynamic, and wave models. For every hurricane scenario used as input, a simulation provides the corresponding storm-surge scenario. The dependence of the storm-surge scenarios on the effects of climate change is captured by considering the effects of climate change on the hurricane scenarios. For each hurricane both track and characteristics are derived under the worst-case climate scenario presented in the IPCC AR5 report (IPCC, 2013). Such scenario is the 8.5 Representative Concentration Pathway (RCP 8.5), which assumes a radiative forcing level of 8.5 W/m^2 in 2100.

Kendra et al. (2017) generated 238 storm-surge scenarios that include 1200 data over the stretch of the Tar and Pamlico River between the cities of Greenville and Washington in North Carolina. To show that the chosen approach does not require a high number of observations to produce accurate results, from the available 238 sets of observations, we randomly select

25 sets of observations for the model selection and calibration, where a set refers to a specific hurricane and associated storm-surge scenario.

Although the data refer to a specific time and region (in 2100 and the Pamlico River area), the probabilistic model presented by Contento et al. (2018) could be used to estimate the quantities of interest also for different times and regions as long as the values of **x** and **s** are within the range of the values used to calibrate the models. Since **x** and **s** might miss some location characteristics and/or hurricane characteristics, the model could also be retrained or updated by Bayesian updating with new data for a specific time and region when available. For example, the model could be updated with the results of new simulations based on new hurricane scenarios (possibly reflecting better predictions on the effects of climate change).

For the model calibration and validation, three groups of observations are created. The first group (Group A) is used for the model selection and calibration. Group B is used for the model validation considering hurricane characteristics used in the model calibration but considering different locations. Group C is used for the model validation considering locations used in the model calibration but considering different hurricane characteristics.

Specifically, Group A contains observations at 60 locations that are the same for each of the 25 scenarios, for a total of 1500 observations. Among these observations, 70% are used for model selection and calibration. The model selection and calibration is repeated resampling the 70% of the observations to check consistency in the results (i.e., that the results do not depend on the selected data). Group B contains observations at 20 new locations that are again the same for each of the 25 scenarios, for a total of 500 observations. Group C contains observations at the same 60 locations considered for the observation of Group A, but coming from four additional scenarios. To avoid extrapolation in the prediction of the storm-surge height, these four scenarios are selected so that the characteristics of the hurricanes falls within the ranges of the characteristics of the hurricanes used for model calibration. Fig. 2.2 shows the locations of the observations in Groups A, B, and C.

2.4.2 Probabilistic Model for Storm Surge

The probabilistic model presented by Contento et al. (2018) combines two different models: a logistic regression model and a nonstationary random field. The logistic regression model predicts if the single location is wet or dry. While, the random field based on the ILSA approach predicts the

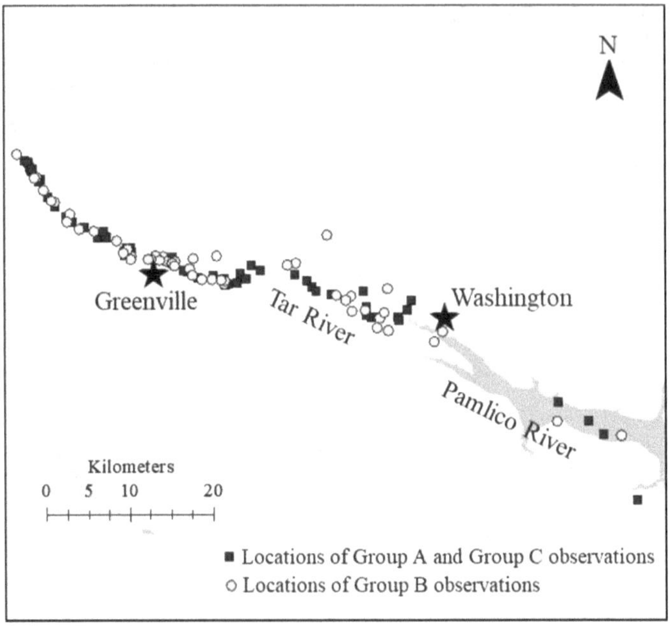

Fig. 2.2 Area of the locations of the storm-surge observations.

storm-surge height, y_{ij}, for the wet locations ($y_{ij} > 0$). Contrary to other models in the literature, the ILSA-based random field model allows for the combined use of observations coming from historical records and high-fidelity model simulations even if they are at different locations. Both the logistic regression model and the ILSA-based random field use explanatory functions to capture the underlying physics of the phenomena. The explanatory functions are selected from an initial set of candidates. Such explanatory functions are combinations of the location and hurricane characteristics **x** and **s**. The candidate explanatory functions are either taken from the literature (Irish et al., 2008; Jia and Taflanidis, 2013) or considered because of their physical meaning. Table 2.2 presents the complete list of the candidate explanatory functions h_{qij}'. Each functions is then standardised to be dimensionless as follows $h_{qij} = [h_{qij}' - E(h_{qij}')]/\mathrm{Var}(h_{qij}')$ where $E(h_{qij}')$ is the sample mean of h_{qij}' and $\mathrm{Var}(h_{qij}')$ is the sample variance of h_{qij}' obtained considering all i and j. Fig. 2.3 shows the physical meaning of the h_{qij}' with geographic meaning (i.e., directions, locations, and distances).

Since locations with altitude $x_{3j_i} \leq 0$ (i.e., locations below the sea level) are more likely to be wet, both the logistic regression and the random field model are calibrated and validated separately for locations with $x_{3j_i} > 0$ and

Table 2.2 Explanatory Functions That Capture the Geomorphological Features of the Landing Site and the Characteristics of the Hurricane

Function	Description
$h'_{0ij} = 1$	Intercept
$h'_{1ij} = g(x_{1j_i}, s_{1i}, s_{2i})$	Longitudinal distance between observation location and landing point
$h'_{2ij} = g(x_{2j_i}, s_{1i}, s_{2i})$	Latitudinal distance between observation location and landing point
$h'_{3ij} = x_{3j_i}$	Altitude of the observation location
$h'_{4ij} = g(x_{1j_i}, x_{2j_i}, s_{1i}, s_{2i})$	Distance between observation location and landing point
$h'_{5ij} = s_{1i}$	Longitude of the landing point
$h'_{6ij} = s_{2i}$	Latitude of the landing point
$h'_{7ij} = s_{3i}$	Heading angle of the hurricane at landing
$h'_{8ij} = s_{4i}$	Central pressure
$h'_{9ij} = s_{5i}$	Forward velocity
$h'_{10ij} = s_{6i}$	Radius of maximum wind speed
$h'_{11ij} = \cos[g(x_{1j_i}, x_{2j_i}, s_{1i}, s_{2i})]$	Cosine of the angular distance between the direction of the heading angle and the direction connecting the observation point to the landing point (cosine of the angle $h^{'*}_{11ij}$ in Fig. 2.3)
$h'_{12ij} = s_{4i}^2$	Second power of the central pressure
$h'_{13ij} = s_{4i} \cdot s_{6i}$	Cross product of central pressure and radius of maximum wind speed
$h'_{14ij} = s_{4i}/s_{6i}$	Ratio of central pressure and radius of maximum wind speed

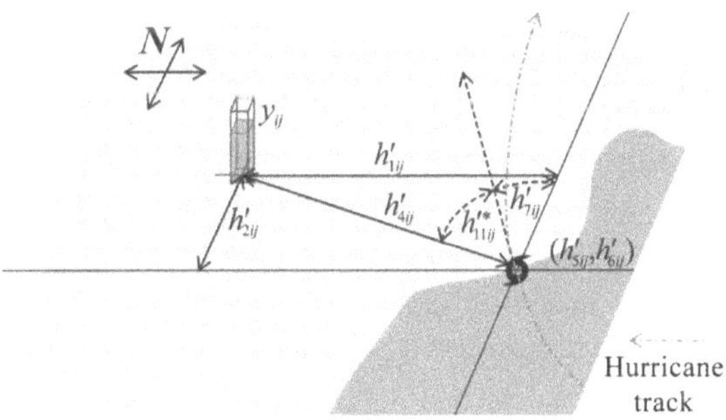

Fig. 2.3 Graphic representation of selected explanatory functions.

$x_{3j_i} \leq 0$. In order to do so, the observations in Groups A–C are divided into two groups. Group 1 contains the observations with $x_{3j_i} > 0$, and Group 2 contains the observations with $x_{3j_i} \leq 0$. The model selection is done without considering this distinction.

2.4.2.1 Logistic Regression Model

First, the logistic regression model is used to estimate the probability p_{ij} that a given location j is wet for a certain hurricane i. The model is written as:

$$p_{ij}\left(\mathbf{x}_{j_i}, \mathbf{s}_i, \boldsymbol{\theta}_l\right) = \frac{\exp\left[\boldsymbol{\theta}_l^{\mathrm{T}} \cdot \mathbf{h}_{ij}^{*}\left(\mathbf{x}_{j_i}, \mathbf{s}_i\right)\right]}{1 + \exp\left[\boldsymbol{\theta}_l^{\mathrm{T}} \cdot \mathbf{h}_{ij}^{*}\left(\mathbf{x}_{j_i}, \mathbf{s}_i\right)\right]} \tag{2.1}$$

where $\boldsymbol{\theta}_l$ is the vector of model parameters, and $\mathbf{h}_{ij}^{*}(\mathbf{x}_{j_i}, \mathbf{s}_i)$ is the vector of the powers of the h_{qij}, from the first to the fourth order. Starting from the full vector, Contento et al. (2018) use a selection process to remove the explanatory functions that are statistically insignificant. The selection process has two steps. The first step uses the Akaike information criterion (AIC) (Akaike, 1974) for a preliminary selection. The second step uses a cross-validation technique to make the final selection. In this second step, the selection process randomly divides the observations used for model selection (Group A) into a calibration set (80% of the observations) and a selection set (20% of the observations). Then, it calibrates all of the competing models using the calibration set, and generates several test sets through resampling of the selection set. The cross-validation technique selects a reduced number of candidates by comparing the models' prediction errors (the total number of wrong predictions for the test sets). Specifically, for each candidate model, the selection process computes the mean value and the 95% confidence interval of the model error. Then, the process chooses as final candidates the models whose 95% confidence interval includes the lowest mean value. Among these candidate models, the selection process finally selects the model that has the least number of explanatory functions. The selected model is then recalibrated using all the data in Group A. Fig. 2.4 shows the selection process for the 11 candidate models with the lowest AIC. For each of the 11 models showed in the example, the dot represents the mean value of the model error while the bar shows the 95% confidence interval. Among the models with a low prediction error, only Models 5–8 have a 95% confidence interval that contains the lowest mean prediction error that is the one of Model 5. Since Model 5 has fewer explanatory functions than any of the Models 5–8, the final model is Model 5. For the selected model, the model parameters, $\boldsymbol{\theta}_l$, are estimated using a Bayesian approach (Gardoni et al., 2002). The form of Model 5 along with the statistics of the model parameters can be found in Contento et al. (2018). Once the posterior distribution of the model parameters, $f''(\boldsymbol{\theta}_l)$, is found, a predictive

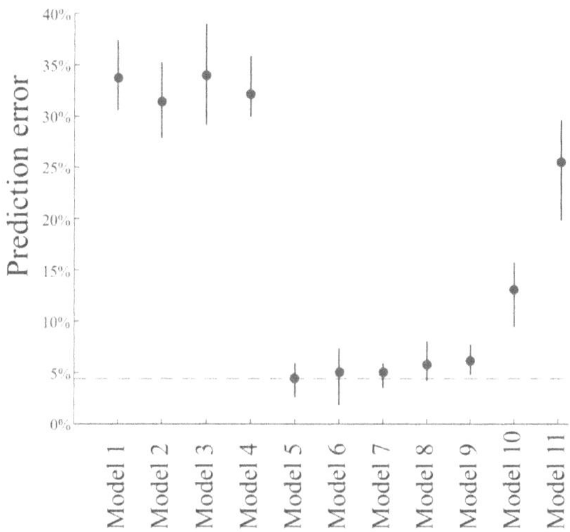

Fig. 2.4 Mean values (*dots*) and 95% confidence intervals (*bars*) of the prediction error for the different models.

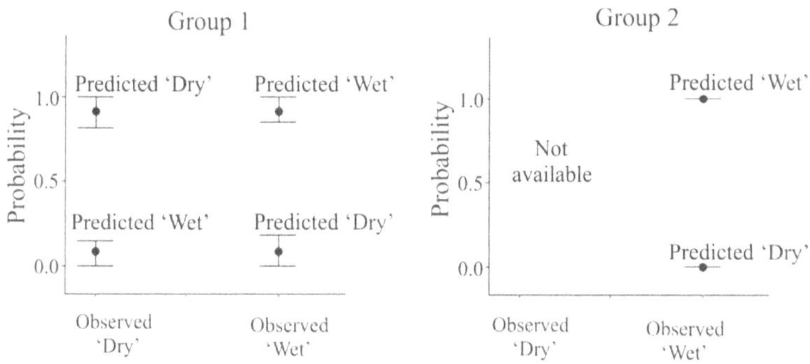

Fig. 2.5 Predicted probabilities of dry and wet observations.

estimate, $\tilde{p}_{ij}(\mathbf{x}_{ji}, \mathbf{s}_i)$, of the probability of the location j being wet is obtained following Gardoni et al. (2002).

Fig. 2.5 shows the comparison between observations and predictions for the data used in the model calibration (80% of Group A) divided into Groups 1 and 2 as described earlier. In each of the two graphs, the dots represent the mean probabilities for wet and dry locations to be predicted as wet or dry. The bars are the one standard deviation confidence intervals. Since all the observation in Group 2 are wet, the model predicts that these locations are wet with probability close to 1 and confidence intervals close to 0 (the small difference from 1 and 0 is due to the numerical approximations in finding $\boldsymbol{\theta}_i$).

2.4.2.2 Random Field Model

A nonstationary spatial random field is used to estimate the storm-surge height for the wet locations. The general formulation of a nonstationary spatial random field can be found in, for example, Schmidt et al. (2011). Generally, a random field is composed by three terms: a mean structure of the field; a zero-mean nonstationary random field; and a white noise. The mean structure describes the mean value of the quantity of interest as it varies spatially. The zero-mean nonstationary random field captures the nonstationary covariance. The white noise captures the micro-scale uncertainty. The mean structure is typically a deterministic function ideally constructed based on the underlying physics of the phenomenon. The covariance generally is constructed using a parametric function. Finally, the white noise is typically modelled with an additive zero mean normal random variables with unknown standard deviation.

The random field used in Contento et al. (2018) is based on the ILSA proposed by Xu and Gardoni (2018). The random field is modelled as a function of explanatory functions, $\mathbf{h}_{ij}(\mathbf{x}_{j_i}, \mathbf{s}_i)$, which are considered as latent dimensions of the random field. For purpose of normalisation, the random field model is developed for the natural logarithm of the storm-surge heights, $\mathbf{z} = \ln(\mathbf{y})$. The random field is written as:

$$z_{ij}\left(\mathbf{x}_{j_i}, \mathbf{s}_i, \boldsymbol{\theta}_{IL}\right) = m\left(\mathbf{x}_{j_i}, \mathbf{s}_i, \boldsymbol{\theta}_m\right) + v\left(\mathbf{x}_{j_i}, \mathbf{s}_i, \boldsymbol{\theta}_\Sigma, \boldsymbol{\theta}_\sigma\right) + \sigma\varepsilon\left(\mathbf{x}_{j_i}\right) \tag{2.2}$$

where the mean field, $m(\mathbf{x}_{j_i}, \mathbf{s}_i, \boldsymbol{\theta}_m)$, the zero mean correlated field, $v(\mathbf{x}_{j_i}, \mathbf{s}_i, \boldsymbol{\theta}_\Sigma, \boldsymbol{\theta}_\sigma)$, and the white noise $\sigma\varepsilon(\mathbf{x}_{j_i})$ contain model parameters, $\boldsymbol{\theta}_{IL} = (\boldsymbol{\theta}_m, \boldsymbol{\theta}_\Sigma, \boldsymbol{\theta}_\sigma, \sigma)$, that are calibrated using the data described in Section 2.4.1. The mean field is a linear combination of the explanatory functions:

$$m\left(\mathbf{x}_{j_i}, \mathbf{s}_i, \boldsymbol{\theta}_m\right) = \sum\nolimits_k \theta_{mk} h_k\left(\mathbf{x}_{j_i}, \mathbf{s}_i\right) \tag{2.3}$$

The variances and spatial correlations in the zero-mean nonstationary random field are modelled as functions of the Euclidian distance, the differences between explanatory functions at two locations, and unknown model parameters. The elements of the covariance matrix given by $v(\mathbf{x}_{j_i}, \mathbf{s}_i, \boldsymbol{\theta}_\Sigma, \boldsymbol{\theta}_\sigma)$ and the white noise can be written as:

$$\begin{aligned}\left[\boldsymbol{\Sigma}_{\mathbf{zz}}(\mathbf{x}, \mathbf{s}, \boldsymbol{\theta}_\Sigma, \boldsymbol{\theta}_\sigma, \sigma)\right]_{rs} &= \left[\boldsymbol{\Sigma}_{\mathbf{vv}}(\mathbf{x}, \mathbf{s}, \boldsymbol{\theta}_\Sigma, \boldsymbol{\theta}_\sigma)\right]_{rs} + \sigma^2 \cdot 1_{\{r=s\}} \\ &= \sigma_v^2(\mathbf{x}, \mathbf{s}, \boldsymbol{\theta}_\sigma) \cdot \exp\left[-Q_{rs}(\mathbf{x}, \mathbf{s}, \boldsymbol{\theta}_\Sigma)\right] + \sigma^2 \cdot 1_{\{r=s\}}\end{aligned} \tag{2.4}$$

where $\sigma_v(\mathbf{x}, \mathbf{s}, \boldsymbol{\theta}_\sigma)$ is the standard deviation, $Q_{rs}(\mathbf{x}, \mathbf{s}, \boldsymbol{\theta}_\Sigma)$ is the generalized distance in the correlation function, and σ^2 is the variance of the white noise, $\mathbf{x} = [\mathbf{x}_{j_i}]\ \forall i, j$, and $\mathbf{s} = [\mathbf{s}_i]\ \forall i$. Details for deriving $\sigma_v(\mathbf{x}, \mathbf{s}, \boldsymbol{\theta}_\sigma)$ and $Q_{rs}(\mathbf{x}, \mathbf{s}, \boldsymbol{\theta}_\Sigma)$ can be found in Xu and Gardoni (2018).

For the ILSA-based random field, the model selection starts from the full vector of the explanatory functions, $\mathbf{h}_{ij}(\mathbf{x}_{j_i}, \mathbf{s}_i)$. To avoid underestimating the storm-surge height, Contento et al. (2018) used only the wet observations in Group A for the model selection of the random field. The model calibration and selection is done as for the logistic regression.

In this case, all the 15 explanatory functions shown in Table 2.2 are retained in the selected model. After the parameter estimation, a predictive joint distribution of the natural logarithms of the storm-surge heights $\widetilde{\mathbf{z}}(\mathbf{x}, \mathbf{s})$ is derived using the posterior distribution of $\boldsymbol{\theta}_{IL}$, $f''(\boldsymbol{\theta}_{IL})$. The median predictions of the storm-surge heights $\widetilde{\mathbf{y}}(\mathbf{x}, \mathbf{s})$ and their confidence intervals are consequently derived from $\widetilde{\mathbf{z}}(\mathbf{x}, \mathbf{s})$. The form of the selected model, along with the statistics of the model parameters, and an exhaustive explanation on the derivation of $\widetilde{\mathbf{y}}(\mathbf{x}, \mathbf{s})$ can be found in Contento et al. (2018).

Fig. 2.6 show the results of the fitting obtained for the observations in Groups 1 and 2. The dots in Fig. 2.6 are the predicted heights versus the actual observed heights, while the diagonal lines are the 1:1 lines that represent perfect predictions. The predictions for both Groups 1 and 2 are unbiased.

2.4.3 Water Depth Predictions

For each set of location and hurricane characteristics, the model provides predictions in terms of the probability of having a wet location, $\widetilde{p}_{ij}(\mathbf{x}_{j_i}, \mathbf{s}_i)$, and the marginal distribution of the storm-surge height

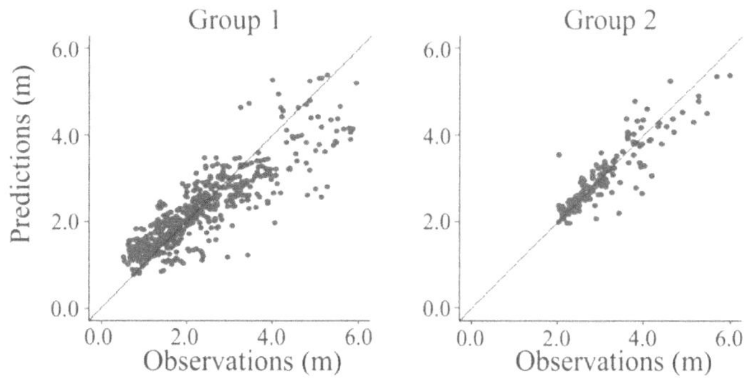

Fig. 2.6 Fitting of the training data for the random field model y_{ij}.

$\widetilde{y}_{ij}(\mathbf{x}_{j_i}, \mathbf{s}_i)$. Since the distribution of the storm-surge height refers to the mean sea level, alone it is not an exhaustive piece of information and, at the same time, does not convey an intuitive representation of the flooding risk. To solve this issue, an option is to provide results in terms of water depth, $k_{ij}(\mathbf{x}_{j_i}, \mathbf{s}_i)$, at the site. The prediction for the water depth can be obtained by subtracting the altitude of the prediction location x_{3j_i} from the prediction of the storm-surge height as:

$$\widetilde{k}_{ij}(\mathbf{x}_{j_i}, \mathbf{s}_i) = \widetilde{y}_{ij}(\mathbf{x}_{j_i}, \mathbf{s}_i) - x_{3j_i} \tag{2.5}$$

Consequently, with this transformation, a prediction is completely defined by the two values $\left[\widetilde{p}_{ij}(\mathbf{x}_{j_i}, \mathbf{s}_i), \widetilde{k}_{ij}(\mathbf{x}_{j_i}, \mathbf{s}_i)\right]$.

We can now look at the accuracy of the calibrated models in predicting the likelihood that a location is wet and the corresponding water depth (if a location is wet) for different locations than then ones used in the models calibration (but for the same hurricanes), and for different hurricanes. The first set of predictions is made considering the observations in Group B (described in Section 2.4.1). In this case, the predictions are made for hurricane characteristics taken from the initial set of 25 hurricanes chosen for the model calibration, but for locations different from those used to develop the model. The second set of predictions is made considering the observations in Group C. In this second case, the hurricane characteristics are different from those in the data used for the model calibration.

2.4.3.1 Predictions at Different Locations

Fig. 2.7 shows the results for different locations (but the same hurricanes). The dots show the median predictions versus the observations, while the bars are the corresponding one standard deviation confidence intervals. The diagonal lines represent the perfect correspondence between the predictions and observations. For each prediction, the lower is the probability that the location is wet, the dimmer is the colour of the dot and corresponding bar. In this way, it is possible to combine the information from the logistic regression model and the random field.

Fig. 2.7 shows the predictions corresponding to the data in Groups 1 and 2. In general, the values of the observations and predictions for the locations in Group 2 are higher than those in Group 1. The reason is that these locations are those on the riverbed, and the initial water height is summed to the storm surge when considering the water depth (the altitude is negative).

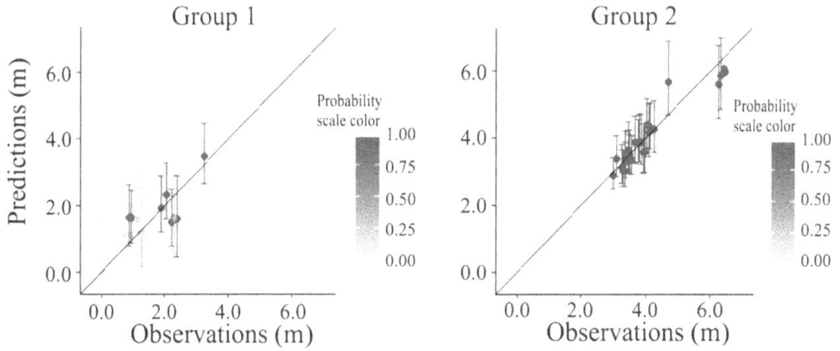

Fig. 2.7 Predictions of water depth k_{ij} due to hurricanes used to calibrate the model (different locations).

In Fig. 2.7, the darker colour of the markers show that generally for Group 1 the model predicts higher values of the water depth with higher confidence. This tendency of the model means that the locations with higher values of water depth, for which the risk of damage is higher, are more likely correctly identified. Fig. 2.7 also shows that the locations of the observations in Group 2 are all correctly predicted as wet (consistently with the results shown in Fig. 2.5). The mean values of the ILSA-based random field predictions are generally more accurate (closer to the 1:1 line) than those for the locations in Group 1.

2.4.3.2 Predictions for Different Hurricanes

The predictions for hurricane scenarios not used in the model calibration are made using the four hurricanes with characteristics shown in Table 2.3.

Figs. 2.8 and 2.9 show the predictions of the water depth obtained for the four additional hurricane scenarios. The meaning of markers, bars, and colours is the same explained in Fig. 2.7.

Generally, we can make similar observations as those made for Fig. 2.7. The highest values of water depth are predicted with good accuracy and, at the same time, the related locations have a high probability of being wet. The accuracy in the prediction increases for the observations belonging to Group 2. The predictions presented in Section 2.4.3 do not aim to exemplify an exhaustive study on the effects of climate change on storm surge but they show that the probabilistic model presented in Contento et al. (2018) is a suitable tool that can be used for probabilistic analyses on storm surge considering or not the effects of climate change.

Table 2.3 Hurricane Characteristics

	Hurricane characteristics					
Hurricane	s_1 (°lon)	s_2 (°lat)	s_3 (rad)	s_4 (hpa)	s_5 (m/s)	s_6 (km)
1	−76.162	34.856	5.16	989.81	5.02	52.80
2	−75.518	35.112	5.75	986.83	3.13	47.77
3	−75.736	35.142	5.22	988.57	5.22	38.85
4	−75.695	35.156	5.16	994.82	5.94	36.29

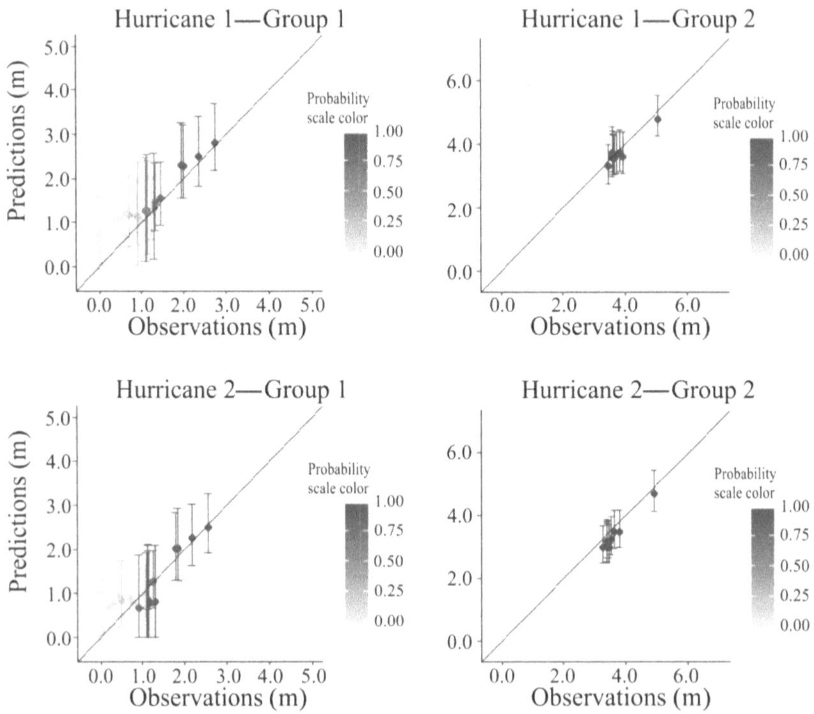

Fig. 2.8 Predictions of water depth k_{ij} for different hurricanes characteristics (Hurricane 1 and 2).

2.5 Development of Mitigation and Adaptation Strategies

This section describes in more detail the use of the results from the storm-surge model (described in Section 2.4 and part of Step 2 of the general framework for hurricane risk analysis) and of the results from the loss estimates from Step 3 of the framework.

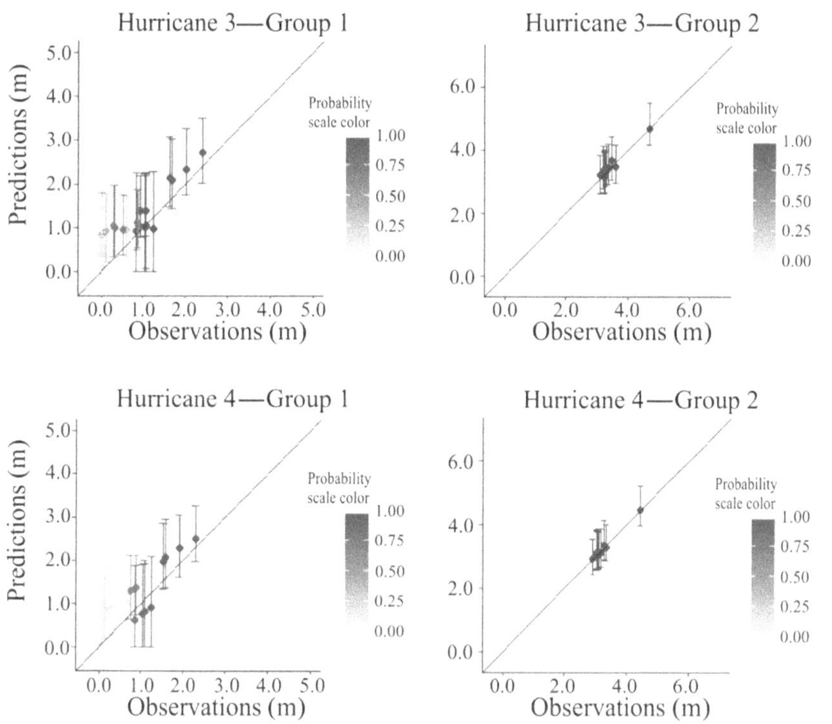

Fig. 2.9 Predictions of water depth k_{ij} for different hurricanes characteristics (Hurricane 3 and 4).

2.5.1 Use of Storm-Surge Predictions

Storm-surge predictions can be used in the form of hazard maps that provide information (e.g., the base flood elevation) for the choice of mitigation and adaptation strategies. At the level of the built or modified natural environment, hazard maps can be used to guide in the development of adaptation strategies. Such strategies could be both interventions on structural elements (e.g., elevating buildings on piles, installing breakaway walls, improving pile foundations, installing opening in foundation walls) and on nonstructural elements (e.g., elevating equipment, constructing flood-resistant equipment enclosures) (FEMA, 2005). The development of accurate hazard maps that account for the effects of climate change play an important role in defining the specific design characteristics of such adaptation strategies (i.e., the height of the elevated floors, and the characteristics of breakaway walls).

At the community level, hazard maps can inform on the effectiveness of mitigation strategies in terms of their ability to reduce the effects of climate change on the characteristics of the hazard. Hazard maps can also be used as a planning tool for environmental and human adaptation strategies. For the environmental adaptation strategies, hazard maps may help identify the most effective locations and characteristics of structures such as seawalls, levees and storm-surge barriers (FEMA, 2005), or green infrastructure such as wetland. For the human adaptation strategies, hazard maps may help establish evacuation routes, define high-risk areas where development should be avoided by enforcing specific zoning policies, and define the most suitable areas for relocation that could be a successful way to adapt to both sea-level rise and storm surge (Olshansky, 2018).

2.5.2 Use of Damage and Loss Estimations

Damage and loss distributions obtained from the third step of the framework can be used for the comparison of mitigation and adaptation strategies through the evaluation of their cost-effectiveness (Bjarnadottir et al., 2011), in decision models for life-cycle risk assessment of structures and infrastructure (Lee and Ellingwood, 2017; Gardoni et al., 2016), and in financial models used to device strategies to edge the risks (Grace et al., 2005; Jaffee et al., 2010; Hofer et al., 2018). In this section, we focus specifically on the use of the damage and loss estimations in financial models and specifically on insurance premium and CAT bond pricing. Insurance premium and CAT bond pricing, although different, are both directly affected by the risk associated to the policy (Grace et al., 2005) and the uncertainties in their estimation (Hofer et al., 2018). Modern pricing techniques can be used to devise effective risk hedging opportunities. To be effective, such techniques need accurate loss estimations that account for the underlying uncertainties.

The pricing of insurance premiums depends on several factors such as the required deductible, the desired coverage amount, and the risk associated to the policy. The deductible and the coverage amount are set a priori. On the other hand, the risk associated to the policy is more difficult to define because it depends on several aspects such as the kind of hazard and the characteristics of the insured property (e.g., type and location). Different insurance companies adopt different formulations for pricing the premiums. Using a generic formulation, Contento et al. (2017) compared the expected hurricane insurance premiums considering the present climate conditions as well as the RCP 8.5 climate change scenario. The results suggested an increase in the premium due to the effects of climate change that ranges from 5% to 20% depending from the type of building to be insured. There are several strategies to avoid excessive increases in premium. The most common strategies consist in acting directly on the premium by adding or increasing the deductible, transferring part of the insurance exposure to reinsurance companies, or transferring part of the insurance risks to the capital market using insurance-linked securities like CAT bonds (Ou Yang, 2010; Hofer et al., 2018).

CAT bonds are risk-linked securities that transfer a specified set of risks from a sponsor to investors. CAT bonds are structured as coupon-paying bonds with default linked to the occurrence of a trigger event. There are different types of trigger events for hurricane CAT bonds. Some of them are related to the occurrence and characteristics of the individual event (i.e., the occurrence of a hurricane with intensity higher than a predefined threshold), others are related to the consequences of the event (i.e., the exceedance of a set amount of losses for a single insurance or in a specific area of interest). In both cases the trigger event may be affected by the effects of climate change. For the first type of trigger events, the effects would be captured by the results of Steps 1 and 2. For the second type, the effects would be captured by the results of Step 3.

2.6 Summary

The terms 'mitigation' and 'adaptation' strategies are typically used with different meanings in the contexts of natural hazards and climate change. The chapter started by providing clear definitions of 'mitigation' and 'adaptation' strategies, and proposing a consistent use of these terms with dealing with natural hazards affected by climate change.

Then, the chapter presented a general framework for hurricane risk analysis. The framework has four steps: the modelling of the hurricane activity, the modelling of the hazard scenarios, the modelling of damage and losses, and the development of mitigation and adaptation strategies. This chapter presented a probabilistic model for storm-surge predictions that can be used in the second step of the framework. The chapter also showed the possible use of the information obtained by a risk analysis in the development of mitigation and adaptation strategies.

The probabilistic model for storm-surge predictions combines a logistic regression model and random field. The logistic regression model estimates the probability that a location is wet. The random field estimates the corresponding distribution of water depth. The main advantages of this model compared to those available in the literature are its computational efficiency, the possibility to be calibrated with data coming from both high-fidelity simulations and historical records, and its ability to provide predictions at locations different from those for which the model was calibrated. The chapter showed examples of predictions that have been obtained calibrating the model with data from high-fidelity model simulations. Such simulations were performed assuming the effects of the climate change scenario described by the RCP 8.5.

The last part of the chapter showed the possible uses of hazard maps (obtained in the second step of the framework), and of damage and loss estimations (obtained in the third step). Hazard maps are an important instrument for the choice and design of mitigation and adaptation strategies both at the level of the built or modified natural environment and at the community level. Damage and loss estimates can be used for the comparison of mitigation and adaptation strategies, in decision models for life-cycle risk assessment of structures and infrastructure, and in financial models. Specifically considering financial models, the chapter showed how the results of the risk analysis can inform the insurance premiums and the pricing of CAT bonds.

2.7 Potential Design and Practice Evolutions

The use of risk analyses could help the design process both at the level of the built or modified environment and the community. At the level of the built or modified environment level, it could help define specific design characteristics. For example, hazard maps providing the base flood elevation derived based on a selected climate change scenario could be adopted in the common practice. At the community level, risk analyses could help in land planning processes. However, current state-of-the-art tools (i.e., software) for hurricane risk analyses are largely research tools rather than design tools. The development of new design-oriented tools could help the evolution of the engineering practice.

2.8 Open Research Questions

This chapter shows a framework for hurricane risk analysis, as well as some of the most advanced theoretical models that can be used as part of such framework. However, there are several aspects within such framework that could be improved. One aspect is that the predictions of damages and losses carry a high uncertainty deriving from the choice of the climate change scenario. A higher control of greenhouse gasses emissions on one side and the improvement of climatological models on the other could reduce the uncertainty in the choice of the scenario to adopt and in the projections for a given scenario. The reduction of the uncertainty in the hazard description would consequently be reflected in the uncertainty in the damage and loss estimations. Another aspect is related to the models available in the literature and consequently included in the available software. In most of the cases, the dependence of the models in Steps 1 and 2 on climate change is restricted to the use of projections of SSTs. The reason is that changes in SSTs are projected with higher confidence than changes in most of the other climatological variables. Adding the dependence of the models on other climatological variables affected by climate change may help improve the accuracy of the predictions.

Acknowledgements

The research presented in this chapter was supported in part by the Centre for Risk-Based Community Resilience Planning funded by the US National Institute of Standards and Technology (NIST Financial Assistance Award Number: 70NANB15H044). The views expressed are those of the authors, and may not represent the official position of the sponsor.

References

Akaike, H., 1974. A new look at the statistical model identification. IEEE Trans. Autom. Control AC-19, 716–723.

Asher, M.J., Croke, B.F.W., Jakeman, A.J., Peeters, L.J.M., 2015. A review of surrogate models and their application to groundwater modeling. Water Resour. Res. 51, 5957–5973.

Batts, M.E., Cordes, M.R., Russell, L.R., Shaver, J.R., Simiu, E., 1980. Hurricane wind speeds in the United States. National Bureau of Standards, Report Number BSS-124, US Department of Commerce.

Bjarnadottir, S., Li, Y., Stewart, M.G., 2011. A probabilistic-based framework for impact and adaptation assessment of climate change on hurricane damage risks and costs. Struct. Saf. 33 (3), 173–185.

Booij, N., Holthuijsen, L.H., Ris, R.C., 1996. The SWAN wave model for shallow water. In: Proceedings of the 25th International Conference on Coastal Engineering, Orlando, USA. vol. 1. pp. 668–676.

Bretschneider, C.L., 1972. A non-dimensional stationary hurricane wave model. In: Proceedings of 1972 Offshore Technology Conference, Houston, Texas.

Bunya, S., Dietrich, J.C., Westerink, J.J., Ebersole, B.A., Smith, J.M., Atkinson, H., Jensen, R., Resio, R.A., Luettich, D.T., Dawson, C., Cardone, V.J., Cox, A.T., Powell, M.D., Westerink, H.J., Roberts, H.J., 2010. A high-resolution coupled riverine flow, tide, wind, wind wave, and storm surge model for southern Louisiana and Mississippi. Part I: model development and validation. Mon. Weather Rev. 138, 345–377.

Camargo, S.J., 2013. Global and regional aspects of tropical cyclone activity in the CMIP5 models. J. Climate 26, 9880–9902.

Chavas, D.R., Lin, N., Emanuel, K., 2015. A model for the complete radial structure of the tropical cyclone wind field. Part I: comparison with observed structure. J. Atmos. Sci. 72 (9), 3647–3662.

Contento, A., Chhabra, T., Gardoni, P., Dierer, S., 2017. Analysis of the effects of climate change on insurance against hurricanes. In the 12th Int. Conference on Structural Safety & Reliability (ICOSSAR2017), Vienna, Austria, August 2017.

Contento, A., Xu, H., Gardoni, P., 2018. A physics-based transportable probabilistic model for climate change dependent storm surge. In: Gardoni, P. (Ed.), Handbook of Sustainable and Resilient Infrastructure. Routledge.

Cooper, J.A.G., Pile, J., 2013. The adaptation-resistance spectrum: a classification of contemporary adaptation responses to climate-related coastal change. Ocean Coast. Manag. 94, 90–98.

Dietrich, J.C., Zijlema, M., Westerink, J.J., Holthuijsen, L.H., Dawson, C., Luettich, R.A., Jensen, R.E., Smith, J.M., Stelling, G.S., Stone, G.W., 2011. Modeling hurricane waves and storm surge using integrally-coupled, scalable computations. Coast. Eng. 58, 45–65.

Emanuel, K., 2004. Tropical cyclone energetics and structure. In: Fedorovich, E., Rotunno, R., Stevens, B. (Eds.), Atmospheric Turbulence and Mesoscale Meteorology. Cambridge University Press, Cambridge.

Emanuel, K., Rotunno, R., 2011. Self-stratification of tropical cyclone outflow. Part I: implications for storm structure. J. Atmos. Sci. 68, 2236–2249.

Emanuel, K., Ravela, S., Vivant, E., Risi, C., 2006. A statistical deterministic approach to hurricane risk assessment. Bull. Am. Meteorol. Soc. 87 (3), 299–314.

FEMA, 2005. Hurricane mitigation. In: A Handbook for Public Facilities. Department of Homeland Security, Federal Emergency Management Agency, Mitigation Division, Washington, D.C.

FEMA, 2009a. HAZUS-MH MR4 Hurricane Model Technical Manual. Department of Homeland Security, Federal Emergency Management Agency, Mitigation Division, Washington, DC.

FEMA, 2009b. HAZUS-MH MR4 Flood Model Technical Manual. Department of Homeland Security, Federal Emergency Management Agency, Mitigation Division, Washington, DC.

FEMA, 2012. Multi-Hazard Loss Estimation Methodology Hurricane Model Technical Manual. Department of Homeland Security, Federal Emergency Management Agency, Mitigation Division, Washington, DC.

Gardoni, P. (Ed.), 2017. Risk and Reliability Analysis: Theory and Applications. Springer, Cham.

Gardoni, P., Mosalam, K.M., Der Kiureghian, A., 2002. Probabilistic capacity models and fragility estimates for RC columns based on experimental observations. J. Eng. Mech. ASCE 128 (10), 1024–1038.

Gardoni, P., Guevara-Lopez, F., Contento, A., 2016. The life profitability method (LPM): a financial approach to engineering decisions. Struct. Saf. 63, 11–20.

Georgiou, P.N., 1985. Design Windspeeds in Tropical Cyclone-Prone Regions. PhD thesis, Faculty of Engineering Science, University of Western Ontario, London, Ontario, Canada.

Grace, M.F., Klein, R.W., Liu, Z., 2005. Increased hurricane risk and insurance market responses. J. Insur. Regul. 24 (2), 3–32.

Hofer, L., Gardoni, P., Zanini, M.A., 2018. CAT bond pricing and coverage design against natural perils. In: Gardoni, P. (Ed.), Handbook of Sustainable and Resilient Infrastructure. Routledge.

Holland, G.J., 1980. An analytic model of the wind and pressure profiles in hurricanes. Mon. Weather Rev. 108, 1212–1218.

Holland, G.J., Belanger, J.I., Fritz, A., 2010. A revised model for radial profiles of hurricane winds. Mon. Weather Rev. 138, 4393–4401.
Houston, S.H., Shaffer, W.A., Powell, M.D., Chen, J., 1999. Comparisons of HRD and SLOSH surface wind fields in hurricanes: implications for storm surge modeling. Weather Forecast. 14, 671–686.
IPCC, 2007. Climate models and their evaluation. In: Solomon, S., Qin, D., Manning, M., Chen, Z., Marquis, M., Averyt, K.B., Tignor, M., Miller, H.L. (Eds.), Climate Change 2007: The Physical Science Basis. Contribution of Working Group I to the Fourth Assessment Report of the Intergovernmental Panel on Climate Change. Cambridge University Press, Cambridge and New York, NY, pp. 589–662.
IPCC, 2013. Stocker, T.F., Qin, D., Plattner, G.-K., Tignor, M., Allen, S.K., Boschung, J., Nauels, A., Xia, Y., Bex, V., Midgley, P.M. (Eds.), Climate Change 2013: The Physical Science Basis. Contribution of Working Group I to the Fifth Assessment Report of the Intergovernmental Panel on Climate Change. Cambridge University Press, Cambridge and New York, NY.
Irish, J.L., Resio, D.T., Ratcliff, J.J., 2008. The influence of storm size on hurricane surge. J. Phys. Oceanogr. 38, 2003–2013.
Jaffee, D., Kunreuther, H., Michel-Kerjan, E., 2010. Long-term property insurance. J. Insur. Regul. 29, 167–187.
Jarvinen, B., Lawrence, M., 1985. An evaluation of the SLOSH storm surge model. Bull. Am. Meteorol. Soc. 66, 1408–1411.
Jelesnianski, C.P., Chen, J., Shaffer, W.A., 1992. SLOSH: sea, lake, and overland surges from hurricanes. NOAA Tech. Report NWS 48.
Jia, G., Taflanidis, A.A., 2013. Kriging metamodeling for approximation of high-dimensional wave and surge responses in real-time storm/hurricane risk assessment. Comput. Methods Appl. Mech. Eng. 261–262, 24–38.
Jia, G., Taflanidis, A.A., Nadal-Caraballo, N.C., Melby, J.A., Kennedy, A.B., Smith, J.M., 2016. Surrogate modeling for peak or time-dependent storm surge prediction over an extended coastal region using an existing database of synthetic storms. Nat. Hazards 81 (2), 909–938.
Johnson, D.R., Fischbach, J.R., Ortiz, D.S., 2013. Estimating surge-based flood risk with the coastal Louisiana risk assessment model. J. Coast. Res. Spec. Issue 67, 109–126.
Jolliffe, I.T., 2002. Principal Component Analysis, Series: Springer Series in Statistics, Second ed. Springer, NY.
Kendra, M.D., Xue, X., Xu, J., Wang, N., Kolar, R.L., Geoghegan, K.M., 2017. A coupled model system for hurricanes, storm surge and coastal flooding to support community resilience planning under climate change. In: 12th Int. Conference on Structural Safety & Reliability (ICOSSAR2017), Vienna, Austria, August 2017.
Kepert, J.D., 2010. Slab- and height-resolving models of the tropical cyclone boundary layer. Part I: comparing the simulations. Q. J. R. Meteorol. Soc. 136, 1686–1699.
Kim, S.W., Melby, J.A., Nadal-Caraballo, N.C., Ratcliff, J., 2015. A time-dependent surrogate model for storm surge prediction based on an artificial neural network using high-fidelity synthetic hurricane modelling. Nat. Hazards 76 (1), 565–585.
Knutson, T.R., Sirutis, J.J., Zhao, M., Tuleya, R.E., Bender, M., Vecchi, G.A., Villarini, G., Chavas, D., 2015. Global projections of intense tropical cyclone activity for the late twenty-first century from dynamical downscaling of CMIP5/RCP4.5 scenarios. J. Climate 28, 7203–7224.
Kossin, J.P., Emanuel, K.A., Camargo, S.J., 2016. Past and projected changes in western North Pacific tropical cyclone exposure. J. Climate 29, 5725–5739.
Langousis, A., Veneziano, D., 2009. Theoretical model of rainfall in tropical cyclones for the assessment of long-term risk. J. Geophys. Res. 114, D02106.

LaRow, T.E., Lim, Y.-K., Shin, D.W., Chassignet, E.P., Cocke, S., 2008. Atlantic basin seasonal hurricane simulations. J. Climate 21, 3191–3206.
Lee, J.Y., Ellingwood, B.R., 2017. A decision model for intergenerational life-cycle risk assessment of civil infrastructure exposed to hurricanes under climate change. Reliab. Eng. Syst. Saf. 159, 100–107.
Lee, K.H., Rosowsky, D.V., 2007. Synthetic hurricane wind speed records: development of a database for hazard analysis and risk studies. ASCE, Nat. Hazard. Rev. 8 (2), 23–34.
Li, Y., van de Lindt, J., Dao, T., Bjarnadottir, S., Ahuja, A., 2012. Loss analysis for combined wind and surge in hurricanes. Nat. Hazard. Rev. 13, 1–10.
Lin, N., Emanuel, K., Oppenheimer, M., Vanmarcke, E., 2012. Physically based assessment of hurricane surge threat under climate change. Nat. Clim. Chang. 2 (6), 462–467.
Lin, N., Emanuel, K., Vanmarcke, E., 2014. Hurricane risk analysis: a review on the physically-based approach. In: Deodatis, G., Ellingwood, B.R., Frangopol, D.M. (Eds.), Safety, Reliability, Risk and Life-Cycle Performance of Structures and Infrastructures. CRC Press, Boca Raton, FL, pp. 1291–1297.
Lonfat, M., Rogers, R., Marchok, T., Marks, F.D., 2007. A parametric model for predicting hurricane rainfall. Mon. Weather Rev. 135, 3086–3097.
Mattocks, C., Forbes, C., 2008. A real-time, event-triggered storm surge forecasting system for the state of North Carolina. Ocean Model. 25, 95–119.
Mizuta, R., Yoshimura, H., Murakami, H., Matsueda, M., Endo, H., Ose, T., Kamiguchi, K., Hosaka, M., Sugi, M., Yukimoto, S., Kusunoki, S., Kitoh, A., 2012. Climate simulations using MRI-AGCM3.2 with 20-km grid. J. Meteorol. Soc. Jpn. 90A, 233–258.
Mudd, L., Wang, Y., Letchford, C., Rosowsky, D., 2014. Assessing climate change impact on the US east coast hurricane hazard: temperature, frequency, and track. Nat. Hazard. Rev. 15 (3), 1–11.
Mudd, L., Rosowsky, D., Letchford, C., Lombardo, F., 2017. Joint probabilistic wind–rainfall model for tropical cyclone hazard characterization. J. Struct. Eng. 143(3) 04016195.
Murphy, C., Gardoni, P., McKim, R. (Eds.), 2018. Climate Change and Its Impacts. Springer, Cham.
Neumann, C.J., 1991. The National Hurricane Center Risk Analysis Program (HURISK). In: NOAA Technical Memorandum NWS NHC 38. National Oceanic and Atmospheric Administration (NOAA), Washington, DC.
Olshansky, R.B., 2018. Recovery after disasters: how adaptation to climate change will occur. In: Murphy, C., Gardoni, P., McKim, R. (Eds.), Climate Change and Its Impact: Risks and Inequalities. Springer, Cham.
Ou Yang, C., 2010. Managing Catastrophic Risk by Alternative Risk Transfer Instruments. Publicly Accessible Penn Dissertations, p. 220.
Phadke, A.C., Martino, C.D., Cheung, K.F., Houston, S.H., 2003. Modeling of tropical cyclone winds and waves for emergency management. Ocean Eng. 30, 553–578.
Powell, M.D., Vickery, P.J., Reinhold, T.A., 2003. Reduced drag coefficient for high wind speeds in tropical cyclones. Nature 422 (6929), 279–283.
Powell, M., Soukup, G., Cocke, S., Gulati, S., Morisseau-Leroy, N., Hamid, S., Dorst, N., Axe, L., 2005. State of Florida hurricane loss projection model: atmospheric science component. J. Wind Eng. Ind. Aerodyn. 93 (8), 651–674.
Resio, D.T., Westerink, J.J., 2008. Modeling the physics of storm surges. Phys. Today 61 (9), 33–38.
Roland, A., Cucco, A., Ferrarin, C., Hsu, T.-W., Liau, J.-M., Ou, S.-H., Umgiesser, G., Zanke, U., 2009. On the development and verification of a 2-D coupled wave-current model on unstructured meshes. J. Mar. Syst. 78, 244–254. Suppl. 1.
Russell, L.R., 1971. Probability distributions for hurricane effects. J. Waterw. Harb. Coast. Eng. Div. 97, 139–154.

Scawthorn, C., Flores, P., Blais, N., Seligson, H., Tate, E., Chang, S., Mifflin, E., Thomas, W., Murphy, J., Jones, C., Lawrence, M., 2006. Hazus-MH flood loss estimation methodology. II: damage and loss assessment. Nat. Hazard. Rev. 7 (2), 72–81.

Schmidt, A.M., Guttorp, P., O'Hagan, A., 2011. Considering covariates in the covariance structure of spatial processes. Environmetrics 22 (4), 487–500.

Schwerdt, R.W., Ho, F.P., Watkins, R.W., 1979. Meteorological criteria for standard project hurricane and probable maximum hurricane wind fields, Gulf and East Coasts of the United States. NOAA technical report NWS 23, US Department of Commerce.

Stevenson, A. (Ed.), 2010. Oxford Dictionary of English. third ed, Oxford University Press, Oxford.

Stewart, M.G., Val, D., Bastidas-Arteaga, E., O'Connor, A., Wang, X., 2014. Climate adaptation engineering and risk-based design and management of infrastructure. In: Frangopol, D.M., Tsompanakis, Y. (Eds.), Maintenance and Safety of Aging Infrastructure. Taylor and Francis Publishers, pp. 641–684.

Strachan, J., Vidale, P.L., Hodges, K., Roberts, M., Demory, M.-E., 2013. Investigating global tropical cyclone activity with a hierarchy of AGCMs: the role of model resolution. J. Climate 26, 133–151.

Thompson, E.F., Cardone, V.J., 1996. Practical modeling of hurricane surface wind fields. J. Waterw. Port Coast. Ocean Eng. 122 (4), 195–205.

Tuleya, R.E., DeMaria, M., Kuligowski, R.J., 2007. Evaluation of GFDL and simple statistical model rainfall forecasts for U.S. landfalling tropical storms. Weather Forecast. 22, 56–70.

Vickery, P.J., Twisdale, L.A., 1995. Prediction of hurricane wind speeds in the US. J. Struct. Eng. 121, 1691–1699.

Vickery, P.J., Skerlj, P.F., Steckley, A.C., Twisdale, L.A., 2000a. Hurricane wind field model for use in hurricane simulations. ASCE, J. Struct. Eng. 126 (10), 1203–1221.

Vickery, P.J., Skerjl, P.F., Twisdale, L.A., 2000b. Simulation of hurricane risk in the U.S. using empirical track model. J. Struct. Eng. 126, 1222–1237.

Vickery, P.J., Masters, F.J., Powell, M.D., Wadhera, D., 2009a. Hurricane hazard modeling: the past, present, and future. J. Wind Eng. Ind. Aerodyn. 97 (7–8), 392–405.

Vickery, P.J., Wadhera, D., Powell, M.D., Chen, Y., 2009b. A hurricane boundary layer and wind field model for use in engineering applications. J. Appl. Meteorol. Climatol. 48, 381–405.

Westerink, J.J., Luettich, R.A., Blain, C.A., Scheffner, N.W., 1994. ADCIRC: an advanced three-dimensional circulation model for shelves, coasts and estuaries. Report 2: Users Manual for ADCIRC-2DDI. Department of the Army, US Army Corps of Engineers.

Westerink, J.J., Luettich, R.A., Feyen, J.C., Atkinson, J.H., Dawson, C., Roberts, H.J., Powell, M.D., Dunion, J.P., Kubatko, E.J., Pourtaheri, H., 2008. A basin to channel scale unstructured grid hurricane storm surge model applied to southern Louisiana. Mon. Weather Rev. 136 (3), 833–864.

Xu, H., Gardoni, P., 2018. Improved latent space approach for modelling non-stationary spatial–temporal random fields. Spat. Stat. 23, 160–181.

Zhao, M., Held, I.M., Lin, S.-J., Vecchi, G.A., 2009. Simulations of global hurricane climatology, interannual variability, and response to global warming using a 50-km resolution GCM. J. Climate 22, 6653–6678.

Further Reading

Bornn, L., Shaddick, G., Zidek, J.V., 2012. Modeling nonstationary processes through dimension expansion. J. Am. Stat. Assoc. 107 (497), 281–289.

Burnecki, K., Kukla, G., 2003. Pricing of zero-coupon and coupon CAT bonds. Appl. Math. 30, 315–324.

Fouedjio, F., Desassis, N., Romary, T., 2015. Estimation of space deformation model for non-stationary random functions. Spat. Stat. 13, 45–61.
Fuglstad, G.-A., Simpson, D., Lindgren, F., Rue, H., 2015. Does non-stationary spatial data always require non-stationary random fields? Spat. Stat. 14 (Part C), 505–531.
Härdle, W.K., Cabrera, B.L., 2010. Calibrating CAT bonds for Mexican earthquakes. J. Risk Insur. 77 (3), 625–650.
Hubbert, G.D., McInnes, K.L., 1999. A storm surge inundation model for coastal planning and impact studies. J. Coast. Res. 15 (1), 168–185.
Katzfuss, M., 2013. Bayesian nonstationary spatial modeling for very large datasets. Environmetrics 24 (3), 189–200.
Lindgren, F., Rue, H., Lindström, J., 2011. An explicit link between Gaussian fields and Gaussian Markov random fields: the stochastic partial differential equation approach. J. R. Stat. Soc. Series B Stat. Methodology 73 (4), 423–498.
Mateu, J., Fernández-Avilés, G., Montero, J.M., 2013. On a class of non-stationary, compactly supported spatial covariance functions. Stoch. Env. Res. Risk A. 27 (2), 297–309.
Nychka, D., Wikle, C., Royle, J.A., 2002. Multiresolution models for nonstationary spatial covariance functions. Stat. Model. 2 (4), 315–331.
Paciorek, C.J., Schervish, M.J., 2006. Spatial modelling using a new class of nonstationary covariance functions. Environmetrics 17 (5), 483–506.
Reich, B.J., Eidsvik, J., Guindani, M., Nail, A.J., Schmidt, A.M., 2011. A class of covariate-dependent spatiotemporal covariance functions. Ann. Appl. Stat. 5 (4), 2425–2447.
Risser, M.D., Calder, C.A., 2015. Regression-based covariance functions for nonstationary spatial modelling. Environmetrics 26 (4), 284–297.
Sampson, P.D., Guttorp, P., 1992. Nonparametric estimation of nonstationary spatial covariance structure. J. Am. Stat. Assoc. 87 (417), 108–119.
van Vuuren, D.P., Edmonds, J., Kainuma, M., Riahi, K., Thomson, A., Hibbard, K., Hurtt, G.C., Kram, T., Krey, V., Lamarque, J.-F., Masui, T., Meinshausen, M., Nakicenovic, N., Smith, S.J., Rose, S.K., 2011. The representative concentration pathways: an overview. Clim. Change 109, 5–31.
Vanmarcke, E., 2010. Random Fields: Analysis and Synthesis. World Scientific, Singapore.

CHAPTER THREE

Climate Change Impact and Adaptation for Urban Drainage Systems

Yekenalem Abebe, Solomon Tesfamariam
University of British Columbia, School of Engineering, Kelowna, BC, Canada

3.1 Introduction

IPCC's Fifth Assessment Report concluded that, due to global warming, heavy precipitation events are expected to increase in frequency and intensity under climate change (Pachauri et al., 2015). In Canada, annual mean temperatures have increased by 1.6°C over the last 50 years (Environment and Climate Change Canada, 2016). Similarly, annual precipitation has increased from 5% to 35% in southern Canada over the last century (Groisman and Easterling, 1993). Changes in precipitation patterns towards more intense storms could lead to widespread pluvial flooding, aggravated by inappropriate land-use planning, increased paving, and loss of water storage space. Drainage networks that are designed based on historical climate regimes may be defunct in the future. Pluvial flooding occurs when extreme rainfall overwhelms drainage systems and the excess runoff cannot be absorbed (Maksimović et al., 2009). This could result in significant damage costs and environmental impact. The damage cost could further increase due to the synergetic effects of infrastructure ageing, climate change, and urbanisation. In 2016, more than $10 billion worth of stormwater assets was in poor or very poor condition across Canada, and with the current reinvestment level, a further decline in asset condition is expected (Canadian Infrastructure Report Card, 2016). On the other hand, a growing urban population requires new infrastructure development and alteration in land cover towards predominantly impervious surfaces. Such shifts will reduce the amount of water that infiltrates and it can intensify flood hazards (Harbor, 1994; Pauleit et al., 2005). Furthermore, impervious areas collect and accumulate pollutants and particles that are washed off during rainfall and pollutes receiving water bodies.

Climate Adaptation Engineering
https://doi.org/10.1016/B978-0-12-816782-3.00003-6

Restoring flooding and associated damages is expensive for property owners, municipalities, and insurance companies. Hence, there is a critical need to support municipalities in their decision-making processes to effectively manage stormwater infrastructure, prioritise investment decisions, and optimise cost while incorporating climate adaptation strategies in the process. To date, there have been various reported studies on climate change impact and adaptation of urban drainage systems. Waters et al. (2003) investigated the impact of a 15% increase in extreme rainfalls on a typical urban catchment in southern Ontario and proposed adaptation alternatives. Kirshen et al. (2014) presented a case study on Somerville's urban drainage system in Massachusetts. They used multicriteria scenario analysis to evaluate alternative adaptation strategies, based on the volume of flooding, volume of combined sewage discharged, and peak flow rate. Zhou et al. (2012) investigated three climate adaptation options in Odense, Denmark and presented a risk-based framework to assess the economic feasibilities of each option. Kang et al. (2016) studied alternatives to improve the drainage system in Incheon, Korea and proposed a seven-step decision-making framework. Berggren et al. (2011) investigated the hydraulic performance of urban drainage systems in future climate condition in Kalmar, Sweden. Semadeni-Davies et al. (2008) assessed the potential impacts of climate change on stormwater flows in Helsingborg, South Sweden. They concluded that urbanisation and projected increases in heavy rainfalls are set to increase flood risk and recommended the installation of sustainable urban drainage systems (SUDS) techniques to relieve the adverse impacts. Most of the reported studies provide detailed investigation on a selected section of a drainage network.

In this chapter, we present a three-phase framework that accounts for city-scale flood vulnerability assessment as well as detailed mitigation investigation. Furthermore, important steps in performance, hazard, and risk assessment are outlined. In addition, economic analysis of adaptation options as a way of selecting the most appropriate option is discussed. The proposed framework is described in detail in Section 3.2. In Section 3.3, the framework is applied on a case study area in Vernon, Canada.

3.2 Methodology

The proposed methodology outlines a comprehensive framework to assess climate change impact and plan adaptation alternatives for drainage infrastructure (Fig. 3.1). It follows a top-down approach that accounts for large-scale vulnerability assessment as well as detailed adaptation investigation. In addition,

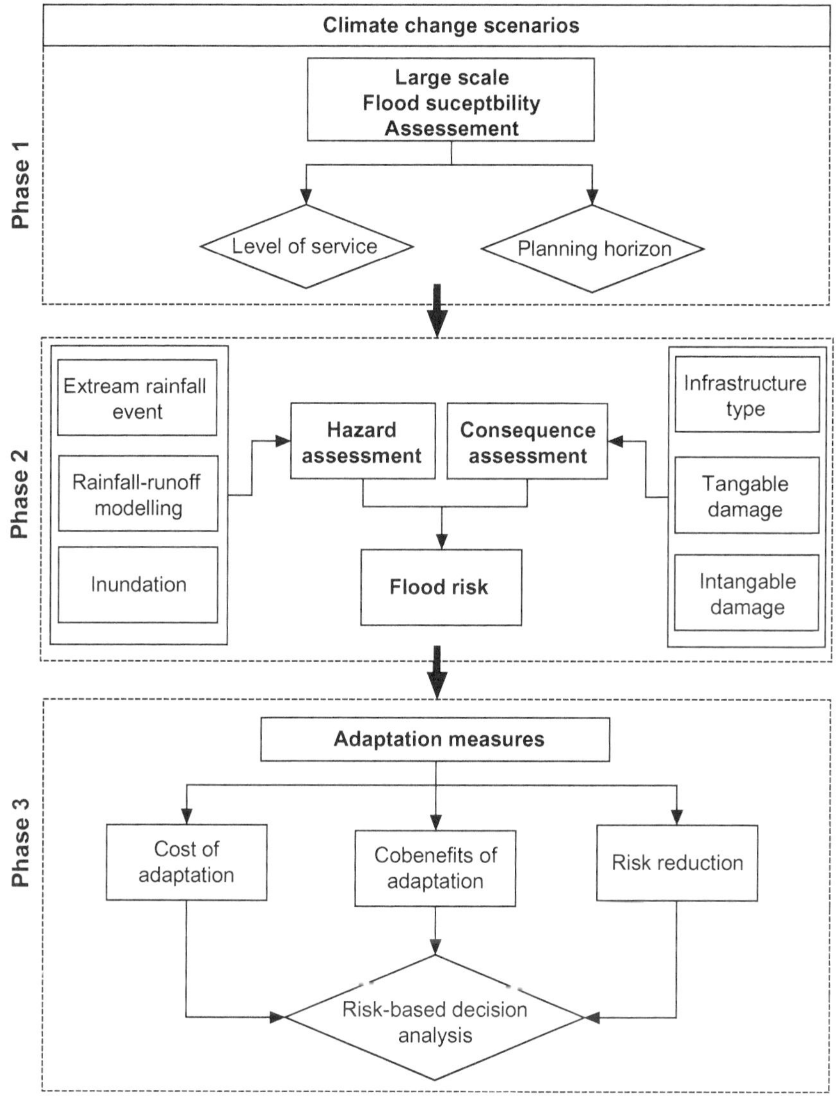

Fig. 3.1 Climate change impact and adaptation decision-making framework for urban drainage systems.

important steps in performance, hazard, and risk assessment are outlined and economic analysis of adaptation alternatives is discussed. The framework is subdivided into three phases:

- Phase 1: city-scale pluvial flood vulnerability assessment is carried out to identifying critical locations for further investigation. This step is crucial to focus resources and carry out detail studies only on critical sections of the drainage network. Furthermore, the planning horizon and the required level of service (LOS) need to be defined before proceeding to the next phase.

- Phase 2: detailed climate change impact assessment is carried out. The performance of the drainage system for current and future climate conditions is assessed using one-dimensional-two-dimensional (1D-2D) inundation simulation.
- Phase 3: adaptation measures relevant to urban drainage systems are identified and their respective performance in terms of risk reduction is assessed. Finally, the overall benefit of adaptation options is compared to choose among potential alternatives.

3.2.1 Phase One: City-Scale Vulnerability Assessment

In spite of increasing attention towards sustainable practices for stormwater management, adaption planning to pluvial flooding remains a challenging task due to the complexity of urban environment (Zhou et al., 2012). Large-scale modelling of an urban drainage network is difficult and time-consuming. Urban drainage systems consist of several underground structures (such as pipes and manholes) and overland flow routes (including roadside ditches, roadways, storage ponds, and swells). These interconnected structures often exhibit complex nonlinear dynamics, which makes it difficult to optimise the technical performance of the system and estimate the net benefits of adaptation measures. In addition, urban areas have a complex socioeconomic interaction that makes it difficult to quantify damage cost and risk that is required for adaptation planning.

In the first step of the framework, large-scale flood vulnerability assessment is proposed to identify critical locations that might be highly affected by climate change. This is crucial for municipalities to focus resources and carry out detailed investigations only on the critical neighbourhood. Thus, a flood vulnerability assessment approaches using a coupled GIS and Bayesian belief network (BBN) model that can capture the casual nexus between pluvial flood causative factors is proposed (Fig. 3.2). Details on the methodology and implementation are reported in Abebe et al. (2018). To construct the BBN model, important factors that can contribute to the occurrence of urban flooding are identified. The factors used are land cover, population density, slope, soil drainage class, drainage density, DEM, rainfall, and drainage capacity. The number of reported basement flooding and sewer backup damages can be used to represent the flood vulnerability index (FVI). At the

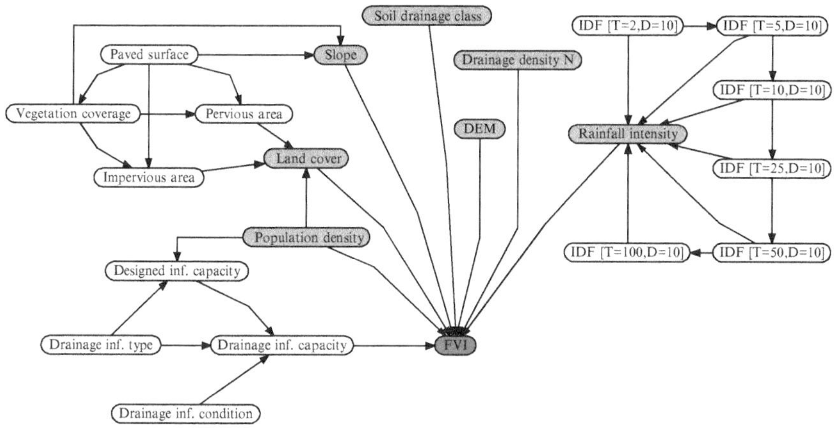

Fig. 3.2 BBN-based pluvial flood vulnerability assessment model (Abebe et al., 2018).

end of this phase, critical locations with high FVI will be identified for further analysis.

3.2.1.1 Level of Service

The required level of flood protection and water quality standards are the main considerations in urban drainage systems design. Stormwater drainage systems are composed of minor and major systems. The minor drainage system is designed to handle storms that are more frequent. Paths taken by runoff beyond the capacity of the minor system are called the major system. The major system is intended for the less frequent storm. It consists of natural waterways, temporary overload relief swales, and green areas. The major and minor systems are closely interrelated and their design and management need to be done in conjunction to achieve an overall stormwater management objective (Maksimović et al., 2009).

The performance of a drainage system can be measured by its hydraulic capacity to safely convey runoff through the catchment without causing flooding. Municipalities usually set a design standard based on the return period that states the required level of protection. For example, the city of Vernon, Canada recommends that the minor system should be capable to handle a five-year storm in low-density residential areas, while, the major system is designed to handle a 100-year storm (Bailey, 1992). For the storm sewer and the major overland flow path, this corresponds to the annual probability of having an overflow is 0.20 and 0.01, respectively. Therefore, the drainage system LOS can be defined by the precipitation return period where flood damage starts (Arnbjerg-Nielsen and Fleischer, 2009).

The ability of urban drainage systems to function adequately in a variety of conditions is an important system characteristic (Nanos and Filion, 2016). Climate change induced upsurge in extreme rainfall events can affect the performance of stormwater systems. To date, the design of stormwater infrastructure is based on an underlying assumption that the probability distribution of precipitation extremes is statistically stationary (Rosenberg et al., 2010). However, this assumption is invalid due to climate change (Srivastav et al., 2014). Therefore, the intensity and frequency of pluvial flood hazard can significantly increase in the future.

3.2.2 Phase Two: Climate Change Impact Assessment

Stormwater drainage infrastructure are planned based on local weather conditions. However, hydrologic data are no longer static over time due to climate change and a new design approach is required. The combination of climate change, land-use change, and infrastructure deterioration could exacerbate urban flooding. When municipalities look to improve existing infrastructure and build new stormwater systems, these changing conditions have to be considered in the planning process. In this section important steps in drainage systems performance, flood risk, and climate change impact assessment are discussed.

3.2.2.1 Hazard Assessment

Urban pluvial flooding process is a function of precipitation, runoff generation rate, overland, and sewer flow capacity of an area. Each of these hydrological and hydraulic processes must be analysed to assess flood hazard. This requires rainfall-runoff modelling, sewer flow modelling (1D simulation), and overland flow routing (1D or 2D simulation).

Extreme Rainfall Events

The first important step in hazard assessment is to determine the system loading or extreme rainfall events. Intensity-duration-frequency (IDF) curves are most commonly used to characterise precipitation events. The IDF curves provide precipitation accumulation depths for various return periods (T) and different durations (D). They are typically developed by fitting extreme value distributions, like Gumbel, generalised extreme value (GEV), or lognormal, to annual maximum precipitation (AMP) time series (Mirhosseini et al., 2013). However, IDF curves prepared based on historical data must be updated to reflect the expected shift in the future climate.

General circulation models (GCMs) and regional climate models (RCMs) provide the dynamics of the climate system and projections of future climate parameters (Eden et al., 2014). These models have low spatial resolution and do not provide the relevant temporal and spatial details required for drainage systems planning. Therefore, downscaling techniques (dynamic or statistical) are employed to prepare localised and high-resolution data. Statistical procedures generally formulate predictor-predicted relationships between GCMs and locally observed data (Herath et al. 2016). Various statistical downscaling procedures have been proposed for updating IDF curves (e.g. Hanssen-Bauer et al., 2005; Jacobeit et al., 2014; Nguyen et al., 2007; Wilby et al., 1998). In this study, equidistance quantile matching (EQM) method proposed by Srivastav et al. (2014) is used. The first step in EQM method is to formulate a statistical relationship between the observed station and the GCM historical (baseline) AMPs. In the second step, it establishes a statistical relationship between the GCM base period and the future projections. Finally, the relationship between the first two steps is used to update the IDF curves for future periods. More details on the methodology can be found in Sandink et al. (2017) and Srivastav et al. (2014).

Rainfall-Runoff Modelling

To quantify flood hazard, first the rainfall-runoff relationship must be established. This is a complex process and many models are available that vary in their accuracy (Lee et al., 2005; Rajurkar et al., 2004). The time-area method is one of the simplest and fastest alternative to estimate total runoff flow to manholes (O'Loughlin et al., 2010). This method is based on a rainfall excess hyetograph and time-area representation of the contributing catchment. Catchments are defined by the percentage of impervious area, time of concentration, and the time-area curve. The excess runoff depends on the initial losses from wetting and depression storage and continuous hydrological losses caused by evapotranspiration and interception.

Inundation Modelling

Inundation modelling can be based on physical-based models resolving continuity and momentum water equations or conceptual models that satisfy only continuity equation and use theoretical relationships among parameters (Ochoa-Rodríguez et al., 2013). Since physical-based models are computationally intensive and require longer processing time, conceptual models using GIS-based surface flow models and digital elevation model (DEM) are an alternative and simplified approach. However, this approach might affect forecast precision and it cannot predict water velocity (Liu and Pender, 2010). Examples of such methods include Rapid Flood Spreading Model (Lhomme et al., 2008; Wang et al., 2010), and Cellular Automata approaches (Guidolin et al., 2016). Van Dijk et al. (2014) compared these techniques with physical-based models and found that surface flow models prove to be an easy and fast alternative to identify vulnerable locations in hilly and flat areas. Whereas coupled 1D-2D simulation models give better results in areas at the transition between hill and flat. Pluvial inundation modelling techniques are evolving towards coupled 1D-2D simulation models, which can simulate the interaction between surface and sewer flows (Van Dijk et al., 2014). In this study, a coupled 1D-2D simulation is used to analyse the inundation extent for different climate change projection scenarios.

3.2.2.2 Consequence (Damage) Assessment

Flooding can have physical, social, or ecologic impacts (Merz et al., 2010). Tangible damages are quantified in monetary terms, whereas, intangible damages like social and ecological impact are difficult to estimate (Natural Resources Canada, 2017). Social impact assessment focuses on victims' psychological and emotional ordeal, while, ecological impact assesses the disruption to the natural ecosystem (Kubal et al., 2009). Methods for quantifying intangible damages are not well developed and most urban flood risk analysis studies consider only tangible damages (Lekuthai and Vongvisessomjai, 2001). Tangible damages can be direct damages that occur immediately or indirect damages that occur over time and can impact activity in surrounding areas (Thieken et al., 2009). Usually, indirect damages are estimated as a percentage of the direct damage estimation. Direct damages can be further grouped into structural, internal, or external (Natural Resources Canada, 2017). These damages can be assessed using depth damage, fragility, or loss curves, which describe the relationship between the depth of inundation and the economic loss. Since, it is impractical to prepare individual curves for all structures in a study area, a classification scheme that groups similarly behaving structures into one class is typically applied. These relationships can be established from historical flood events and damage data or can be synthesised from a detailed loss assessment on a sample structure. Flood damage is influenced by multiple factors, such as flow velocity, duration, contamination, etc., therefore, loss functions based only on inundation depth may have a large uncertainty (Thieken et al., 2009).

3.2.2.3 Flood Risk

Flood risk in a given planning horizon is determined by aggregating the damage costs resulting from various extreme rain events, based on their probability of occurrence (Olsen et al., 2015). High probability events have a lower damage cost, on the other hand, events that cause substantial damage may not occur frequently (see Fig. 3.3A). It is important to consider multiple simulations to cover a large range of return periods. Combination of these events with their corresponding return periods yields the risk-density curve $D(p)$ (Arnbjerg-Nielsen and Fleischer, 2009) (see Fig. 3.3B). The expected annual damage $E(L)$ can be computed by integrating the risk-density curve over all probabilities (p) and catchment area (A) as:

$$E(L) = \iint D(p)dpdA \tag{3.1}$$

where $D(p)$ is the damage for a given return period ($T = 1/p$). The integral is solved in the range between zero and infinite, $T \in [0, \infty]$. However, it is not practical to simulate all possible return periods. Consequently, various approximation techniques can be used to estimate $E(L)$ (Olsen et al., 2015). Numerical integration over all probabilities is one option, which requires the calculation of the associated risk for n relevant return periods. It is also possible to find analytical solution by assuming a log-linear relationship between damage cost and return period T. Simulated time series is another method to estimates the average expected damage costs for a given observation period (Olsen et al., 2015). Applying a log-linear relationship between damage cost and return period, it is possible to calculate $E(L)$ with only a few hazard simulations (Arnbjerg-Nielsen and Fleischer, 2009). In this study, the simulated time series method is used. This method estimates

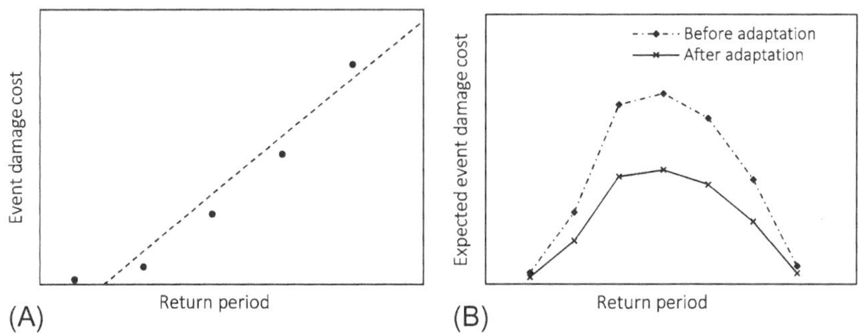

Fig. 3.3 Flood risk assessment (A) return period and damage cost relationship and (B) risk-density curve.

Table 3.1 Expected Annual Damage Calculation

Rank	**Return Period**	**Damage Cost for Event**
1	T_1	C_1
2	T_2	C_2
...	..	..
M	T_m	C_m
		$E(L)=\sum_{i=1}^{m}\frac{C_i}{n}$

the number of exceedance events with in a given period and aggregates the cost (Table 3.1). The rank and return periods of the events can be determined by using the median plotting position using Eq. (3.2):

$$T=\frac{1}{p}=\frac{n+0.4}{m-0.3} \tag{3.2}$$

where n is the duration of the simulated time series, T and p are the return period and the exceedance probability, respectively, m is the rank of the event. For example, when simulating a total study period of 100 years, and assuming the LOS for the stormwater drainage system to be $T=5$ years; there will be 18 flood events with the first ranked event having 129.19 return period and the last ranked event with 5.11 return period. Assuming a log-linear relationship between flood damage cost and return period all the flood events damage cost can be estimated as a multiple of the 100-year flood event. The 100-year and 5-year return period events are considered to have 100% and 0% damage cost, respectively. More details on the methodology can be found in Olsen et al. (2015) and Arnbjerg-Nielsen and Fleischer (2009).

3.2.3 Phase Three: Climate Change Adaptation

Although municipalities are facing severe and multifaceted challenges, several options can help to minimise climate change impacts. Climate change adaptation involves a variety of measures to reduce risk. The area between the risk-density curve before and after the adaptation measures represents the risk reduction due to the adaptation measures (Fig. 3.3B). This can be achieved by managing hazard to minimise the frequency and intensity of flood events or by improving the resiliency of vulnerable structures. The appropriate approach depends on the specific context of the study area. In general, adaptation strategies can be understood in four dimensions based on the intent, timing, spatial, and temporal scope (Fig. 3.4). The intent of the

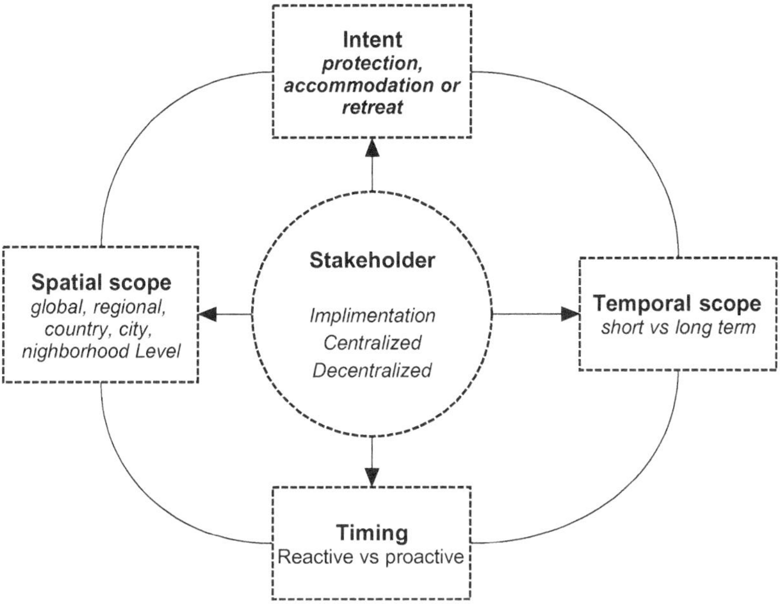

Fig. 3.4 Four dimensions of adaptation strategies.

adaptation could be protection, accommodation, or retreat (Zhou et al., 2011). Protection referrers to the construction of a barrier lessen the impacts of climate change, such as a seawall to protect against coastal flooding. Accommodation allows the flooding to occur in a managed manner. Examples of accommodation actions could include developing evacuation plans, building temporary shelters, and purchasing insurance. The last option is to retreat or move away from the hazard, for example, relocating from flood-prone areas.

Furthermore, climate change adaptation can be reactive or proactive (anticipatory) according to the timing of the action (Smit et al., 1999). Proactive adaptations are implemented before the impact is observed while reactive adaptation actions take place after. Reactive adaptation is often needed to address current impacts, while proactive adaptation is planned for a comprehensive long-term response. In most cases, proactive adaptations are more cost-effective than reactive adaptations. In addition, the scope of adaptation plans widely varies in temporal and spatial scale. Adaptation actions can be large scale or focused on a specific location. It is often desired to have adaptation actions with longer and wider effects. Furthermore, adaptation strategies should be flexible and resilient in terms of both temporal and spatial scope.

Available adaptation measures for urban drainage systems can be discussed under three groups: source control, system control, and downstream measures (Zhou, 2014). Source control helps minimise excess runoff upstream. This can be achieved by facilitating local infiltrations and storage systems. Typical examples include permeable pavements, green roofs, blue roofs, ponds, dry wells, and so on. Downstream control measures maximise the conveyance capacity and efficiency of the drainage system. Examples include pipe enlargement, detention ponds, or relief channels. System controls aim to reduce hazard impact on vulnerable structures, such measures can include topographic or land-use changes and flood defence structures. These adaptation options can be hard (structural) or soft (non-structural) measures depending on their physical impact on the environment.

Cost-benefit analysis is an important tool to understand the relative economic advantages of different adaptation alternatives (Bastidas-Arteaga and Stewart, 2015; see also Chapter 1). It is a systematic process for calculating the net benefits of a decision, policy or a project. Net present value (NPV) is one of the most popular analysis metrics. It computes the difference between the present value of cash inflows and outflows. A positive net present value indicates that the projected generated investment exceeds the anticipated costs. In this case, the risk reduction due to adaptation measures is compared against the investment needed to implement the adaptation measures. An investment with a larger NPV will always be preferred and if NPV is negative, it will result in a net loss. NPV is calculated as follows:

$$\mathrm{NPV} = \sum_{t_o}^{t_e} \frac{E(L)\Delta R - C_{\mathrm{adapt}}}{(1+r)^t} \tag{3.3}$$

where $E(L)$ is the expected annual damage, ΔR is the risk reduction due to adaptation measures, C_{adapt} is the investment cost of adaptation measures, r is the discount rate, and t_o and t_e are the start and end time of the analysis period, respectively.

3.3 Case Study

The city of Vernon is found in the Thompson Okanagan region of British Columbia, Canada (Fig. 3.5). It has a total population of around 40,000 and an area of 95 km^2. The storm drainage pipe network contains 3979 mains and 7414 service lines. Development is mainly concentrated

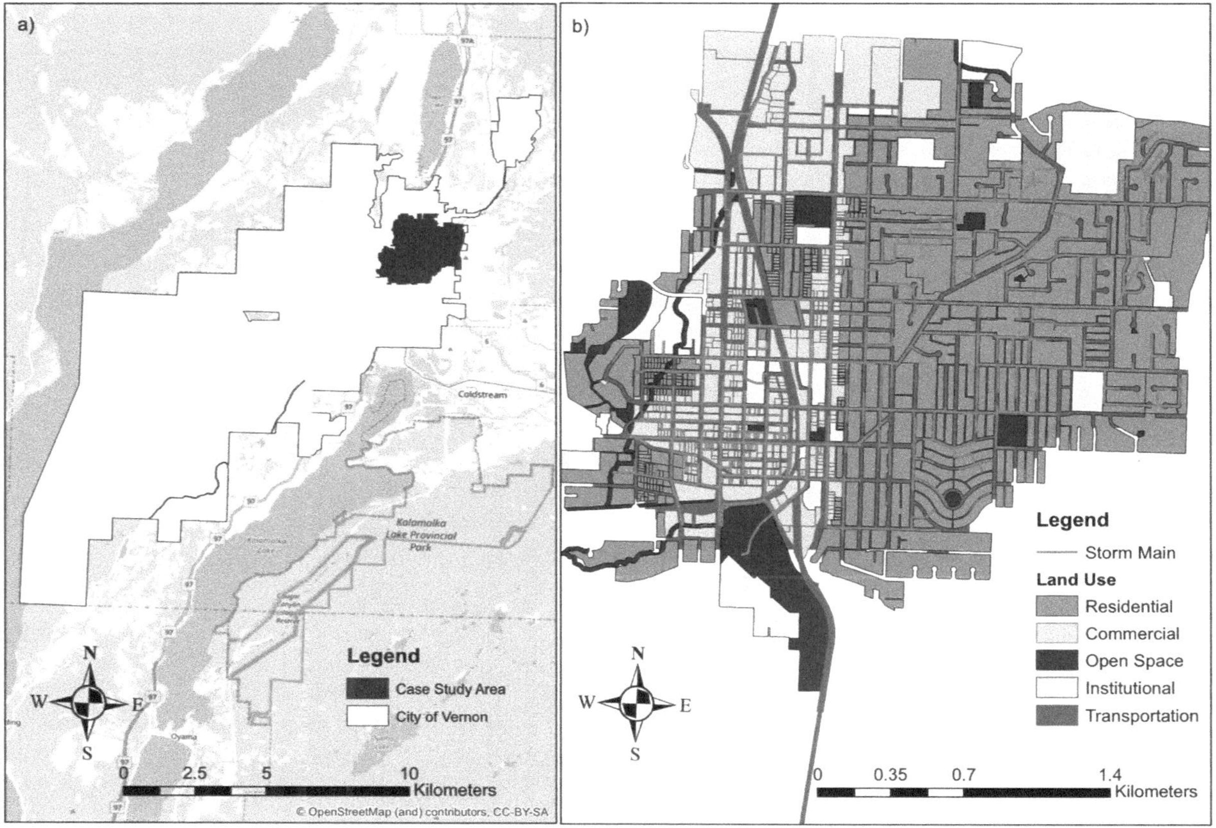

Fig. 3.5 Case study area (A) City of Vernon boundary and (B) close-up view of the case study area.

in the city centre and along the Okanagan lake coastlines. The city centre is located far from the lake and Vernon creek drains the city's storm network into the lake. For this case study, we have zoomed into the central part of the city, where a comprehensive storm network data is available (Fig. 3.5). Our study is limited to pluvial flooding in relation to stormwater drainage system capacity. The total catchment area considered in the study is 5.5 km^2. The catchment has mainly residential detached houses and commercial areas.

3.3.1 Stormwater Drainage System

Pipes, manholes, inlet, and outlet structures are the main components of storm sewer network. Pipes can be grouped into the trunk (main) and service pipes. A service line is smaller in diameter and is tributary to the trunk line. The study catchment contains 15 km of service pipes and 61 km of storm main. Manholes provide access to underground storm sewers for inspection and cleanout. They are provided where there is an abrupt change in direction, gradient, or pipe diameter and at major junctions where multiple storm drains converge. They can be made of either masonry or concrete, usually with cast iron or concrete covers. There are 836 manholes in the case study area. Storm drain inlets are used to collect urban runoff and discharge it to an underground sewer system. Inadequate attention to inlet capacity can cause an undue hazard to the security of many urban activities. There are three main types of inlet structures. Grate inlets consist of an opening in the gutter covered by screens. Curb-opening inlets are vertical openings in the curb covered by a top slab. Slotted drain inlets consist of a slotted opening with bars perpendicular to the opening. Slotted inlets function as weirs with flow entering from the side. Furthermore, there are 30 outfalls in the case study area that are located adjacent to Vernon Creek. Roadside gutters, ditches, and other overland open channel flow route are considered part of the major drainage system. Furthermore, stormwater best management practices (BMPs) are important components of a drainage system. BMPs provide water quality enhancement as well as improve the overall hydraulic capacity of the system. In the study area, there are 1545 catch basins and 9 drywells.

The performance of the stormwater drainage network is modelled using the commercially available software package MIKE FLOOD. MIKE FLOOD is a professional flood modelling application with 1D and 2D flood simulation engines. The core packages of the applications used in this study are MIKE URBAN for collection systems modelling and MIKE 21 for 2D surface flow. The sewer flow and overland flow are coupled at every manhole location. The drainage system performance and pluvial flood hazard are analysed for various climate change scenarios.

3.3.2 IDF Curves

Environment and Climate Change Canada (2017) prepares IDF curves for all weather stations across Canada based on historical rain records (Fig. 3.6). However, these curves need to be updated to reflect future changes in precipitation pattern. Pacific Climate Impact Consortium (PCIC) offers statistically downscaled daily Canada-wide climate scenarios, at a gridded resolution of roughly 10 km for the simulated period of 1950–2100 for three Representative Concentration Pathways-RCP2.6, RCP4.5, and RCP8.5 (PCIC 2013). Based on these downscaled models, existing IDF curves can be updated and used to investigate climate change impact on the study area's drainage system.

In this study, we used the web-based IDF updating tool for Canada developed by Simonovic et al. (2016). The new version of the tool utilised the spatially downscaled models developed by the PCIC and applied EQM method for temporal downscaling. They used GEV distribution to fit the AMP series and update the IDF curves. The updated IDF curves for the Vernon weather station are shown in Fig. 3.7. For example, a 2-year return period and 5-min rainfall intensity are 3.24, 3.46, 3.51, and 3.61 mm/h for historical, RCP2.6, RCP4.5, and RCP8.5 scenarios, respectively. This means that the rainfall intensity expected for the same return period will increase in the future. Similarly, the return period expected for a given rainfall intensity becomes shorter (Table 3.2). Interpolating between the historical and updated IDF curves, we can estimate that a 100-years storm event will occur every 43.8, 30.8, and 15.5 years for RCP2.6, RCP4.5, and RCP8.5 scenarios, respectively. This will have a significant impact on the intensity and frequency of pluvial flood hazard in the study area.

3.3.3 Inundation

Based on the IDF curves, a 1-h triangular hyetograph is constructed to represent the design storm for the different climate scenarios. The study catchment is subdivided into parcel polygons and the percentage of impervious area is estimated for each parcel. Each parcel is assumed to be connected to the nearest manhole in the network. The time of concentration for each parcel is calculated based on the shape of the subcatchments. Assuming a

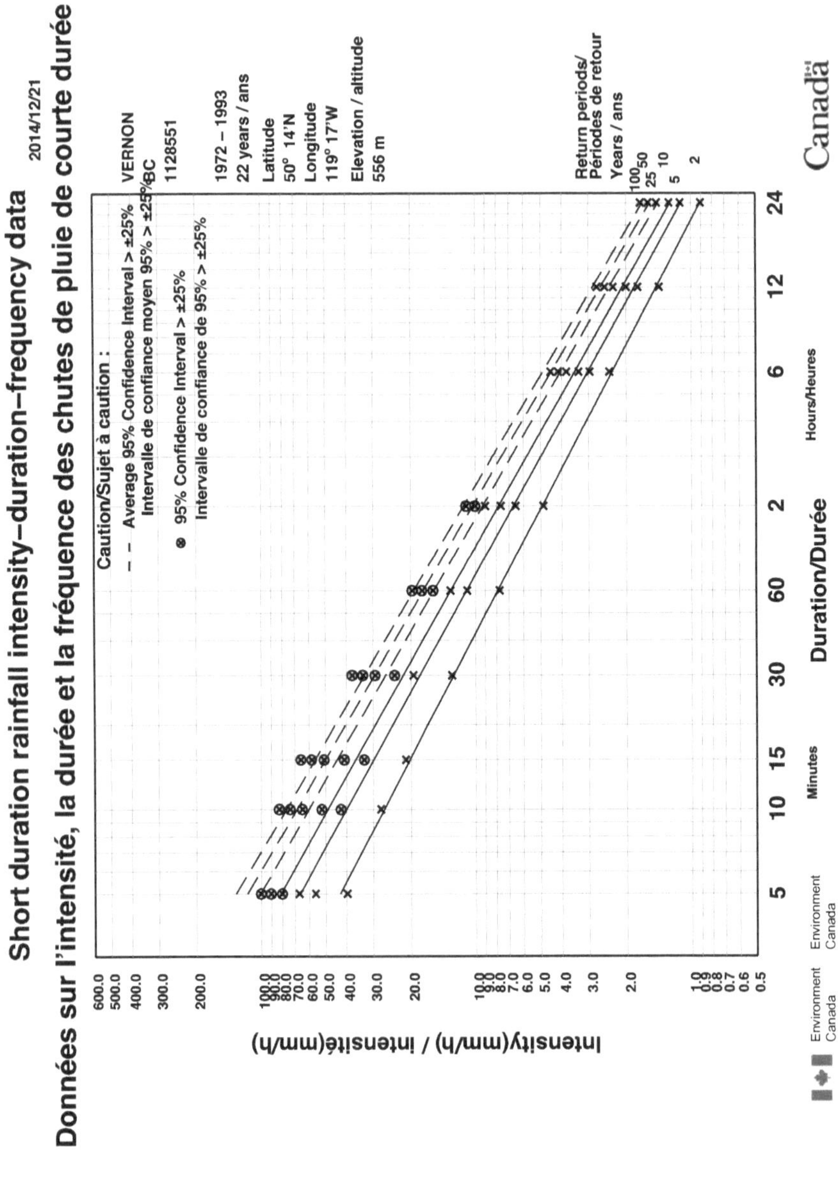

Fig. 3.6 IDF curves for Vernon, Canada (Environment and Climate Change Canada, 2017).

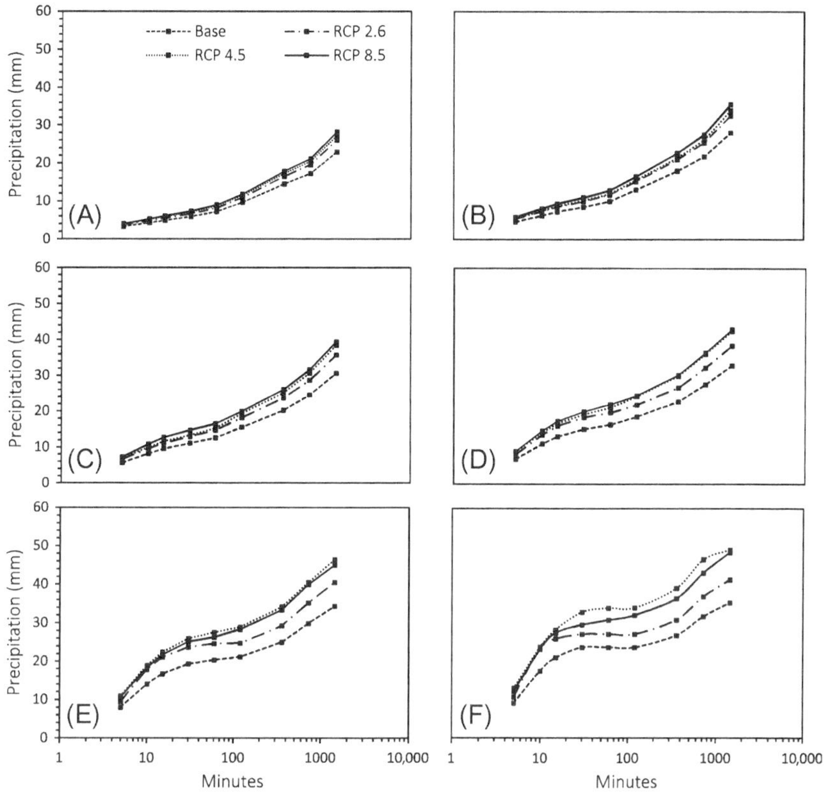

Fig. 3.7 Updated IDF curves for Vernon, Canada for return periods (A) 2 years, (B) 5 years, (C) 10 years, (D) 25 years, (E) 50 years, and (F) 100 years.

Table 3.2 Climate Change Impact on Return Period

Return Period	Today Base	Year 2096 RCP 2.5	RCP 4.5	RCP 8.5
T	100	43.8	30.8	15.5
LOS T	10	5.4	4.5	3.4

linear time-area curve representing the contributing part of the catchment surface as a function of time, the total runoff load to each manhole in the network is calculated. The overland flow from each overflowing manhole is routed using a 5-m resolution DEM and MIKE 21 Flood Screening Tool. Sample results for 50- and 100-year storms in base case scenario and for RCP 8.5 are plotted in Fig. 3.8. From Fig. 3.8, it can be seen that pluvial flood hazard will increase in future climate conditions.

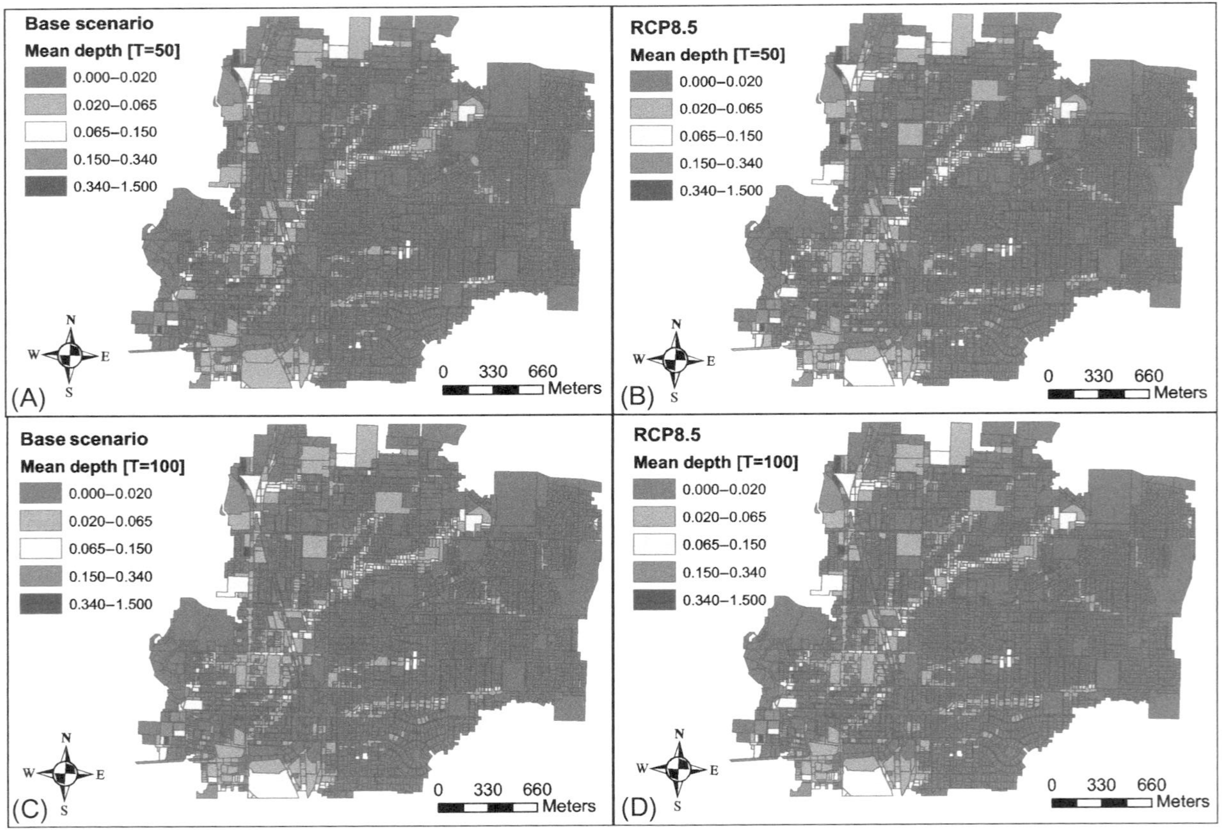

Fig. 3.8 Pluvial flood hazard for deferent scenarios (A) base scenario [$T=50$], (B) RCP 8.5 [$T=50$], (C) base scenario [$T=100$], and (D) risk-density curve.

Table 3.3 Expected Annual Damage Calculation for Base Case Scenario

Rank	*T*	% Cost	Expected Annual Damage *E*(*L*) [CAD]
1	129.1	111.1	$540,696
2	53.1	72.5	$353,167
3	33.4	52.4	$255,393
4	24.4	38.7	$188,801
5	19.2	28.40	$138,241
6	15.8	20.02	$97,471
7	13.4	13.00	$63,309
8	11.7	6.96	$33,908
9	10.3	1.66	$8,102

3.3.4 Cost of Climate Change

Assuming a 10-years LOS and planning horizon of 100 years, in the base case scenario, it is expected to have eight flood events in the next century, with the biggest event resulting in around $500,000 in damage cost and the minor event less than $10,000 (see Table 3.3).

Considering climate change, the number of flood event, as well as the damage cost, will increase significantly. The shift in return period based on the updated IDF curves is summarised in Table 3.2. Accordingly, the LOS will decrease in the future and more flooding events are expected. For example, considering RCP 2.5 scenario, there will be 17 flood events with the maximum damage cost of around $820,000. The event damage cost and expected total damage cost for the various climate scenarios considered are shown in Figs. 3.9 and 3.10.

3.3.5 Adaptation

Available adaptation methods can be discussed under the general concept of SUDS or more traditional sewer system expansion options, like pipe enlargement and underground storage tanks. In this study, we have considered two-adaptation options:

- *Option A*: it is assumed that individual households and commercial property owners will implement LID technologies to facilitate onsite infiltration and reduce peak runoff. LID technologies considered are drywells; green roofs, and permeable pavements. It is assumed that the total impervious area in the study area catchment will reduce by half.

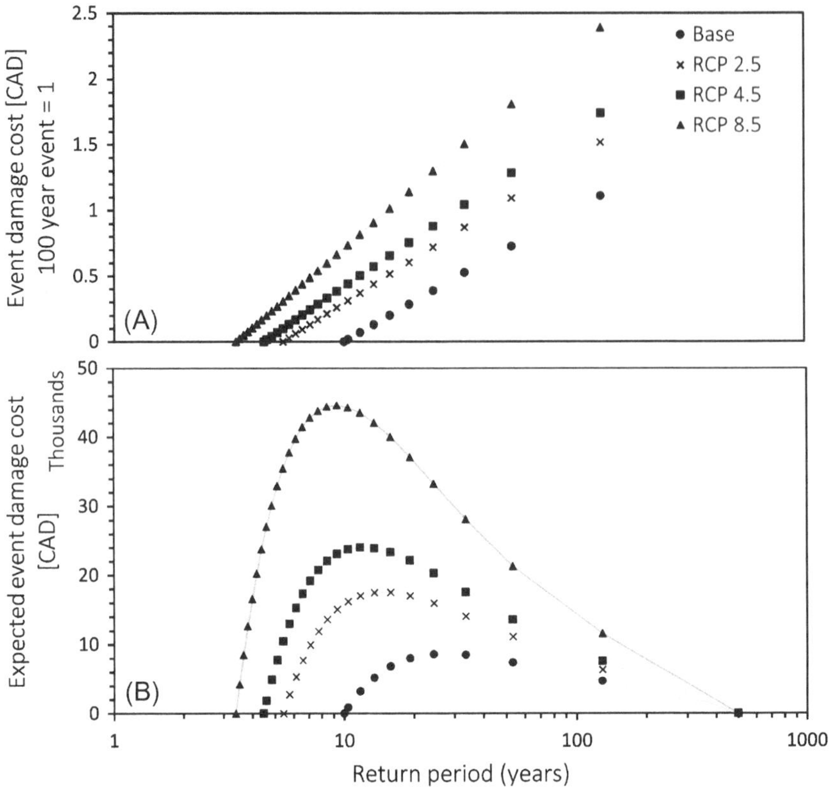

Fig. 3.9 Pluvial flood risk assessment (A) event damage cost and return periods, (B) risk-density curve.

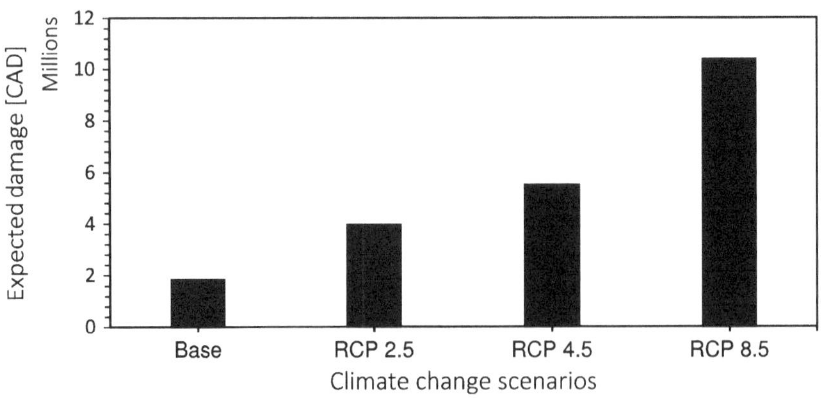

Fig. 3.10 Expected damage cost for different climate change projection scenarios.

- *Option B*: the municipality will construct underground storage tanks to relief the surcharge from critical nodes.

A NPV analysis is carried out for the two-adaptation options considered based on the RCP 4.5 scenario. For option A, the investment cost for each household and commercial site is estimated based on an average initial cost required to install green roofs, drywells, and permeable pavement. For option B, the municipality must upgrade the drainage system by installing storage tanks to relieve overflow runoff from critical manholes. Both adaptation options will help reduce the number of flood events, expected damage per event, and in general flood risk. For example, the damage cost for the first ranked flood event will reduce by 90% and 95% for adaptation options A and B, respectively. However, the NPV for option B is lower and option A provides the most benefit relative to the initial capital cost (Table 3.4). The event damage cost and expected annual damage for the various adaptation alternatives are shown in Fig. 3.11. The area between the risk-density curve before and after the two-adaptation alternatives represents the risk reduction due to the adaptation measures (Fig. 3.11B).

Table 3.4 NPV for Adaptation Measures

		NPV, in Million [CAD]	
Adaptation Options	**Investment Cost, in Million [CAD]**	**Discount Rate (1%)**	**Discount Rate (0.75%)**
(A) Infiltration	$51	$3.0	$19.4
(B) Storage tank	$60	−$8.2	$7.3

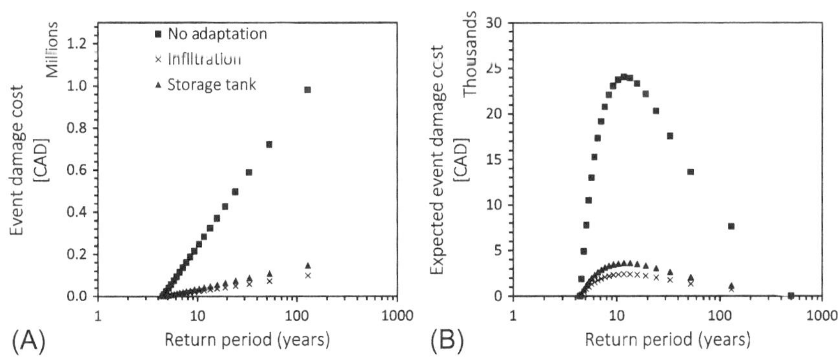

Fig. 3.11 Adaptation alternatives (A) event damage cost and return periods (B) risk-density curve.

3.4 Summary

There is a consensus regarding climate change and its impact on precipitation patterns. The expected increase in intensity and frequency of precipitation events can intensify urban flooding and sewer backup in the future. Urban drainage systems are the first line of defence against pluvial flooding. However, their design has been widely based on the exceedance probability of historical rainfall events. The underlying assumption that the probability distribution of precipitation extremes is statistically stationary is invalid. Therefore, municipalities need to investigate the potential impact of climate change and plan appropriate adaptation alternatives. In this chapter, we presented a generic framework for climate change impact and adaptation planning of stormwater drainage infrastructure. Important steps in performance assessment, hazard, and risk quantification are discussed. The impacts of climate change on the stormwater system is assessed using 1D-2D simulations for present conditions as well as three climate projection scenarios. There are a wide range of adaptation measures available to reduce the expected impact of climate change, including SUDS—also referred to as BMP or LID. SUDS are gaining wide acceptance due to their natural approach to managing stormwater and effectiveness in improving water pollution and reducing flood susceptibility. Other structural adaptation options, like constructing underground storage tanks or enlargement of pipe size are also viable alternatives. The decision regarding the type of adaptation measure to implement should be based on economic analysis of all feasible options.

3.5 Potential Design and Practice Evolution

Climate change is a big challenge for municipalities across Canada and adapting stormwater systems to cope with the increasing precipitation is crucial to building a resilient community. Effective adaptation planning will require detail investigation on the magnitude and location of the expected impacts. Furthermore, accurate information on the cost and performance of adaptation alternatives like green infrastructure is needed. However, the process of incorporating climate change into existing planning processes can be a difficult task for local decision-makers. Municipalities have a constrained budget and may lack the technical expertise to deal with climate change. Uncertainty in future climate projections and the absence of compressive data on the costs and benefits of adaptation alternatives further complicate the process. For example, the biggest challenge in the case study discussed in this chapter was the lack of accurate and verifiable data to validate the model outputs. Consequently, various assumptions have to be made for the model development and cost-benefit analysis. To overcome such challenges better and more accessible information on the performance of green infrastructure and other adaptation strategies is important. Furthermore, municipal capacity building and training on the full value of green infrastructure and how to integrate it into other projects, such as highway improvements, infrastructure rehabilitation, and urban redevelopment programs is crucial.

Considering future climate conditions in the planning and management process of urban drainage systems is important. Most design guidelines currently stipulate the use of existing IDF curves that are prepared from historical rainfall data. It will be difficult to completely base the planning procedures on predicted future precipitation patterns due to the uncertainties in climate modelling. However, both historical and update IDF curves can be used together to make more robust decisions. For new development sites, municipalities should consider climate change impact assessment and adaptation options as one part of the plan approval process.

Stormwater infrastructure designs in the face of climate change can generally follow three main principles. They can be designed using historic flows, without any adjustment to design flows to consider climate change and accept a gradual decrease in the LOS over the planning horizon. This approach can be acceptable when the expected flood damage is minor. On the other hand, the design can be based on predicted future flows, adjusted from climate change projection scenarios. This can provide enough hydraulic capacity throughout the design life of the infrastructure, however, it could require high sunk cost and financial burden to the developer early on. Another alternative is to design for the present climate with sufficient flexibility to allow the incorporation of additional capacity in the future. Planning for a shorter service life and retrofit when conditions necessitate provides better flexibility to manage uncertainty. However, upgrade costs can still be substantial. Therefore, using a combination of the three approaches for different drainage components can provide optimal results.

3.6 Open Research Questions

Further work in the area of stormwater drainage system adaptation should focus on:

- Performance assessment of adaptation measures and BMPs
- Economic evaluation and cobenefits of adaptation measures like green roofs, pervious pavements, and other infiltration systems
- Improvement in flood loss estimation and fragility curves
- Downscaling of climate data and updating IDF curves at the urban catchment levels
- Risk of storm systems failure considering both structural deterioration and hydraulic capacity reduction due to climate change
- Stormwater infrastructure (pipes, manholes, etc.) deterioration modelling considering climate variables and other environmental factors
- Climate change impact on operation and maintenance of stormwater infrastructure such as storm drain flushing/cleaning frequency
- Design optimisation of drainage systems taking into account specific environmental conditions and potential climate change scenarios

Acknowledgements

The financial support through MITACS and Natural Sciences and Engineering Research Council of Canada (NSERC) Discovery Grant Program (RGPIN-2014-05013) is acknowledged.

References

Abebe, Y., Kabir, G., Tesfamariam, S., 2018. Assessing urban areas vulnerability to pluvial flooding using GIS applications and Bayesian belief network model. J. Clean. Prod., Elsevier 174, 1629–1641.

Arnbjerg-Nielsen, K., Fleischer, H., 2009. Feasible adaptation strategies for increased risk of flooding in cities due to climate change. Water Sci. Technol. 2 (60), 273–281.

Bailey, M., 1992. Regulations, standards and specifications for the design and installation of drainage systems. https://www.vernon.ca/sites/default/files/docs/bylaws/schedule_f.pdf%3e. (Oct. 20, 2017).

Bastidas-Arteaga, E., Stewart, M.G., 2015. Damage risks and economic assessment of climate adaptation strategies for design of new concrete structures subject to chloride-induced corrosion. Struct. Saf., Elsevier 52, 40–53.

Berggren, K., Olofsson, M., Viklander, M., Svensson, G., Gustafsson, A.-M., 2011. Hydraulic impacts on urban drainage systems due to changes in rainfall caused by climatic change. J. Hydrol. Eng. 17 (1), 92–98.

Canadian Infrastructure Report Card, 2016. Stormwater Infrastructure. http://www.canadainfrastructure.ca/en/%3e. (Oct. 12, 2017).

Eden, J.M., Widmann, M., Maraun, D., Vrac, M., 2014. Comparison of GCM- and RCM-simulated precipitation following stochastic postprocessing. J. Geophys. Res. Atmos. 119 (19), 11,040–11,053.

Environment and Climate Change Canada, 2016. Canadian Environmental Sustainability Indicators. https://www.canada.ca/en/environment-climate-change/services/environmental-indicators/temperature-change.html%3e. (Oct. 30, 2017).
Environment and Climate Change Canada, 2017. Engineering Climate Datasets. http://climate.weather.gc.ca/prods_servs/engineering_e.html%3e. (Apr. 19, 2017).
Groisman, P.Y., Easterling, D.R., 1993. Variability and trends of total precipitation and snowfall over the united state and Canada. J. Climate 7 (1), 184–204.
Guidolin, M., et al., 2016. A weighted cellular automata 2D inundation model for rapid flood analysis. Environ. Model. Softw. 84, 378–394.
Hanssen-Bauer, I., Achberger, C., Benestad, R., Chen, D., Førland, E., 2005. Statistical downscaling of climate scenarios over Scandinavia. Climate Res. 29 (3), 255–268.
Harbor, J.M., 1994. A practical method for estimating the impact of land-use change on surface runoff, groundwater recharge and wetland hydrology. J. Am. Plan. Assoc., Taylor & Francis 60 (1), 95–108.
Jacobeit, J., Hertig, E., Seubert, S., Lutz, K., 2014. Statistical downscaling for climate change projections in the Mediterranean region: methods and results. Reg. Environ. Change, Springer 14 (5), 1891–1906.
Kang, N., Kim, S., Kim, Y., Noh, H., Hong, S., Kim, H., 2016. Urban drainage system improvement for climate change adaptation. Water 8 (7), 268.
Kirshen, P., Caputo, L., Vogel, R., Mathisen, P., Rosner, A., Renaud, T., 2014. Adapting urban infrastructure to climate change: a drainage case study. J. Water Resour. Plan. Manag. 141 (4), 1–11.
Kubal, C., Haase, D., Meyer, V., Scheuer, S., 2009. Integrated urban flood risk assessment—adapting a multicriteria approach to a city. Nat. Hazards Earth Syst. Sci. 9, 1881–1895.
Lee, H., McIntyre, N., Wheater, H., Young, A., 2005. Selection of conceptual models for regionalisation of the rainfall-runoff relationship. J. Hydrol. 312 (1–4), 125–147.
Lekuthai, A., Vongvisessomjai, S., 2001. Intangible flood damage quantification. Water Resour. Manag. 15, 343–362.
Lhomme, J., Sayers, P.B., Gouldby, B.P., Samuels, P.G., Wills, M., Mulet-Marti, J., 2008. Recent development and application of a rapid flood spreading method. In: Food Risk Management: Research and Practice, Oxford, UK.
Liu, Y., Pender, G., 2010. A new rapid flood inundation model. In: Proceedings of the first IAHR European Congress, 4–6.
Maksimović, Č., Prodanović, D., Boonya-Aroonnet, S., Leitão, J.P., Djordjević, S., Allitt, R., Allit, R., 2009. Overland flow and pathway analysis for modelling of urban pluvial flooding. J. Hydraul. Res. 47 (4), 512–523.
Merz, B., Kreibich, H., Schwarze, R., Thieken, A., 2010. Assessment of economic flood damage. Nat. Hazards Earth Syst. Sci. 10 (8), 1697–1724.
Mirhosseini, G., Srivastava, P., Stefanova, L., 2013. The impact of climate change on rainfall intensity-duration-frequency (IDF) curves in Alabama. Reg. Environ. Chang. 13 (1), 25–33.
Nanos, M.G., Filion, Y.R., 2016. Risk-based performance assessment of stormwater drainage networks under climate change: a case study in the city of Kingston. In: World Environmental and Water Resources Congress, pp. 73–81.
Natural Resources Canada, 2017. Canadian Guidelines and Database of Flood Vulnerability Functions. http://hazuscanada.ca/sites/all/files/nrc-canadianguidelines-final_2017-03-30_draft.pdf%3e. (Jun. 22, 2017).
Nguyen, V.-T.-V., Nguyen, T.-D., Cung, A., 2007. A statistical approach to downscaling of sub-daily extreme rainfall processes for climate-related impact studies in urban areas. Water Sci. Technol. Water Supply, IWA Publishing 7 (2), 183.
O'Loughlin, G., Chocat, B., Huber, W., 2010. Rainfall-runoff processes and modelling. J. Hydraul. Res. 34 (2015), 733–751.
Ochoa-Rodríguez, S., Onof, C., Maksimović, Č., Wang, L.-P., Willems, P., Van Assel, J., Gires, A., Ichiba, A., Bruni, G., Ten Veldhuis, M.-C., 2013. Urban Pluvial Flood

Modelling Current Theory and Practice. http://www.raingain.eu/sites/default/files/wp3_review_document.pdf%3e. (Feb. 15, 2017).
Olsen, A., Zhou, Q., Linde, J., Arnbjerg-Nielsen, K., 2015. Comparing methods of calculating expected annual damage in urban pluvial flood risk assessments. Water 7 (1), 255–270.
Pachauri, R.K., Meyer, L., Van Ypersele, J.-P., Brinkman, S., Van Kesteren, L., Leprince-Ringuet, N., Van Boxmeer, F., 2015. Climate Change 2014: Synthesis Report. In: Contribution of Working Groups I, II and III to the fifth assessment report of the Intergovernmental Panel on Climate Change, IPCC.
Pauleit, S., Ennos, R., Golding, Y., 2005. Modelling the environmental impacts of urban land use and land cover change - a study in Merseyside, UK. Landsc. Urban Plan., Elsevier 71 (2–4), 295–310.
Rajurkar, M.P., Kothyari, U.C., Chaube, U.C., 2004. Modelling of the daily rainfall-runoff relationship with artificial neural network. J. Hydrol. 285 (1–4), 96–113.
Rosenberg, E.A., Keys, P.W., Booth, D.B., Hartley, D., Burkey, J., Steinemann, A.C., Lettenmaier, D.P., 2010. Precipitation extremes and the impacts of climate change on stormwater infrastructure in Washington state. Clim. Change 102 (1–2), 319–349.
Sandink, D., Schardong, A., Simonovic, S.P., 2017. Computerized tool for the development of intensity-duration-frequency curves under changing climate—Technical Manual. Available from: https://www.idf-cc-uwo.ca/Manuals. (accessed 18 November 2018).
Semadeni-Davies, A., Hernebring, C., Svensson, G., Gustafsson, L.G., 2008. The impacts of climate change and urbanisation on drainage in Helsingborg, Sweden: Combined sewer system. J. Hydrol. 350 (1–2), 100–113.
Simonovic, S.P., Schardong, A., Sandink, D., Srivastav, R., 2016. A web-based tool for the development of intensity duration frequency curves under changing climate. Environ. Model. Softw., Elsevier Ltd 81, 136–153.
Smit, B., et al., 1999. The science of adaptation: a framework for assessment. Mitig. Adapt. Strat. Gl. 4.3-4 199–213.
Srivastav, R.K., Schardong, A., Simonovic, S.P., 2014. Equidistance quantile matching method for updating IDF curves under climate change. Water Resour. Manag. 28 (9), 2539–2562.
Thieken, A.H., Ackermann, V., Elmer, F., Kreibich, H., Kuhlmann, B., Kunert, U., Maiwald, H., Merz, B., Müller, M., Piroth, K., Schwarz, J., Schwarze, R., Seifert, I., Seifert, J., 2009. Methods for the evaluation of direct and indirect flood losses. In: 4th International Symposium on Flood Defence, Toronto, Canada, 6–8.
Van Dijk, E., Van der Meulen, J., Kluck, J., Straatman, J.H.M., 2014. Comparing modelling techniques for analysing urban pluvial flooding. Water Sci. Technol. 69 (2), 305–311.
Wang, C., Wan, T.R., Palmer, I.J., 2010. Urban flood risk analysis for determining optimal flood protection levels based on digital terrain model and flood spreading model. Visual Comput., Springer-Verlag 26 (11), 1369–1381.
Waters, D., Watt, E.W., Marsalek, J., Anderson, B.C., 2003. Adaptation of a storm drainage system to accommodate increased rainfall resulting from climate change. J. Environ. Plan. Manag. 46 (5), 755–770.
Wilby, R.L., Wigley, T.M.L., Conway, D., Jones, P.D., Hewitson, B.C., Main, J., Wilks, D.S., 1998. Statistical downscaling of general circulation model output: a comparison of methods. Water Resour. Res. 34 (11), 2995–3008.
Zhou, Q., 2014. A review of sustainable urban drainage systems considering the climate change and urbanization impacts. Water 6 (4), 976–992.
Zhou, Q., Arnbjerg-Nielsen, K., Mikkelsen, P.S., Nielsen, S.B., Halsnæs, K., 2011. Urban drainage design and climate change adaptation decision making (Ph.D. dissertation). DTU Environment.
Zhou, Q., Halsnaes, K., Arnbjerg-Nielsen, K., 2012. Economic assessment of climate adaptation options for urban drainage design in Odense, Denmark. Water Sci. Technol. 66 (8), 1812–1820.

CHAPTER FOUR

Corrosion of Concrete and Steel Structures in a Changing Climate

Mark G. Stewart*, Emilio Bastidas-Arteaga†
*Centre for Infrastructure Performance and Reliability, School of Engineering, The University of Newcastle, Callaghan, NSW, Australia
†Research Institute in Civil and Mechanical Engineering, UMR CNRS 6183, Université de Nantes, Nantes Cedex, France

4.1 Introduction

A significant proportion of the world's infrastructure uses steel and concrete as the main construction materials. This infrastructure was valued in 2005 at over $200 trillion, and is expected to exceed $300 trillion in value by 2025 (ARCADIS, 2015). Steel and reinforced concrete (RC) are highly susceptible to corrosion. Corrosion can cause loss of strength and stiffness of concrete and steel structures, can cause leaks or ruptures to occur in pipes, storage tanks, and ships, and cause mechanical and electrical equipment to malfunction. Although many facilities are protected by measures such as protective coatings, cathodic protection, etc., it is widely recognised that these are not always effective or even technically or commercially feasible (Paik and Melchers, 2014). Hence, it is not surprising that the National Association of Corrosion Engineers estimated that the global cost of corrosion is about $3 trillion per year, or 3.4% of global GDP (NACE, 2016). If climate change is expected to increase damage risks by a very modest 1% per year, this leads to additional losses of $30 billion per year.

Climate change may be a potentially important factor for corroding structures. Because climate change may influence the environment to which infrastructure is exposed, therefore, it may alter the factors known to affect the corrosion or deterioration of steel and concrete structures, including CO_2 level, temperature, relative humidity, sea-water level, nutrient concentration, ocean acidification, time of wetness, airborne salinity, and airborne pollutants. Each of these can influence initiation or progression of the corrosion process and thus have an effect on remaining service life. Therefore, assuming a constant climate condition for the life period of infrastructure may lead to inaccurate or nonconservative predictions.

Climate Adaptation Engineering
https://doi.org/10.1016/B978-0-12-816782-3.00004-8

This chapter focuses on corrosion of (i) RC beams, columns, and slabs in a marine environment and (ii) steel sheet piling for port infrastructure. It first describes the deterioration process, their modelling, their effect on structural performance and reliability, and adaptation measures that may be undertaken to ameliorate corrosion-induced damage to infrastructure under a changing climate. The chapter also provides guidance as to relevant literature in this field.

4.2 Corrosion Processes Under a Changing Climate

4.2.1 RC Structures

Deterioration is a common phenomenon for many ageing RC structures, which may arise from many causes and take on many forms like alkali-aggregate reaction, sulphate attack, and freeze-thaw cycles. However, in many cases the damage of most interest to infrastructure owners, asset managers, and engineers is carbonation and chloride-induced reinforcement corrosion. The assessment of corrosion effects on RC structures is a difficult task because several deterioration mechanisms interact in the process; ingress of the corroding agent—that is, chlorides or carbon dioxide, corrosion of reinforcing steel, and concrete cover cracking (Bastidas-Arteaga et al., 2013). The ingress of the corroding agent induces corrosion of the reinforcing bars. Corrosion reduces the structural capacity and the accumulation of corrosion products in the steel/concrete interface generates concrete cover cracking and reduction of bond. The corrosion process is divided into two stages namely 'corrosion initiation' and 'corrosion propagation'.

Thus, the deterioration process will generally comprise of:

- corrosion initiation:
  - diffusion of aggressive agents through protective cover, and/or
  - direct ingress of aggressive agents through cracks
- corrosion propagation:
  - loss of cross-sectional area of reinforcing or prestressing steel,
  - changes in ductility and mechanical properties of reinforcing or prestressing steel,
  - reduction of bond, and
  - crack initiation and propagation (spalling, delamination) caused by expansive rust products.

These deteriorating processes can interact to produce varying reductions in serviceability and strength performance. Fig. 4.1 shows a schematic of a typical deterioration process for a RC structural element. Before corrosion

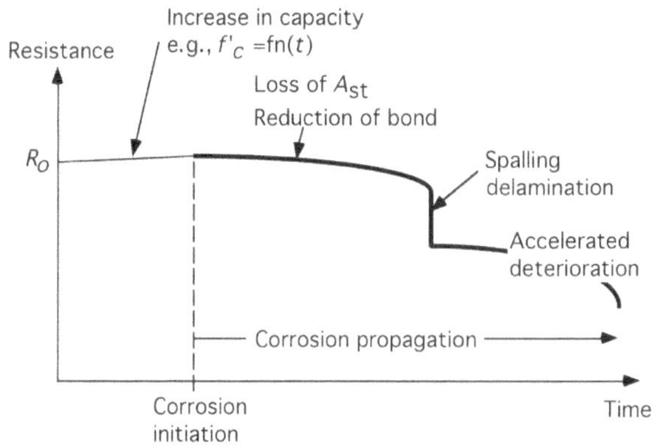

Fig. 4.1 Schematic representation of deterioration process for a RC structural member.

initiation, the increase in capacity could be related with the increase of concrete strength with time. Once corrosion starts, the decrease of resistance is linked to the loss of cross-section of rebars. For more details see Val and Stewart (2009) and Bastidas-Arteaga and Stewart (2017).

Experimental evidence indicates that carbonation and chloride ingress are highly influenced by environmental and climatic conditions of the surrounding environment—that is, atmospheric CO_2 concentration, temperature, and humidity (Saetta et al., 1993, de Larrard et al., 2014). According to the International Panel of Climate Change (IPCC), the environmental CO_2 concentration could increase from 379 ppm in 2005 to nearly 1000 ppm by 2100 (IPCC, 2013 and Chapter 1). The changes in environmental temperature, relative humidity, and carbon dioxide concentration can increase corrosion risks resulting in more widespread corrosion damage and loss of structural safety. Consequently, the effect of atmospheric CO_2 concentration change and global warming on both chloride ingress and carbonation should be considered for long-term sustainable and resilient management of RC structures.

Risk-based methods are highly suited to assess the cost-effectiveness of climate adaptation measures (e.g., Stewart and Deng, 2015). For instance, Bastidas-Arteaga et al. (2010) proposed a stochastic approach to study the influence of global warming on chloride ingress and its effect on corrosion initiation. Stewart and Peng (2010) carried out a preliminary risk and cost-benefit study on adaptation measures to mitigate the effects of carbonation of RC structures. Other studies also focused on the assessment of climate change on the durability of concrete structures in specific locations. Stewart et al. (2011, 2012), Wang et al. (2012), and Peng and Stewart

(2014, 2016) studied the impact of climatic change on corrosion-induced damage in Australia and China. They proposed a probabilistic approach to assess corrosion damage (cracking and spalling) taking into account the influence of climate change on regions characterised by different geographical conditions. Bastidas-Arteaga and Stewart (2015, 2016) then assessed the cost-effectiveness of adaptation measures for new and existing RC infrastructure in France, and Peng and Stewart (2015) assessed cost-effectiveness of adaptation measures for new RC structures in Australia and China.

4.2.2 Steel Structures

The deterioration of steel infrastructure caused by corrosion has been a constant concern for infrastructure owners, asset managers, and structural and mechanical engineers. Therefore, an increasing interest has been focused on predicting the rate and progression of longer-term corrosion of steel infrastructure which takes into consideration of the effects of protective measures or when such measures have become ineffective.

A key issue for asset management is the development of mathematical models for the corrosion process, particularly for long-term exposures. A brief review of such models is available (Melchers, 2008a). Unfortunately, the models presently available in the structural steel corrosion literature either are mainly based on laboratory experiments or are entirely empirical. These may lead to high levels of uncertainty and fail to extrapolate to the long-term structural life estimation, respectively. Melchers (2003) used real-world data for model development together with the corrosion fundamental principles and proposed models for general corrosion and later for pitting (Melchers, 2008b). The model is summarised in Fig. 4.2 and shows the sequential phases (0–4) that are governed by characteristic corrosion-rate controlling processes. The effect of temperature and oxygen availability is also illustrated in Fig. 4.2. For more details see Melchers and Jeffrey (2008).

Climate change is an important factor for corroding steel structures because it may alter temperature, sea-water level, nutrient concentration, ocean acidification, time of wetness, airborne salinity, airborne pollutants, etc. (Melchers, 2014). In addition, pollution of sea water may lead to a rise in dissolved inorganic nitrogen (DIN) concentration which will accelerate corrosion process.

If time-dependent corrosion rates are known, the remaining capacity of steel structures can be estimated. For example, this may be the reliability of corroded structural steel bridges (e.g., Kayser and Nowak, 1989;

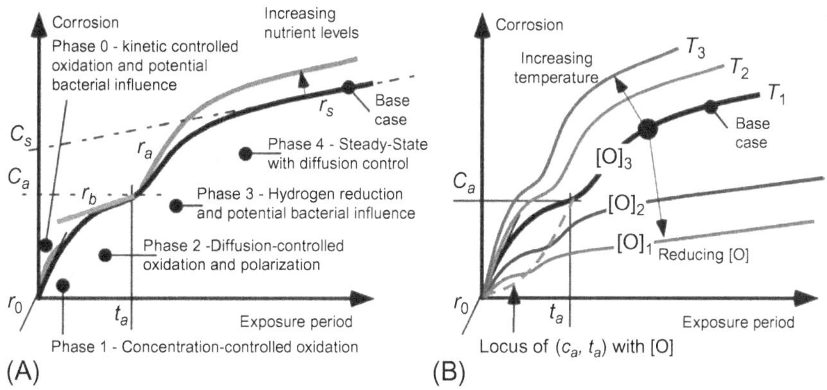

Fig. 4.2 Model for corrosion showing (A) sequential phases, model parameters (r_0, c_a, t_a, c_s, r_s) and the early and later influence of nutrient availability levels and (B) effect of increasing water temperature and reducing dissolved oxygen concentration (Melchers and Jeffrey, 2008).

Hosseini et al., 2013), the corrosion and reliability of steel offshore structures (Melchers, 2005), or the reliability of corroding steel sheet piling (Peng et al., 2017). While there is some research about corrosion effects on capacity of structural steel structures, there is relatively little research which takes account of the effects of a changing environment. Nguyen et al. (2013) conducted a preliminary study of climate change effects on corrosion rate of steel structures. Chaves et al. (2016) and Peng et al. (2017) included the effects of temperature increase and sea level rise produced by climate change and DIN concentration increase caused by pollution on the probabilistic corrosion process. Kallias and Imam (2013) studied the effect that climate change has on corrosion of steel railway bridges. However, the time-dependent effects of a changing environment on steel structures remains relatively unknown and more detailed work needs to be done, particularly from the perspective of probabilistic considerations, modelling, and structural reliability.

4.3 Cost-Effectiveness Assessment of Adaptation Strategies

As discussed in Chapter 1, the economic risk is:

$$E(L) = \Pr(C)\Pr(H|C)\Pr(D|H)\Pr(L|D)L \tag{4.1}$$

where $\Pr(C)$ is the annual probability that a specific climate scenario will occur, $\Pr(H|C)$ is the annual probability of a climate hazard (heat, humidity, etc.)

conditional on the climate scenario, $\Pr(D|H)$ is the probability of corrosion damage for the baseline case of no extra protection (i.e., 'business as usual') for a known level of temperature, humidity, etc., $\Pr(L|D)$ is the conditional probability of economic loss given occurrence of the damage, and L is the loss or consequence if full damage occurs.

The expected loss after climate adaptation is derived from Eq. (4.1) as:

$$E_{\text{adapt}}(L) = \sum (1 - \Delta R) E(L) - \Delta B \tag{4.2}$$

where ΔR is the reduction in risk caused by climate adaptation measure, $E(L)$ is the 'business as usual' risk given by Eq. (4.1), and ΔB is the cobenefit of adaptation such as reduced losses to other hazards, increased energy efficiency of new materials, etc. The net benefit or net present value (NPV) is equal to benefit minus the cost. The decision problem is to maximise NPV or benefit-to-cost ratio (BCR):

$$\text{NPV} = \sum E(L)\Delta R + \Delta B - C_{\text{adapt}} \quad \text{BCR} = \frac{\sum E(L)\Delta R + \Delta B}{C_{\text{adapt}}} \tag{4.3}$$

where C_{adapt} is the cost of adaptation measures including opportunity costs that reduce risk by ΔR. In this chapter, cobenefit is assumed as $\Delta B = 0$. If, for example, an adaptation measure for RC is an increase in concrete strength (improved durability), then a cobenefit is that the capacity of a RC beam will increase if concrete strength is upgraded. In this case, considering cobenefit in the analysis can result in adaptation measures transitioning from not cost-effective to becoming cost-effective (Peng and Stewart, 2015).

4.4 Case Studies

4.4.1 RC Beams, Slabs, and Columns in a Marine Environment

4.4.1.1 Adaptation Measures

There is a wide range of existing and 'low-tech' options that can enhance the durability of RC structures and these can be applied to reduce the adverse effects of climate change. The design options generally include the selection of cover, concrete mix, surface coating barriers, extraction, and cathodic protection. In addition to reducing environmental exposure as much as possible, practical adaptation solutions in a new design may come from increasing cover and strength grade, or any approaches that reduce material diffusion coefficient without compromising the reliability and serviceability of concrete. Adaptation measures for new and existing concrete structures may include:

- surface treatments,
- realkanization,
- extra design cover,
- increase concrete durability,
- stainless or galvanised steel reinforcement,
- corrosion inhibitors,
- cathodic protection, and
- replacement of existing cover with new concrete.

The time of adaptation is highly variable and dependent on extent and location of corrosion damage. Some adaptation strategies could be applied at the time of construction (coatings/surface treatments, reinforcement), and others at time of corrosion initiation (realkanization, chloride extraction). Clearly, it is preferable to use adaptation strategies that are implemented during design and construction rather than in-service (e.g., when corrosion damage occurs) as the latter will be much more costly in terms of direct costs and inconvenience/user delays and other indirect costs (Stewart et al., 2012).

Adaptation strategies will have varying degrees of effectiveness and cost. Some will require regular maintenance over the life of the structure, such as surface treatments, which will increase their life cycle cost. Given that there are many millions of new and existing RC infrastructure in many countries, the cost of adaptation can be immense. For this reason, a risk-based approach is needed to assess the optimal level, if any, of adaptation measures. This includes the cost, location, timing, and extent of adaptation measures.

An increment in cover thickness can increase the time of carbonation and chloride ingress to reach concrete reinforcement and in turn delay corrosion initiation. It is therefore one of the most obvious and simplest adaptation options in the design of RC infrastructure under a changing climate to maintain structural durability and serviceability. While the change of cover is considered as the most straightforward design approach to reduce the impact of a changing climate, other options may also include the selection or design of concrete materials to reduce the diffusion coefficient of deleterious substances, that is, slow the ingress of those substances, and hence delay corrosion of concrete reinforcement. In practice, selection of a higher strength grade of concrete is one approach to reduce the diffusion coefficient, in addition to enhancing its load-carrying properties—see Peng and Stewart (2015) and Bastidas-Arteaga and Stewart (2015) for further details of this adaptation strategy.

What is important is the need to estimate time-dependent changes in damage risks, the effectiveness and cost of one or more adaptation measures, costs of repairing damage, and other criteria needed to assess the optimal level of adaptation measures both now and into the future.

4.4.1.2 Proposed Adaptation Framework

The main practical problem concerning adaptation of structures lies in evaluating its costs and effectiveness. Fig. 4.3 shows the proposed framework for dealing with this problem that combines:

- general methods that could be applied to any structure (deterioration models, stochastic approaches and cost-benefit analysis), with
- information specific to each structure (climate change predictions and structural characteristics).

The following sections describe the general methods that are employed in the proposed framework. Information specific to each structure will be detailed in the numerical application.

4.4.1.3 Deterioration Modelling

Deterioration modelling allows estimating the effects of chloride ingress with regard to serviceability or ultimate limit states. Ultimate limit states are highly dependent on both, geometrical characteristics (cross-sectional dimensions, span length, etc.) and loading (dead, live, seismic, etc.). Therefore, to generalise the results, this section focuses on a serviceability limit state in which the cost-effectiveness of adaptation measures is evaluated in terms of its effect on the time to corrosion damage of the concrete cover

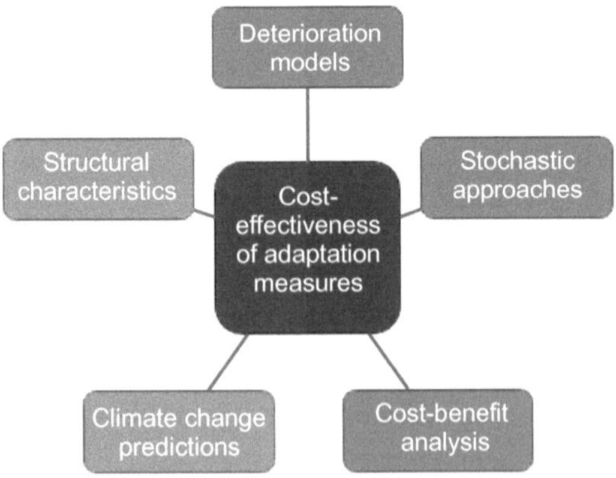

Fig. 4.3 Proposed framework for determining the cost-effectiveness of adaptation measures.

(severe cracking or spalling). Corrosion-induced cover cracking and damage occurs on the concrete surface above and parallel to the rebars. The time to corrosion damage (severe cracking or spalling), T_{sp} is thus obtained as the sum of three stages: (i) corrosion initiation (T_i); (ii) crack initiation (T_{1st}, time to first cracking–hairline crack of 0.05 mm width); and (iii) crack propagation (T_{sev}, time for crack to develop from crack initiation to a limit crack width), that is, $T_{sp} = T_i + T_{1st} + T_{sev}$. After corrosion initiation, the kinematics of T_{1st} and T_{sev} is controlled by corrosion propagation.

The time to corrosion initiation, T_i, is estimated by comparing the chloride concentration at the cover depth, c_t, with a threshold concentration for corrosion initiation C_{th}. The adopted chloride ingress model considers the interaction between three physical processes: chloride ingress, moisture diffusion, and heat transfer (Nguyen et al., 2017; Bastidas-Arteaga, 2018).

Chloride-induced corrosion is characterised by pitting corrosion with a time-variant corrosion rate $i_{corr}(t)$ ($\mu A/cm^2$). Given the complexity of the corrosion process, i_{corr} depends on many factors such as concrete pH and availability of oxygen, and water in the corrosion cell. For instance, the optimum relative humidity for corrosion is 70%–80%. This study considers a time-variant corrosion rate model that takes into account the effect of temperature change (DuraCrete, 2000).

The time to crack initiation, T_{1st}, is estimated based on the model by El Maaddawy and Soudki (2007). The time to severe cracking, T_{ser} (time when concrete cover cracking reaches a limit crack width of 1 mm), is computed by the model by Mullard and Stewart (2011). Mullard and Stewart (2011) have modelled rate of crack propagation which includes a confinement factor (k_c) that represents an increase in crack propagation due to the lack of concrete confinement around external reinforcing bars. If the reinforcing bar is in an internal location then $k_c = 1$, but for rebars located at edges and corners of RC structures then k_c is in the range of 1.2–1.4. For more details see Mullard and Stewart (2011).

4.4.1.4 Repair Strategy and Damage Risks

The probability of first damage at time t for original concrete is:

$$\Pr(D|H) = \Pr\left[t \geq T_{sp}\right] \tag{4.4}$$

where T_{sp} is the time when concrete cover severely cracks, and where the asset owner can specify the limit crack width as the criterion for repair.

A patch repair is the most common technique to repair corrosion damage in RC structures. For a patch repair, the concrete cover is typically removed to approximately 25 mm past the steel bars (which are then cleaned of corrosion products) and a repair material is installed. The maintenance strategy assumes that (Stewart, 2001):

- concrete is inspected at time intervals of 2 years;
- patch repair is carried out immediately after corrosion damage has been discovered–hence, $\Pr(L|D) = 100\%$;
- damage limit state exceedance results in entire RC surface being repaired;
- repair provides no improvement in durability performance of the repaired structure (i.e., it is repaired with the same cover and concrete quality as the original design specification); and
- damage may re-occur during the remaining service life of the structure, that is, multiple repairs may be needed.

In addition, the time-dependent damage risks of the repaired material will not be the same as for the original material due to changed temperature and humidity at the time of repair (i.e., when the concrete is new). Hence, the damage risk for repaired (new) concrete exposed to the environment for the first time at time of repair (t_{rep}) will change depending on the new climatic conditions and time of repairs. Eq. (4.4) is a simplified version of a complex time-dependent formulation of damage fragility, for more details see Bastidas-Arteaga and Stewart (2015, 2016).

Other research has modelled the spatial variability of damage on RC surfaces, which enables the likelihood and extent of damage to be predicted. Repair strategies can then be developed based on the extent and location of the corrosion damage. For more details see Mullard and Stewart (2009, 2012) and Peng and Stewart (2014, 2016).

This case study illustrates the assessment of time-dependent damage risks for new RC structures placed in a chloride-contaminated environment under various exposures and climate change scenarios. The structural components are subjected to atmospheric exposure to salt spray (XS1 exposure according to the European Norm EN-206, 2000). The climatic conditions are defined by an oceanic environment placed at a middle latitude (i.e., Europe) where the yearly mean temperature and relative humidity vary between the intervals [5°C and 25°C] and [60% and 80%], respectively. The EN-206 (2000) durability design requirements for a structural lifetime of 100 years and a rebar diameter of 16 mm are (i) 55 mm cover and (ii) 30 MPa concrete compressive strength for these exposure conditions.

The probabilistic models used to estimate damage probabilities are presented in Table 4.1. It is assumed that all the random variables are statistically independent. Monte-Carlo simulation analysis is used as the computational tool to propagate uncertainties through the analysis, although analytical methods could also be used–for example, Stewart and Melchers (1997).

Fig. 4.4 presents the time-dependent probability of severe cracking for various climate change scenarios—that is, $\Pr(D|H)$ where damage represents severe cracking. Fig. 4.4 clearly shows that the rate of damage risk is highly dependent on climate change effects and environmental exposure. If there is no climate change, the probability of severe cracking increases with time and remains constant irrespective of time of repair. However, if climate change reduces the environmental relative humidity, that is, $\Delta RH = -10\%$ in 100 years, the chloride ingress mechanism slows down, and consequently, the probability of damage decreases. For instance, for a structure with no repairs, the probability of damage decreases from 20% to 8% after 100 years of service. In this case, climate change has a 'positive

Table 4.1 Probabilistic Models of the Random Variables (Bastidas-Arteaga and Stewart 2015)

Variable	Units	Mean	COV	Distribution
Reference chloride diffusion coefficient, $D_{c,ref}$	m^2/s	3×10^{-11}	0.20	Log-normal
Environmental chloride concentration, C_{env}	kg/m^3	7.35	0.20	Log-normal
Concentration threshold for corrosion initiation, C_{th}	wt% cem.	0.5	0.20	Normal[a]
Cover thickness, c_t	mm	55	0.25	Normal[b]
Reference humidity diffusion coefficient, $D_{h,ref}$	m^2/s	3×10^{-10}	0.20	Log-normal
Thermal conductivity of concrete, λ	W/(m°C)	2.5	0.20	Beta on [1.4;3.6]
Concrete specific heat capacity, c_q	J/(kg°C)	1000	0.10	Beta on [840;1170]
Density of concrete, ρ_c	kg/m^3	2400	0.04	Normal[a]
Corrosion rate, $i_{corr-20}$	$\mu A/cm^2$	2.586	0.66	Log-normal
28 day concrete compressive strength, $f_c(28)$	MPa	$1.3(f'_{ck})$	0.18	Normal[a]
Concrete tensile strength, f_t	MPa	$0.53(f_c)^{0.5}$	0.13	Normal[a]
Concrete elastic modulus, E_c	MPa	$4600(f_c)^{0.5}$	0.12	Normal[a]

[a]Truncated at 0 mm.
[b]Truncated at 10 mm.

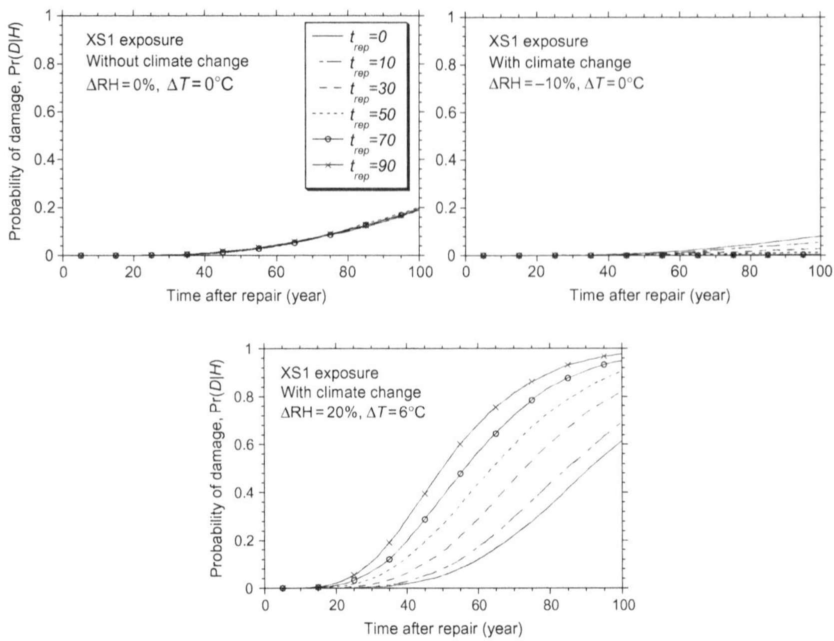

Fig. 4.4 Probability of damage $\Pr(D|H)$ for various climate change scenarios.

effect' on RC durability reducing by 60% the corrosion damage risk. An opposite behaviour is observed when climate change increases the temperature and relative humidity. For the same conditions, the probability of corrosion damage increases from 20% to 98%.

For all adaptation options construction and repair cost data are needed, and such cost data are country, site, and structure specific and so it is difficult to make generalisations about these costs. In this case study, costs are expressed in 2012 US dollars. It is assumed that design and inspection costs are similar for different adaptation measures and so they are not needed for this comparative analysis. Hence, adaptation strategies will only affect the expected damage costs. As we are concerned about outdoor exposures then the external RC structural elements of interest are slabs, beams, and columns. Corrosion damage is assumed to occur on one (exposed) face of a slab and beam, and all faces of a column.

4.4.1.5 Risk Reduction and Cost-Effectiveness of Adaptation Strategy

The cost of repair or replacement and associated user losses, etc. is considerable and for some structures user losses are often much greater than direct repair, replacement, and maintenance costs. To allow for a minor user disruption cost the total loss is assumed as $L=\$500$ per m^2 of exposed concrete, and $Pr(L|D)=100\%$.

It is assumed that an increase in design cover would increase cost of forms, concrete, reinforcement, finishing, and labour by an amount proportional to the extra volume of concrete needed. Table 4.2 presents the adaptation costs (C_{adapt}) for 5 and 10mm increase in extra cover for various structural elements. Clearly, adaptation costs are higher for a square column if cover is increased on all four faces, and damage can also occur on all four faces.

Fig. 4.5 shows the economic risks $E(L)$ for existing cover and $E_{adapt}(L)$ for two extra cover designs and various climate change scenarios, where for this scenario-based analysis $Pr(C)=100\%$, $Pr(H|C)=100\%$, and the hazard H is represented by assumed changes in ΔRH and ΔT produced by climate change over 100 years. The case without a changing climate, $\Delta RH=0\%$ and

Table 4.2 Adaptation Costs (C_{adapt}) for Various Structural Elements

Structural Element	Element Size D_e (mm)	Adaptation Cost ($/m^2) 5 mm Increase in Cover	Adaptation Cost ($/m^2) 10 mm Increase in Cover
Slabs	100	6.50	13.00
Slabs	300	3.75	7.50
Beams	200–800	7.50	15.00
Sq. columns	300	12.00	24.00
Sq. columns	600	7.70	15.00

Note: Slab depth or beam or column width.

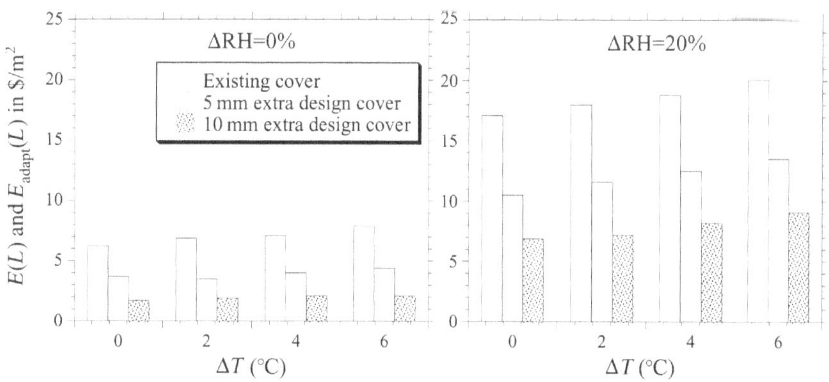

Fig. 4.5 Economic risks $E(L)$ and $E_{adapt}(L)$ for 100 years, for $\Delta RH=0\%$ and $\Delta RH=20\%$.

$\Delta T = 0°C$, is also presented. The analyses assume a discount rate $r = 4\%$, a structural lifetime of 100 years, and is calculated using Eqs. (4.1) and (4.2). The risk reduction ($\Delta R = (E(L) - E_{adapt}(L))/E(L)$) varies from 30% to 50% for 5 mm increases in cover, and 60% to 80% for a 10 mm increase in cover.

It is observed in Fig. 4.5 that economic risks are larger when increases in temperature and relative humidity are considered. This is explained by the rise of both chloride ingress and corrosion rate when the structure is exposed to higher temperature and relative humidity due to climate change (Bastidas-Arteaga et al., 2010). Therefore, the acceleration of these deterioration mechanisms increases the number of repairs and damage costs during the structural lifetime. It is also noted that adaptation strategies reduce the mean damage costs because the number of repairs is reduced and/or the time to repair is longer when there is an increase of the concrete cover. As expected, a 10 mm increase of the design cover is the more effective adaptation strategy. However, these results cannot be used to compare the cost-effectiveness of an adaptation strategy because they do not include the adaptation costs.

Tables 4.3 and 4.4 summarise the mean BCR and Pr(BCR > 1) for various climate change scenarios and two structural components ($D_e = 300$ mm slabs and columns). Fig. 4.6 and Tables 4.3 and 4.4 show that the mean BCR is highly dependent on both the exposure and the type of structural component. In some cases the mean BCR is lower than one indicating that the adaptation strategy is not cost-effective for some structural components under given climate change scenarios. Similar behaviour is observed in Fig. 4.6 for Pr(BCR > 1) where the probabilities of cost-effectiveness are lower for small structural elements for which the adaptation cost is higher.

Higher temperature and relative humidity accelerate the deterioration processes by increasing the cost-effectiveness of the implementation of an adaptation measure. For RC slabs (see Table 4.3) the mean BCR is only higher than one when climate change could induce increases of relative humidity equal or higher than 10% in 100 years. However, even under these scenarios, the Pr(BCR > 1) indicates that the risks associated to a bad investment are higher. This behaviour is similar for columns (see Table 4.4) but mean BCR and Pr(BCR > 1) are smaller because the adaptation cost is higher (see Table 4.2). It is noted that an increase of 5 mm cover provides higher estimates of BCR and Pr(BCR > 1). However, the likelihood that BCR > 1 is less than 60% even for a pessimistic (worst-case) climate change scenario of $\Delta RH = 20\%$ and $\Delta T = 6°C$.

Table 4.3 Mean BCR and Pr(BCR > 1) (Shown in Italics) for Slabs ($D_e = 300$ mm)

	5 mm Increase in Design Cover				10 mm Increase in Design Cover			
ΔRH	$\Delta T = 0°C$	$\Delta T = 2°C$	$\Delta T = 4°C$	$\Delta T = 6°C$	$\Delta T = 0°C$	$\Delta T = 2°C$	$\Delta T = 4°C$	$\Delta T = 6°C$
−10%	0.33 *(8%)*	0.53 *(8%)*	0.57 *(9%)*	0.57 *(9%)*	0.29 *(8%)*	0.39 *(8%)*	0.39 *(9%)*	0.40 *(9%)*
0%	0.68 *(18%)*	0.91 *(20%)*	0.94 *(22%)*	0.95 *(23%)*	0.60 *(18%)*	0.67 *(20%)*	0.69 *(22%)*	0.78 *(23%)*
10%	1.17 *(34%)*	1.27 *(37%)*	1.35 *(38%)*	1.32 *(40%)*	0.92 *(34%)*	1.03 *(37%)*	1.07 *(38%)*	1.07 *(40%)*
20%	1.65 *(50%)*	1.70 *(53%)*	1.72 *(55%)*	1.76 *(59%)*	1.36 *(50%)*	1.42 *(54%)*	1.45 *(56%)*	1.47 *(59%)*

Table 4.4 Mean BCR and Pr(BCR > 1) (Shown in Italics) for Columns ($D_e = 300$ mm)

	5 mm Increase in Design Cover				10 mm Increase in Design Cover			
ΔRH	$\Delta T = 0°C$	$\Delta T = 2°C$	$\Delta T = 4°C$	$\Delta T = 6°C$	$\Delta T = 0°C$	$\Delta T = 2°C$	$\Delta T = 4°C$	$\Delta T = 6°C$
−10%	0.10 *(5%)*	0.16 *(7%)*	0.18 *(8%)*	0.19 *(8%)*	0.09 *(4%)*	0.12 *(4%)*	0.12 *(5%)*	0.13 *(5%)*
0%	0.21 *(7%)*	0.29 *(8%)*	0.30 *(9%)*	0.31 *(11%)*	0.19 *(5%)*	0.21 *(5%)*	0.21 *(6%)*	0.24 *(6%)*
10%	0.36 *(9%)*	0.40 *(11%)*	0.42 *(12%)*	0.41 *(12%)*	0.29 *(5%)*	0.32 *(6%)*	0.34 *(7%)*	0.34 *(7%)*
20%	0.45 *(13%)*	0.50 *(13%)*	0.52 *(14%)*	0.55 *(14%)*	0.43 *(7%)*	0.45 *(8%)*	0.44 *(9%)*	0.46 *(9%)*

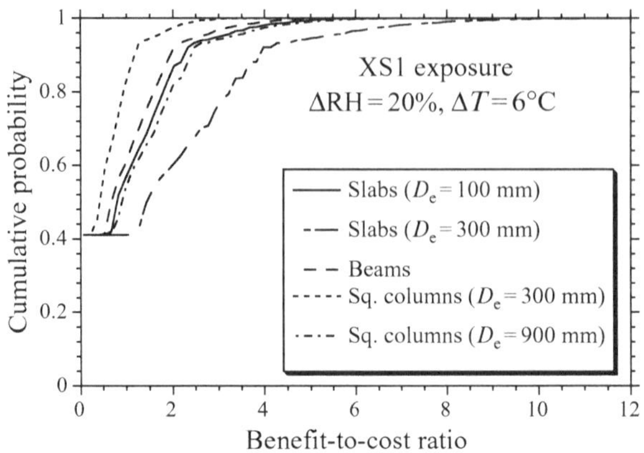

Fig. 4.6 Pr(BCR) for all structural components for a 5 mm increase in cover.

On the other hand, as presented in Fig. 4.4, some 'positive' effects of climate change on concrete durability could be attended if RH decreases with time. These positive effects will therefore reduce the cost-effectiveness of adaptation measures. For instance, if the relative humidity decreases (i.e., $\Delta RH = -10\%$), the chloride ingress rate will also decrease diminishing the number of repairs and consequently repair costs. In such a case, Tables 4.3 and 4.4 indicate that the mean BCRs computed for this climate scenario are generally lower than those of the case when $\Delta RH = 0\%$. This means that the benefits of the adaptation measures could be lower under some climate change conditions. Consequently, the effects of climate adaptation measures should be carefully evaluated in order to decide if they provide benefits of losses with respect to the existing design.

NPV and BCR are influenced by the policy horizon or time period of the analysis, and discount rates. For example, the mean BCR and Pr (BCR > 1) are very sensitive to discount rate and both decision metrics are larger for low discount rates (see Bastidas-Arteaga and Stewart (2016, 2017) for more details). This is explained by the fact that small discount rates imply that future costs are larger at present cost by increasing the cost-effectiveness of adaptation measures for repairs close to the end of the structural lifetime. As discussed in Chapter 1, various governments recommend lower discount rates of about 2% for long-term investments. The probabilistic BCR analysis therefore shows that the adaptation strategies are more cost-effective according to these recommendations.

4.4.2 Deterioration of Structural Steel Port Infrastructure

4.4.2.1 Problem Description

Steel sheet piling is commonly used in many ports and harbours worldwide. However, corrosion of steel sheet piling can result in metal loss and reduced structural capacity, which can then lead to failure (see Fig. 4.7). Corrosion results from a chemical reaction, so an increase in sea-water temperature can accelerate the corrosion process. The Fourth Assessment Report of the Intergovernmental Panel on Climate Change predicts that average seawater surface temperature is 'likely' to increase by 6°C over the next 100 years (IPCC, 2007).

The thinning of piles will cause loss of pile cross-section, and then may lead to structural safety concerns. This is because steel profile in the vertical direction determines the (vertical) bending strength of piles, their soil retaining capacity, and operational capacity of the sheet piling system. In addition, perforation of sheet piling may result in loss of containment fill behind the piles or stability problems. For these reasons the main focus of this section is the vertical corrosion effect (thinning), vertical capacity, and pitting perforation of steel sheet piling.

Osorio et al. (2011) investigate the safety of a steel sheet pile using simplified corrosion models by reliability analyses. Boéro et al. (2012) predicted the impact of corrosion on the mechanical behaviour of a quay using a reliability analysis, based on statistical analysis of collected corrosion data in France. Silva et al. (2014) assessed the reliability of a steel plate subject to distributed and localised corrosion. This case study expands this work considerably mainly focusing on the assessment of the cost-effectiveness of an adaptation measure.

Fig. 4.7 Example of failure of a sheet pile retaining wall. *(Photo courtesy R Jeffrey.)*

The corrosion of concern for this type of coastal infrastructure is a phenomenon known as accelerated low water corrosion (ALWC) (Melchers and Jeffrey, 2013; Melchers et al., 2014). The data suggest that the long-term trend in uniform corrosion and also in maximum pit depth tends to follow an anaerobic steady-state linear trend as a function of time t that can be described, for exposure periods greater than about 5 years where c_s is the 'y-intercept' at time $t=0$ and r_s is the slope of the long-term trend, see Fig. 4.8. This model, driven by the parameters c_s and r_s, is used in this section to estimate losses due to ALWC.

The cross section of the sheet pile retaining wall considered in this study case is shown in Fig. 4.9, and the statistical parameters are given in Table 4.5. The fragility $\Pr(D|H)$ of sheet piling to ALWC is obtained from a time-dependent structural reliability analysis (Chaves et al., 2016). It is assumed, as is reasonable in practice, that the piles are unprotected, having no protective paint coatings or cathodic protection. Damage to the retaining wall will halt all dock works and associated services. The probability of sheet piling damage is defined as:

$$\Pr(D|H) = \Pr[G_1(t) < 0 \cup G_2(t) < 0] \tag{4.5}$$

where the time-dependent limit state functions are:

$$G_1(t) = f_y - \sigma_{\max}(t) \quad \text{or} \quad G_2(t) = s - c(t) \tag{4.6}$$

where failure (damage) occurs if (i) the maximum stress on the critically loaded cross-section of the pile ($\sigma_{\max}(t)$) exceeds the steel pile yield stress capacity (f_y) or (ii) the corrosion depth $c(t)$ causes full penetration by exceeding the thickness of the pile (s).

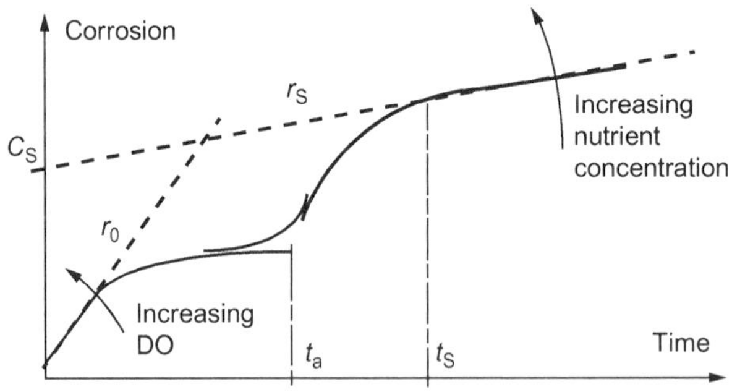

Fig. 4.8 Simplified (engineering) model for marine immersion corrosion loss (Melchers, 2014).

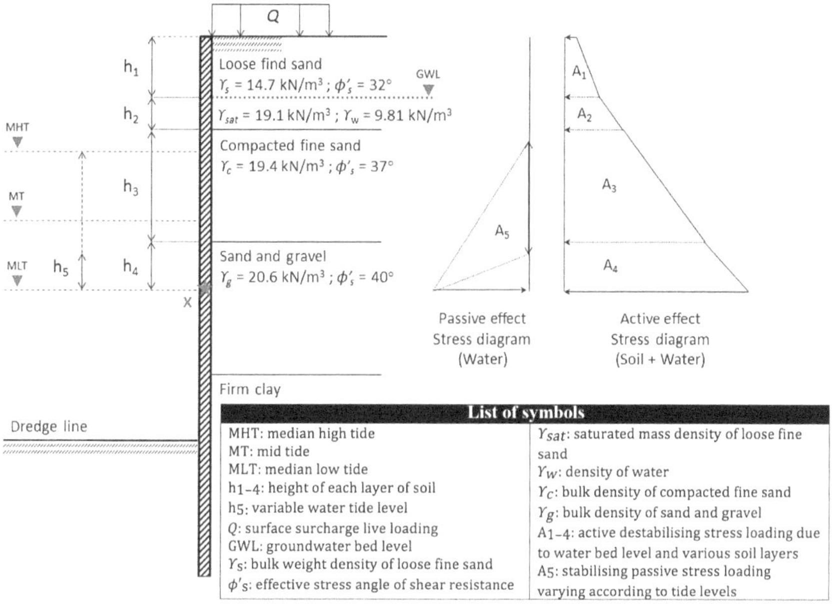

Fig. 4.9 Cross-section of a steel pile retaining wall, showing applied loadings, soil support conditions, and the resulting structural active and passive soil and other pressures (Chaves et al., 2016).

Table 4.5 Basic Design Random Variables and Statistical Properties Assumed

Parameter	Definition/ illustration	Unit	Mean	COV	Distribution
f_y	Eq. (4.6)	N/mm^2	275	0.07	Log-normal
c_S	Fig. 4.8	mm	0.502	0.07	Normal
r_S	Fig. 4.8	mm/year	0.254	0.58	Log-normal
h_1	Fig. 4.9	m	0.4	–	–
h_2	Fig. 4.9	m	0.2	–	–
h_3	Fig. 4.9	m	1.4	–	–
h_4	Fig. 4.9	m	0.2		–
h_5	Fig. 4.9	m	1.5	0.67	Normal
Q	Fig. 4.9	kN/m^2	2.931	1.05	Gumbel
γ_s	Fig. 4.9	kN/m^3	14.7	0.02	Normal
γ_{sat}	Fig. 4.9	kN/m^3	19.1	0.02	Normal
γ_c	Fig. 4.9	kN/m^3	19.4	0.02	Normal
γ_w	Fig. 4.9	kN/m^3	9.81	–	–
γ_g	Fig. 4.9	kN/m^3	20.6	0.02	Normal
d	Table 4.6	mm	14.5	0.05	Normal
s	Fig. 4.10	mm	10.2	0.06	Normal
h	Fig. 4.10	mm	225.0	0.02	Normal
b	Fig. 4.10	mm	750.0	0.02	Normal

Adapted from Chaves, I.A., Melchers, R.E., Peng, L., Stewart, M.G., 2016. Probabilistic remaining life estimation for deteriorating steel marine infrastructure under global warming and nutrient pollution. *Ocean Eng.* 126, 129–137

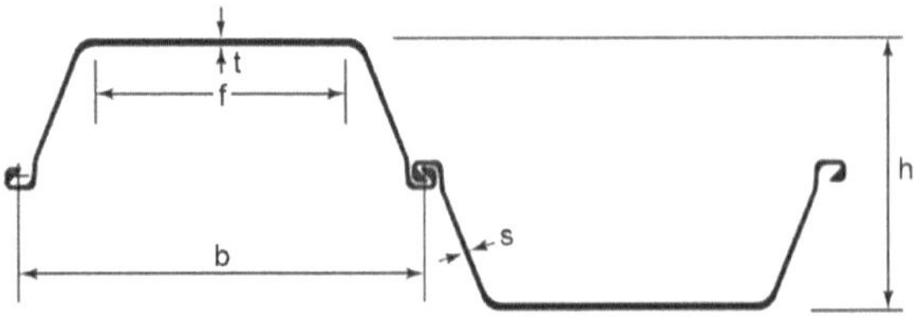

Fig. 4.10 Section of U-profile piles when interlocked together to form a wall.

Table 4.6 AU 25 and AU 26 Commercial Steel Sheet Piling Nominal Dimensions (mm)

Product Section	Width (*b*, mm)	Height (*h*, mm)	Top Flange Thickness (*d*, mm)	Angled Flange Thickness (*s*, mm)	Single Pile Mass (kg/m)	Moment of Inertia (cm^4/m)
AU 25	750	450	14.5	10.2	110.4	56,240
AU 26	750	451	15.0	10.5	113.2	58,140

Current design practice results in the installation of AU 25 U-profile sheet piles, see Fig. 4.10 and Table 4.6. However, if corrosion loss is expected to accelerate due to a changing climate, then a climate adaptation measure may be to select a stronger sheet pile with a larger thickness. In this case, an AU 26 sheet pile is 0.3–0.5 mm thicker and 3% stronger than the AU 25 sheet pile.

4.4.2.2 Results and Discussion

The structural reliability analysis includes the stochastic variability of loads, soil properties, steel material properties, dimensions, and corrosion processes. For more details see Chaves et al. (2016). The fragility $\Pr(D|H)$ for the existing AU 25 and proposed AU 26 sheet piles allowing for a 6°C sea-water temperature increase over the next 100 years is shown in Fig. 4.11. The risk reduction arising from using the higher capacity AU 26 steel pile is shown in Fig. 4.12. Clearly, even though the adaptation measure is the installation of slightly larger (3%) piles, the risk reduction reaches 20% early in the service life of the sheet piles.

It can be assumed that damage shown in Fig. 4.7 will lead to 100% likelihood of loss, hence, $\Pr(L|D)=100\%$. The economic loss (L) from damage of sheet piling can be considerable. The cost to repair damage is likely to be at least $1 million, and repair time at least 1 month. To assess the indirect loss to the owner of a port, the economics of The Port of Botany container terminal in Sydney is used as an illustrative example. The economic activity of the 12 shipping container berths runs to over $2 billion per year

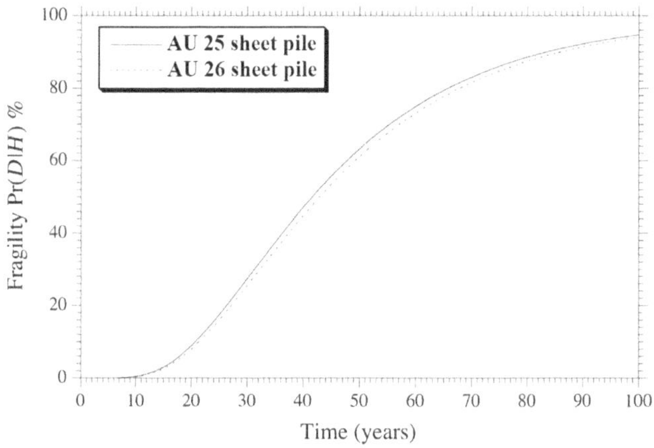

Fig. 4.11 Time-dependent fragility for sheet piles.

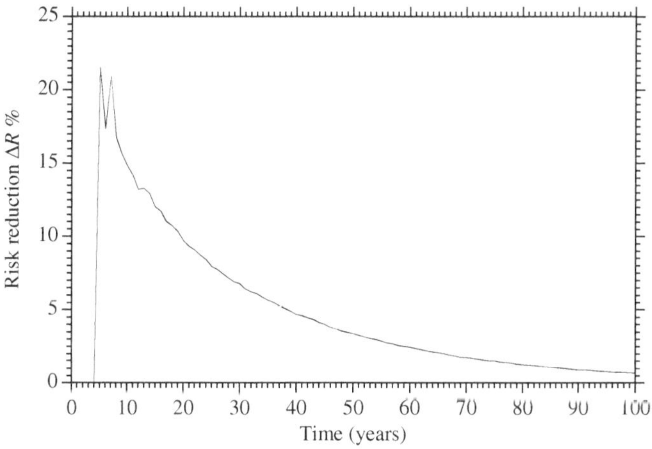

Fig. 4.12 Risk reduction from adaptation.

(Sydney Ports, 2008). This includes costs to the asset owner, trucking costs, worker wages, and economic gains from the efficient import and exports of good in Australia. If one of these 12 berths is unavailable due to sheet piling damage then shipping may be diverted to other berths. However, if all berths are busy then delays can be expected, at a pro-rata cost of $14 million for loss of one berth for 1 month. An upper bound of economic loss, when also considering direct repair costs, is \$15 million. A lower bound is \$1 million assuming loss of one berth for 1 month does not disrupt normal shipping. A mid-estimate of $L = \$8$ million is thus reasonable.

The adaptation cost (C_{adapt}) is based on the additional cost of purchasing larger AU 26 sheet piles. The AU 26 sheet piles are 2.5% heavier than AU 25 piles. The additional material cost for a 200-m-long dock using 30-m deep piles is approximately $C_{adapt} = \$10,000$.

The existing present value risk calculated from Eq. (4.1) for a scenario-based analysis ($\Pr(C) = \Pr(H|C) = 100\%$) of a 6°C increases in sea-water temperature in 100 years and 7% discount rate is $E(L) = \$745,830$. The average risk reduction over 100 years is 5.4%. Assuming no cobenefits, the NPV (or net benefit) of this adaptation measure is NPV = \$30,500. The benefit-to-cost ratio is BCR = 4.05. The use of larger AU 26 sheet piles is cost-effective for this climate scenario. This adaptation measure remains cost-effective even if adaptation costs double, or economic losses are halved. The NPV will increase for discount rates lower than 7%. Fig. 4.13 shows NPV as a function of time. The pay-back period (when benefit exceeds cost) is only 12 years.

There is likely to be uncertainty about climate scenarios. As such, it is useful to conduct a risk-based cost-benefit assessment for infrastructure assuming the current climatic conditions. The analysis reveals that for no change in sea-water temperature the NPV is \$24,900. Hence, even if there is no change in sea-water temperature, it is cost-effective to increase the size of sheet piling, in the present case, to AU 26.

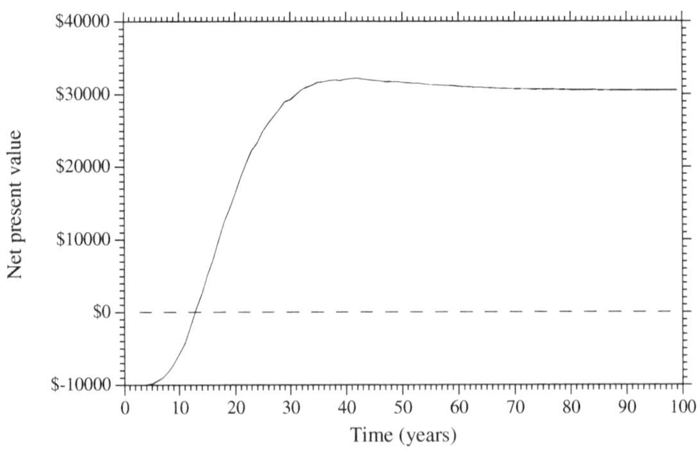

Fig. 4.13 Net present value for adaptation.

4.5 Summary

This chapter has shown that climate change can lead to increased corrosion-induced damage to steel and RC infrastructure. This damage can affect the strength and serviceability of infrastructure, leading to infrastructure owners spending billions of dollars per year to repair damage, or to replace infrastructure. Two case studies were considered: (i) corrosion of RC beams, columns, and slabs exposed to a marine environment and (ii) corrosion of steel sheet piling used for port infrastructure. These case studies calculated damage fragilities and risks, and applied cost-benefit analyses to assess the cost-effectiveness of adaptation measures. The illustrative examples showed that adaptation measures, that is, low cost with a low risk reduction, can still be cost-effective, particularly if the losses from infrastructure damage are relatively high. In other words, modest (or small) reductions in infrastructure vulnerability can be very cost-effective.

Finally, there is no certainty that existing design and construction practices are optimal. Design standards often are based on past experience, as well as new knowledge. However, they are seldom subject to a cost-benefit analysis due to modelling complexity, and more often than not, scarce resources to undertake work of this nature. It is thus desirable to assess the costs and benefits of existing designs. Hence, even if climate projections are overly conservative, adaptation measures still satisfies a 'no regrets' or 'win-win' policy (Susskind, 2010).

4.6 Potential Design and Practice Evolutions

Construction standards provide design recommendations to account, in a simplified way, for uncertainties related to material properties, models, loading, geometry, etc. They consider different mechanical (bending, shear, etc.) or durability (chloride ingress, carbonation, etc.) solicitations for specific conditions (i.e., seismic zones, wind maps, etc.). Modern design standards have evolved to take into account advances in understanding the behaviour of materials and physical phenomena and practical experience. For example, for RC structures subjected to chloride ingress, the first European standards suggested to use a very thin concrete cover (between 15 and 20 mm) at the beginning of the 20th century. At the end of the century, modern design standards include recommendations about the minimal concrete cover, cement content, use of admixtures, use of stainless steel, etc. (Bastidas-Arteaga and Stewart, 2016; EN-206, 2000).

Some studies have found that the durability performance of infrastructure depends on surrounding environmental conditions such as temperature, relative humidity, pH, etc. (de Larrard et al., 2014, Bastidas-Arteaga and Stewart, 2016; Nguyen et al., 2017). These particular conditions could accelerate or slow down the kinematics of deterioration processes. Nevertheless, modern design norms neglect this aspect. Including some durability maps that identify more or less aggressive areas in future standards could help to designers to optimise their solutions respecting to durability performance of constructions. These maps could also include the potential effects of climate change for those areas.

4.7 Open Research Questions

Further work on the area of adaptation of corroding RC and steel infrastructure should focus on:

- the consideration of the effect and the modelling of other environmental actions such as solar radiation, rain, seawater pH, nutriments availability, etc.;
- the downscaling of climate data at the structural and component levels;
- the improvement of the assessment of climate change effects by implementing or developing more realistic deterioration models and considering other environmental actions;
- the consideration of the spatial variability of RC and steel properties, environmental actions, and deterioration processes;
- the consideration of other adaptation strategies such as improvement of concrete quality, cathodic protection, use of coatings, etc.;
- the optimisation of the design of new RC and steel structures taking into account specific environmental conditions and potential climate change scenarios.

Acknowledgements

The authors gratefully acknowledge the support of the Université de Nantes and the Pays de la Loire Regional Council for supporting the project RI-ADAPTCLIM.

References

ARCADIS, 2015. Global Built Asset Wealth Index 2015. https://www.arcadis.com/en/global/our-perspectives/global-built-asset-wealth-index/. (Accessed 15 May 2018).

Bastidas-Arteaga, E., 2018. Reliability of reinforced concrete structures subjected to corrosion-fatigue and climate change. Int. J. Concr. Struct. Mater. 12 (1), 1–10.

Bastidas-Arteaga, E., Stewart, M.G., 2015. Damage risks and economic assessment of climate adaptation strategies for design of new concrete structures subject to chloride-induced corrosion. Struct. Saf. 52, 40–53.

Bastidas-Arteaga, E., Stewart, M.G., 2016. Economic assessment of climate adaptation strategies for existing RC structures subjected to chloride-induced corrosion. Struct. Infrastruct. Eng. 12 (4), 432–449.

Bastidas-Arteaga, E., Stewart, M.G., 2017. Climate change impact on RC structures subjected to chloride ingress and carbonation-induced corrosion. In: Gardoni, P. (Ed.), Handbook of Sustainable and Resilient Infrastructure. Routledge.

Bastidas-Arteaga, E., Chateauneuf, A., Sánchez-Silva, M., Bressolette, P., Schoefs, F., 2010. Influence of weather and global warming in chloride ingress into concrete: a stochastic approach. Struct. Saf. 32 (4), 238–249.

Bastidas-Arteaga, E., Schoefs, F., Stewart, M.G., Wang, X., 2013. Influence of global warming on durability of corroding RC structures: a probabilistic approach. Eng. Struct. 51, 259–266.

Boéro, J., Schoefs, F., Yañez-Godoy, H., Capra, B., 2012. Time-function reliability of harbour infrastructures from stochastic modelling of corrosion. Eur. J. Environ. Civ. Eng. 16 (10), 1187–1201.

Chaves, I.A., Melchers, R.E., Peng, L., Stewart, M.G., 2016. Probabilistic remaining life estimation for deteriorating steel marine infrastructure under global warming and nutrient pollution. Ocean Eng. 126, 129–137.

de Larrard, T., Bastidas-Arteaga, E., Duprat, F., Schoefs, F., 2014. Effects of climate variations and global warming on the durability of RC structures subjected to carbonation. Civ. Eng. Environ. Syst. 31 (2), 153–164.

DuraCrete, 2000. Probabilistic Calculations. DuraCrete—Probabilistic Performance Based Durability Design of Concrete Structures. In: EU—Brite EuRam III. Contract BRPR-CT95-0132 Project BE95-1347/R12-13.

El Maaddawy, T., Soudki, K., 2007. A model for prediction of time from corrosion initiation to corrosion cracking. Cem. Concr. Compos. 29, 168–175.

EN-206. 2000. Concrete—Part 1: Specification, performance, production and conformity.

Hosseini, A., Sahrapeyma, A., Marefat, M., 2013. A reliability-based methodology for considering corrosion effects on fatigue deterioration in steel bridges—part I: methodology. Int. J. Steel Struct. 13 (4), 645–656.

IPCC, 2007. Climate change 2007: synthesis report. In: Pachauari, R.K., Reisinger, A. (Eds.), Contribution of Working Groups I, II and III to the Fourth Assessment Report on Intergovernmental Panel on Climate Change. IPCC, Geneva. Core writing team.

IPCC, 2013. Climate change 2013: the physical science basis. Contribution of Working, In: Stocker, T.F., Qin, D., Plattner, G.-K., Tignor, M., Allen, S.K., Boschung, J., Nauels, A., Xia, Y., Bex, V., Midgley, P.M. (Eds.), Group I to the Fifth Assessment Report of the Intergovernmental Panel on Climate Change. Cambridge University Press, Cambridge and New York, NY. 1535 pp.

Kallias, A.N., Imam, B.M., 2013. Effect of climate change on the deterioration of steel bridges. In: IABSE Symposium Reportvol. 99, pp. 1731–1738. 6.

Kayser, J.R., Nowak, A.S., 1989. Reliability of corroded steel girder bridges. Struct. Saf. 6 (1), 53–63.

Melchers, R.E., 2003. Modeling of marine immersion corrosion for mild and low-alloy steels—part 1: phenomenological model. Corrosion 59 (4), 319–334.

Melchers, R.E., 2005. The effect of corrosion on the structural reliability of steel offshore structures. Corros. Sci. 47 (10), 2391–2410.

Melchers, R.E., 2008a. Development of new applied models for steel corrosion in marine applications including shipping. Ships and Offshore Structures 3 (2), 135–144.

Melchers, R.E., 2008b. Extreme value statistics and long-term marine pitting corrosion of steel. Probab. Eng. Mech. 23 (4), 482–488.

Melchers, R.E., 2014. Using models to interpret data for monitoring and life prediction of deteriorating infrastructure systems. Struct. Infrastruct. Eng. 11 (1), 1–10.

Melchers, R.E., Jeffrey, R.J., 2008. The critical involvement of anaerobic bacterial activity in modelling the corrosion behaviour of mild steel in marine environments. Electrochim. Acta 54 (1), 80–85.

Melchers, R.E., Jeffrey, R., 2013. Accelerated low water corrosion of steel pilling in harbours. Corros. Eng. Sci. Technol. 48, 496–505.

Melchers, R.E., Jeffrey, R.J., Usher, K.M., 2014. Localized corrosion of steel sheet piling. Corros. Sci. 79, 139–147.

Mullard, J.A., Stewart, M.G., 2009. Stochastic assessment of the timing and efficiency of maintenance for corroding RC structures. J. Struct. Eng. ASCE 135 (8), 887–895.

Mullard, J.A., Stewart, M.G., 2011. Corrosion-induced cover cracking: new test data and predictive models. ACI Struct. J. 108 (1), 71–79.

Mullard, J.A., Stewart, M.G., 2012. Life-cycle cost assessment of maintenance strategies for RC structures in chloride environments. J. Bridg. Eng. 17 (2), 353–362.

NACE, 2016. International Measures of Prevention, Application, and Economics of Corrosion Technologies Study. National Association of Corrosion Engineers, Houston, Texas. (1 March).

Nguyen, M.N., Wang, X., Leicester, R.H., 2013. An assessment of climate change effects on atmospheric corrosion rates of steel structures. Corros. Eng. Sci. Technol. 48 (5), 359–369.

Nguyen, P.-T., Bastidas-Arteaga, E., Amiri, O., El Soueidy, C.-P., 2017. An efficient chloride ingress model for long-term lifetime assessment of reinforced concrete structures under realistic climate and exposure conditions. Int. J. Concr. Struct. Mater. 11 (2), 199–213.

Osorio, P., Odenbreit, C., Van Baars, S., Vrouwenvelder, T., 2011. Failure probability of marine steel sheet pile structures with special consideration of the corrosion impact. In: OCEANS 2011. IEEE—Spain, pp. 1–6.

Paik, J.K., Melchers, R.E., 2014. Condition Assessment of Aged Structures. Elsevier, New York.

Peng, L., Stewart, M.G., 2014. Spatial time-dependent reliability analysis of corrosion damage to concrete structures under a changing climate. Mag. Concr. Res. 66 (22), 1154–1169.

Peng, L., Stewart, M.G., 2015. Damage risks and cost-benefit analysis of climate adaptation strategies of RC structures in Australia. Concrete in Australia 41 (3), 36–47.

Peng, L., Stewart, M.G., 2016. Climate change and corrosion damage risks for reinforced concrete infrastructure in China. Struct. Infrastruct. Eng. 12 (4), 499–516.

Peng, L., Stewart, M.G., Melchers, R.E., 2017. Corrosion and capacity prediction of marine steel infrastructure under a changing environment. Struct. Infrastruct. Eng. 13 (8), 988–1001.

Saetta, A., Scotta, R., Vitaliani, R., 1993. Analysis of chloride diffusion into partially saturated concrete. ACI Mater. J. 90 (5), 441–451.

Silva, J.E., Garbatov, Y., Guedes Soares, C., 2014. Reliability assessment of a steel plate subjected to distributed and localized corrosion wastage. Eng. Struct. 59 (0), 13–20.

Stewart, M.G., 2001. Spalling risks, durability and life-cycle costs for RC buildings. In: Int. Conf. Safety, Risk Reliab.—Trends Eng., IABSE, pp. 537–542.

Stewart, M.G., Deng, X., 2015. Climate impact risks and climate adaptation engineering for built infrastructure. J. Risk Uncertain. Eng. Syst. 1(1). 04014001.

Stewart, M.G., Melchers, R.E., 1997. Probabilistic Risk Assessment of Engineering Systems. Chapman and Hall, London.

Stewart, M.G., Peng, J.X., 2010. Life-cycle cost assessment of climate change adaptation measures to minimise carbonation-induced corrosion risks. Int. J. Eng. Under Uncertain. 2, 35–46.

Stewart, M.G., Wang, X., Nguyen, M.N., 2011. Climate change impact and risks of concrete infrastructure deterioration. Eng. Struct. 33 (4), 1326–1337.

Stewart, M.G., Wang, X., Nguyen, M.N., 2012. Climate change adaptation for corrosion control of concrete infrastructure. Struct. Saf. 35 (1), 29–39.

Susskind, L., 2010. Responding to the risks posed by climate change: cities have no choice but to adapt. Town Plan. Rev. 81 (10), 217–235.

Sydney Ports, 2008. Port Botany Container Terminal Expansion Overview. (March 2008).

Val, D., Stewart, M.G., 2009. Reliability assessment of ageing reinforced concrete structures—current situation and future challenges. Struct. Eng. Int. 19 (2), 211–219.

Wang, X., Stewart, M.G., Nguyen, M., 2012. Impact of climate change on corrosion and damage to concrete infrastructure in Australia. Clim. Chang. 110 (3–4), 941–957.

CHAPTER FIVE

Timber Power Pole Network Management in a Changing Climate

Paraic C. Ryan*,†, Mark G. Stewart‡
*Civil, Structural and Environmental Engineering, University College Cork, Cork, Ireland
†Environmental Research Institute, University College Cork, Cork, Ireland
‡Centre for Infrastructure Performance and Reliability, School of Engineering, The University of Newcastle, Callaghan, NSW, Australia

5.1 Introduction

Critical infrastructure provides the vital services that allow advanced societies and economies to operate effectively and efficiently. Without it, many attributes of modern societies could not exist, such as high-quality health care, globalised supply chains, and worldwide communications (Yates, 2014). The reliance of other forms of critical infrastructure on energy infrastructure makes it perhaps the most important component of our modern critical infrastructure networks. The 1998 ice storms in the United States exemplified this, when power outages resulted in cascading failures across a range of other critical infrastructure sectors, namely; telecommunications, transportation, drinking water supply, banking and financial systems (Auld et al., 2006). More recently, power outages resulting from the 2017 storm Ophelia in Ireland resulted in cascading failures in the wastewater treatment, water treatment, and telecommunications sectors (Independent, 2017). It is thus vital that the power infrastructure we build today is capable of reliably serving our needs under future climatic conditions, especially given our ever-increasing dependency on power and power-based IT management and operational systems.

This chapter explores the vulnerability of power infrastructure to climate change, and examines the feasibility of utilising climate change adaptation to reduce potential climate change risks for this infrastructure. A single aspect of the power infrastructure is considered, namely, power distribution poles. These poles undergo far less stringent design and maintenance than other aspects of power infrastructure, and consequently are perhaps the most

Climate Adaptation Engineering
https://doi.org/10.1016/B978-0-12-816782-3.00005-X

vulnerable aspect of the network. The power pole networks themselves also constitute significant national assets across the globe, that is, there are five million timber power poles in Australia worth over $10 billion (Crews and Horrigan, 2000) and over 200 million timber power poles in the United States (Bolin and Smith, 2011). Another important consideration is the consequence of power pole failures, which are primarily caused by wind loading of deteriorated poles (Winkler et al., 2010). These consequences range from loss of power to business, to catastrophic bushfire (wildfire) events with significant loss of life and infrastructure (Ryan et al., 2014).

Climate change adaptation for power pole networks is explored in this chapter using probabilistic modelling techniques, which provide risk-based decision support. The use of probabilistic analysis allows for incorporation of the considerable uncertainty associated with future climate projections, and the inherent variability of assets across an infrastructure network. Uncertainties associated with structural capacity, structural loading, and deterioration of assets with time are also incorporated using probabilistic methods. The approach constitutes a cross-disciplinary framework, which brings together the latest climate projections from the scientific community, and established structural reliability tools from engineering research, to assess and develop practical solutions to climate change. Engineering-based frameworks, like those developed herein for timber power poles, can play a key role in bridging gaps between the climate change scientific community and infrastructure owners and policy makers, resulting in more sustainable and resilient critical infrastructure.

The framework developed in this chapter utilises a three-step process to develop the robust risk-based decision support for climate adaptation. As can be seen from Fig. 5.1, the first requisite step in this process is understanding

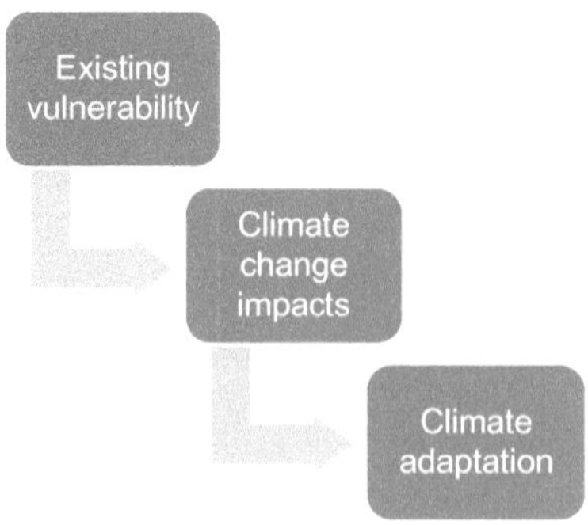

Fig. 5.1 Framework for assessment of climate adaptation for timber power poles.

existing vulnerability. We must develop this clear engineering-based understanding of the vulnerability of the network under today's conditions, before we can begin to predict how the network might preform under conditions in the future. Having established existing vulnerability, the second step is to evaluate climate change impacts for a range of climate change scenarios. The examination of impacts across different scenarios ensures we avoid the worst-case thinking pitfalls highlighted in Chapter 1. The impact analysis will indicate if climate adaptation is necessary. With this knowledge, the feasibility of different climate adaptation strategies can be explored where appropriate, using tools such as probabilistic cost-benefit analysis. The main body of this chapter is broken into three sections in line with this three stage process, that is, Sections 5.2–5.4, respectively, with a chapter summary provided in Section 5.5. The case studies presented in each of the main sections are conducted in an Australian context, however, the framework used can be applied to any country, and indeed to any infrastructure type, that is, transport, water treatment, etc.

5.2 Existing Pole Vulnerability

As discussed in the introduction to this chapter, the requisite first step in assessing climate change impacts and climate adaptation for power infrastructure is understanding the existing vulnerability. This section presents a model framework developed to examine the vulnerability of a notional power pole network under existing environmental conditions, considering the impact of extreme events and network deterioration. This work is discussed in more detail in Ryan et al. (2014). Having developed this first-principals mathematical understanding of the network, adjustments to the model can be used to assess climate change impacts and the effectiveness of climate adaptation options. These adjustments will be discussed further in Sections 5.3 and 5.4. In addition to presenting the model framework, some results of vulnerability under existing conditions are also presented in this chapter for the Australian case study.

5.2.1 Probabilistic Methodology

An assessment of vulnerability or structural reliability requires consideration of structural resistance of the element being considered (R), and the load the element is subjected to (S). In the context of examination of a network of structural systems such as power pole networks, bridge networks, road networks, etc., asset deterioration and network maintenance must also be considered. Thus, for such a network, the relationship between load (S) and resistance (R) is time dependent. The objective of structural reliability analysis is to evaluate the probability of occurrence of a given failure event at a given time through calculation of the failure probability ($P_f(t)$):

$$P_f(t) = P[(R(t) - S(t)) \leq 0] \tag{5.1}$$

where $R(t)$ and $S(t)$ are resistance and load effects at time t, respectively (Barone and Frangopol, 2014; Stewart and Val, 1999). The $R(t)$ term in the limit state function in Eq. (5.1) will be the bending resistance of a power pole incorporating deterioration, while the $S(t)$ term relates to the annual maximum wind load. This limit state was selected based on the most common failure mode of timber power poles (Winkler et al., 2010).

A probabilistic event-based time-dependent methodology was used to incorporate load, resistance, deterioration, and network maintenance into the infrastructure reliability assessment. A Monte Carlo simulation approach was employed due to the recognised flexibility and robustness of this method in assessing reliability of both structural components and systems (Johansson et al., 2013; Li and Li, 2002; Stewart and Val, 1999). The time-dependent nature of the model refers to the fact that each Monte Carlo simulation runs on a year-by-year basis. For instance, in the case study described in this section, a simulation period of 100 years was selected for the power pole network. Thus, for each Monte Carlo simulation, the model steps from 0 to 100 years, with each yearly step including; (a) calculation of the resistance of the pole at year t, accounting for deterioration, (b) calculation of a load at year t derived from a distribution of maximum annual loads, and (c) infrastructure element inspection and network maintenance where applicable. This time-dependent modelling approach simulates the actual stochastic behaviour of the system overtime, thus creating "an artificial history" of infrastructure network performance. It is also referred to as a sequential modelling approach in the literature (Johansson et al., 2013; Billinton and Wang, 1999).

The event-based aspect of the probabilistic model developed herein refers to the fact that the occurrence of certain events over the simulation period can influence the course of a given sequential Monte Carlo simulation. The two key events which can occur are: (a) violation of the limit state, whereby the annual wind load exceeds the deteriorated pole capacity, and (b) the condemning of a pole as a result of the network inspections and maintenance programme. Upon occurrence of a wind failure or the condemning of a pole, the pole in question is replaced by a new pole in the Monte Carlo simulation. This pole is assigned new properties generated from the appropriate distributions, that is, pole diameter, pole bending strength, sapwood depth, etc. The process of deterioration then restarts for this new pole, and the sequential Monte Carlo process continues for the iteration in question up to the end of the simulation period selected. This time-dependent event-based modelling approach means that, in effect, each Monte Carlo iteration represents one pole location in a network of poles, whereby if this pole fails at a given location it must be replaced to prevent power supply interruption.

It is noted at this point that the case studies presented in this chapter focus on Australia, and thus the inspection of the poles in the network will be modelled herein based on current standards (Standards Australia/New Zealand, 2013) and existing pole inspection and maintenance replacement practice employed in the power industry in Australia. Inspection intervals were set in the model at 5 years, with first inspection at 10 years for untreated poles and 20 years for treated poles (Standards Australia/New Zealand, 2013). In accordance with common industry practice in Australia, inspection failure or pole condemning criteria was set at 50% of original pole capacity based on loss of section modulus, meaning if inspection revealed that the pole moment capacity was less than 50% of the original pole moment capacity the pole failed the inspection and was condemned.

5.2.1.1 Modelling Time-dependent Resistance and Load

The initial bending resistance ($R(0)$) of a timber power pole at time $t=0$ can be represented based on established bending theory as follows:

$$R(0) = f_b \cdot \frac{\pi \cdot D^3}{32} \tag{5.2}$$

where f_b is the bending strength of the timber and D is the ground-line diameter of the distribution pole. In order to model Eq. (5.2) probabilistically, statistical parameters are required for: (a) the pole ground-line diameter, (b) the bending strength of the timber species in question, and (c) the model error associated with Eq. (5.2). The statistical parameters for D are obtained from power pole design, according to the existing Australian standards (Standards Australia/New Zealand, 2011) (Standards Australia/New Zealand, 2010), and subsequent supplier sizing range. These sizing ranges stem from the fact that, unlike steel and concrete, timber is a natural material, and thus exact diameters cannot be specified. Logic dictates that pole ground-line diameters for sized poles will be uniformly distributed from the minimum value to the

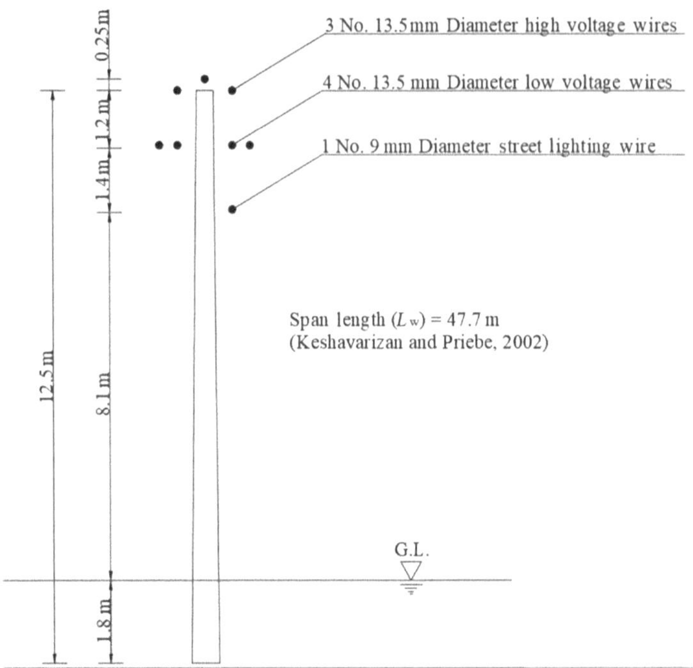

Fig. 5.2 Distribution pole layout.

maximum value within the applicable sizing range. The power pole design was carried out for a typical Australian power setup developed through consultation with the power industry (Fig. 5.2). Information on a mean value for f_b and a joint coefficient of variation (COV) value combining model uncertainty, testing uncertainty, and f_b variation was obtained from experimental test results of power poles (Spencer, 2013).

It is noted, however, that this initial bending resistance ($R(0)$) will tell us little about the performance of timber power poles in service, which undergo significant deterioration over their service life. Thus, for timber power pole networks, as with any infrastructure network analysis, infrastructure element deterioration must be incorporated into the probabilistic model. Timber deterioration was included herein based on the work of Wang et al. (2008a,b), who developed a timber decay model based on 35 years of field data for 77 timber species. The Wang et al. model represents decay progress as an idealised bilinear relationship, characterised by a decay lag time, t_{lag} (years), and a decay rate, r (mm/yr). These r and t_{lag} values can be calculated as:

$$r = k_{wood} \cdot k_{climate} \tag{5.3}$$

$$t_{lag} = 5.5r^{-0.95} \tag{5.4}$$

where k_{wood} is the wood parameter and $k_{climate}$ is the climate parameter. The Wang et al. (2008b) publication provides formulas for k_{wood} which account for timber type, treatment, etc. The uncertainty associated with the decay model was incorporated into the framework using a model error factor for decay, ME_{decay}. According to Wang et al. (2008a), uncertainty increases with increasing timber durability class, with class 1, 2, 3, and 4 timbers having model error COV values of 0.85, 0.9, 1.1, and 1.2, respectively. Thus, depth of decay, d, at a given time, t, incorporating uncertainty and variability can be calculated as:

$$d(t) = ME_{decay} \cdot r \cdot (t - t_{lag}) = ME_{decay} \cdot r \cdot \left(t - 5.5 \cdot ME_{decay} \cdot r^{-0.95}\right) \quad (5.5)$$

The model represented by Eq. (5.5) is, however, for a homogenous timber cross-section. As recognised by Wang et al. (2008b), a timber pole's cross section is far from homogenous. Instead the pole consists of zones of sapwood, heartwood (also termed outerwood), and corewood (Pfautsch et al., 2012; Walker, 2006), all of which will have different rates of deterioration and lag times (Wang et al., 2008b). Wang et al. (2008b) provide t_{lag} equations and values for k_{wood} for sapwood, heartwood, and corewood for both internal and external decay. This information was utilised together with the time-dependent event-based approach (i.e., decay rates change overtime when decay progresses into different sections of the pole) to implement a multilayer timber deterioration model to represent both internal and external decay for timber poles. The nature of this multilayered model for external decay of a treated timber pole is illustrated in Fig. 5.3. The implementation of the model is discussed in more detail in Ryan et al. (2014).

Having used Eq. (5.5) within the multilayered model framework to calculate; (i) the diameter of the internally decayed timber at time, t, $D_1(t)$ and (ii) the external diameter remaining after external decay at time t, $D_2(t)$, the following modification of Eq. (5.2) was used to calculate the time-dependent bending resistance, $R(t)$, of a power pole:

$$R(t) = f_b \cdot \frac{\pi \cdot \left[D_1(t)^4 - D_2(t)^4\right]}{32 \cdot D_1(t)} \quad (5.6)$$

Having developed an expression for $R(t)$, it is necessary to consider the load side of the equation $S(t)$. To capture the time-dependent nature of

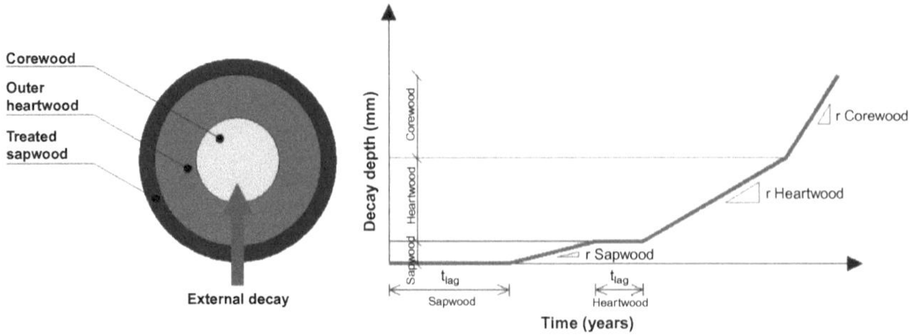

Fig. 5.3 Schematic of multilayered-treated pole decay model adapted from Wang et al. (2008b).

probabilistic wind load, mathematical expressions incorporating wind speed variability, and wind load uncertainty were required. Statistical data for the wind field, v, used in this study were adopted from Wang et al. (2013), who developed statistical distributions for the annual maximum gust wind speeds for a range of locations in Australia. The wind field for the case study locations examined in this chapter is dominated by non-cyclonic synoptic weather systems. These non-cyclonic gust wind speeds are modelled using the Gumbel distribution, with the annual maximum non-cyclonic peak gust speed as follows (Wang et al., 2013):

$$F(v) = e^{-A} \text{ where } A = e^{-\left(\frac{v - v_g}{\sigma_g}\right)} \tag{5.7}$$

where v_g and σ_g are the location and scale parameters for the Gumbel distribution. The statistical parameters for each of the five Australian cities considered in the case study herein are provided in Appendix 5A (Wang et al., 2013).

Having established an expression which probabilistically represents the time-dependent wind speed, the work of Henderson and Ginger (2007) was used to calculate the time-dependent wind load $S(t)$ as follows:

$$S(t) = B \cdot v(t)^2 \tag{5.8}$$

where:

$$B = ME_w \cdot A \cdot \left(C \cdot E^2 \cdot D_w^2 \cdot G \cdot \rho/2\right) \tag{5.9}$$

The terms in the brackets in Eq. (5.9) are related to the nominal values given in the Australian wind loading standards (Standards Australia/New Zealand, 2011) where C is the quasi-steady pressure coefficient related to the C_d parameter in AS/NZS 1170.2, E is the velocity height multiplier that accounts for exposure and height related to $M_{z,cat}$ in AS/NZS 1170.2, D_w is the wind directional effects related to M_d in AS/NZS 1170.2, G is a factor that accounts for gusting effects and is related to K_a and K_l in AS/NZS 1170.2, ρ is the density of air, and ME_w is the model error for the wind load model. The term $v(t)$ in Eq. (5.8) is the time-dependent 0.2-s gust wind speed. This model allows uncertainty and variability to be incorporated into predictions for the wind load on the power poles, conductors, etc., for a given wind speed v, at a given time t. Details of the statistical properties used for the variables in Eq. (5.9) are provided in Appendix 5A.

5.2.2 Existing Vulnerability Findings

The focus of this chapter is on the climate change impacts and climate adaptation. Consequently, only limited results will be presented on existing vulnerability findings, with readers referred to Ryan et al. (2014) for further findings on existing vulnerability. The first of the results presented relates to the effect of incorporating network maintenance into the vulnerability assessment. Only a handful of studies in the existing literature consider the effect of network maintenance actions on the structural reliability of pole networks (Bjarnadottir et al., 2014; Gustavsen and Rolfseng, 2004; Salman et al., 2017). However, given that power pole networks throughout the world are inspected and replaced in order to reduce annual wind failure rates, a realistic assessment of the structural reliability of a distribution pole in a working network must incorporate the effect of these inspections and subsequent maintenance replacements.

Fig. 5.4 shows a comparison of wind vulnerability at $t = 20$ years for analysis with and without consideration of network maintenance for a location in New South Wales in Australia. It is clear that the wind vulnerability of the unmaintained pole is higher, particularly at lower wind speeds. These higher probabilities of wind failure at lower wind speeds represent the portion of the unmaintained pole population which have suffered significant or even complete decay at the 20-year point in time. However, the more realistic maintained network vulnerability curve reflects the fact that many poles which have deteriorated to a large extent by 20 years are likely to have been replaced as a result of network maintenance or pole failures. As mean annual maximum wind speed for the pole design location (New South Wales, Australia) tends to be around 30 m/s, the most influential zone of the vulnerability curve in dictating failure occurrences lies around 40–50 m/s. These

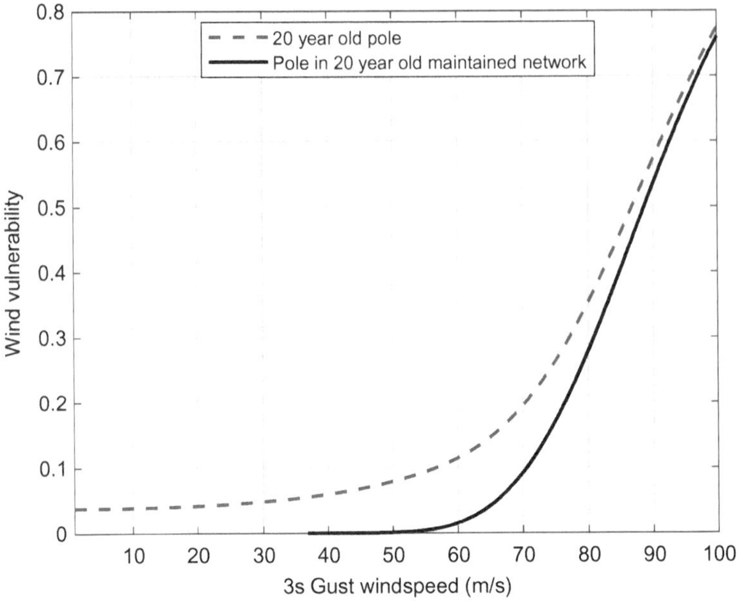

Fig. 5.4 Vulnerability curves for 20 year old pole and pole in a 20 year old maintained network.

results show that in order to get a better idea of the vulnerability of power poles in service, network maintenance must be incorporated into time-dependent structural reliability analysis.

Fig. 5.5 shows the annual wind failure rates and condemning rates for copper chrome arsenate (CCA) treated poles obtained from the one million Monte Carlo simulations. First, considering the condemning rates; from initial inspection at 20 years, to the 100th-year point, condemning rates range from 0.7% to 9.5%, with condemning only occurring every 5 years in accordance with the inspection intervals. The average annual pole condemning rate for the treated poles over the 100-year monitoring period is 1.01%, meaning on average 1% of the treated pole network fail inspection, are condemned and are subsequently replaced each year. This figure compares reasonably well with the pole replacement rate experienced in the Essential Energy's network of one million power distribution poles in rural New South Wales in 2007 which was found to be 1.66% (Spencer and Elder, 2009). It is noted that while obtaining a pole condemning rate of the same order of magnitude as that currently experienced in industry is desirable, the aim of this assessment is not to simulate the existing network, which comprises of a mix of treated and untreated poles of various ages designed

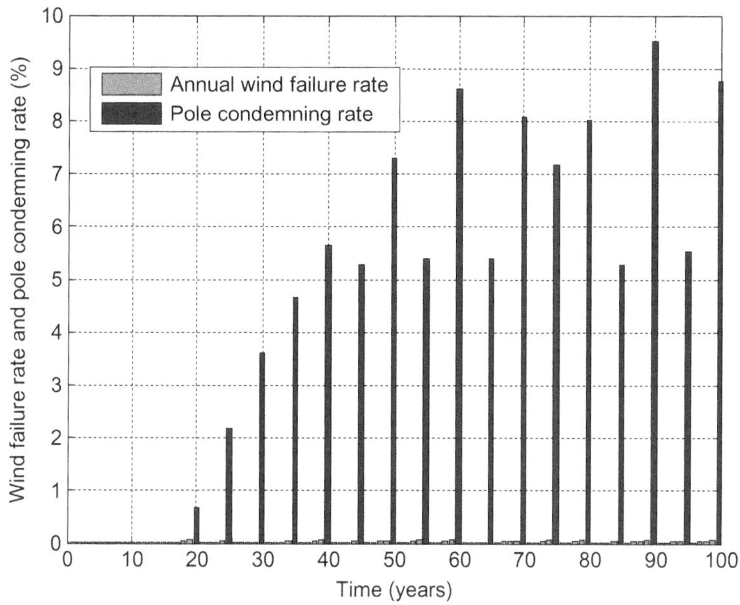

Fig. 5.5 Condemning rates and annual wind failure rates for treated poles.

using either the older allowable stress design methods (Electricity Supply Association of Australia, 1991) or the current AS/NZS standards.

The annual wind failure rates in Fig. 5.5 are a fraction of the 5-yearly pole condemning rates. Consequently, the annual wind failure rates are plotted separately on a more appropriate scale in the bar chart in Fig. 5.6. This figure shows low failure rates in the first 1–17 years, followed by a steep rise from 17 to 20 years. After this point, the annual failure rate initially drops and then rises to a peak every 5 years. This cyclical nature of wind failures is due to the influence of the maintenance inspection intervals. The inspections occur at the end of every 5th year. These inspections lead to the replacement of the most decayed poles, which are most vulnerable to wind failure. Thus, in the year following the inspection, the network is most resilient to wind failure and the failure rates are low, while just before the inspection wind vulnerability is increased meaning failure rates are highest. The average annual wind failure rate for the treated poles over the 100-year monitoring period was 0.036%. Due to restricted access to pole failure records and varying accuracy of existing records, it is difficult to obtain an exact figure for annual wind failure rates for timber power distribution poles in Australia. However, industry experience indicates that annual failure rates in New South Wales

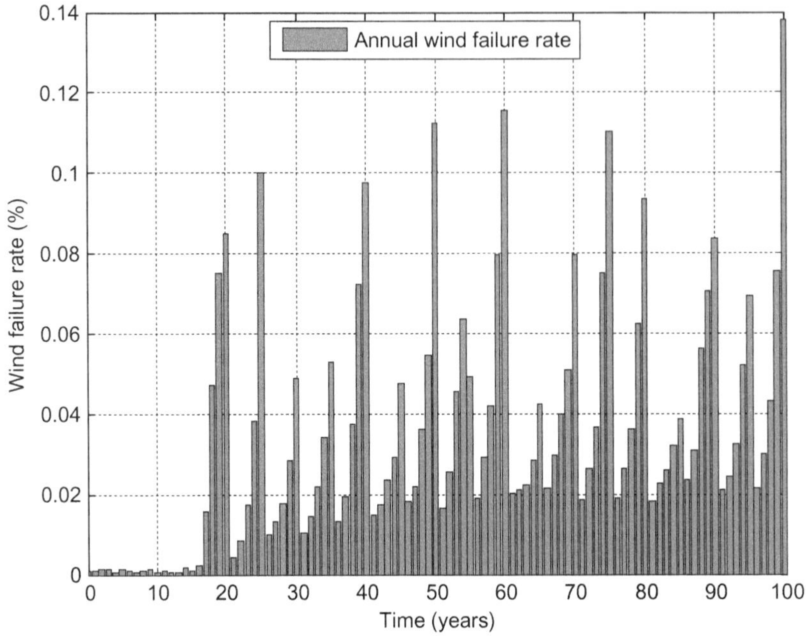

Fig. 5.6 Annual wind failure rates from 0 to 100 years for treated poles.

are currently in the region of 0.01%–0.04%. Thus, the model output for treated poles designed and maintained to existing standards is reasonably close to the wind reliability of the existing network. Again it is noted that while this agreement is encouraging, the aim of this assessment was to assess treated poles designed and maintained in accordance with the recently developed AS/NZS 7000:2010 standard, and not to attempt to emulate the existing network.

An analysis was also conducted to compare the performance of treated and untreated timber poles. This analysis found that under wind load, treated and untreated poles designed and maintained in accordance with existing Australian standards were found to have similar failure rates (0.036% and 0.041%, respectively). This is due to the fact that the untreated pole has a greater original diameter than the treated pole (mean treated pole diameter = 268 mm and mean untreated pole diameter = 313 mm). However, untreated pole networks, which experience more rapid deterioration, required approximately twice as many maintenance-based pole replacements to sustain this level of reliability.

5.3 Climate Change Impact and Risk

The importance of implementing effective climate change adaptation strategies is beginning to penetrate into infrastructure management and policy across the globe (IPCC, 2014). For instance, the US military have recently announced that the Pentagon is undertaking sweeping changes to operation systems and installations to adapt to climate change impacts now and in the future (Goldenberg, 2014). However, in order to implement effective climate change adaptation, it is essential to first understand the nature and magnitude of predicted climate change impacts. Without this understanding of the climate change impacts, we are susceptible to the pitfalls discussed in Chapter 1 such as worst-case thinking and probability neglect. Understanding the extent of climate impacts and climate risk is not straightforward, and requires detailed analysis for two main reasons; (1) the influence of climate change on infrastructure performance is normally dependent on a relatively complex interaction of a number of different climatic effects (i.e., temperature, rainfall, wind speeds, etc.) manifesting in changes in deterioration rates and loading conditions, etc. and (2) there is considerable uncertainty associated with future climate change projections and long-term infrastructure performance.

The case study presented in this chapter examines the climate change impacts on power distribution pole networks across five regions in Australia. Again, probabilistic methods are used to facilitate incorporation of uncertainty at various levels. The methodology below presents how the interactive impact of temperature, rainfall, and wind speed are incorporated into the climate change impact analysis. Section 5.3.2 presents the findings of the case study assessment, illustrating the regional variability of climate change impact. Further details pertaining to this case study can be found in Ryan et al. (2016).

5.3.1 Methodology

5.3.1.1 Modelling Climate Change Impacts

The theoretical model developed in Section 5.2 allows prediction of the impacts of climate change on power distribution infrastructure. Climate change impact on both timber deterioration and wind speed is incorporated into the study. First, considering timber deterioration, the impact of climate change is incorporated into the Wang et al. model (2008a,b) through the

k_{climate} parameter, mentioned in Section 5.3. The k_{climate} parameter is calculated as follows:

$$k_{\text{climate}} = f(RN)^{0.3} \cdot g(T)^{0.2} \tag{5.10}$$

$$\begin{aligned} &g(T) = 0, \quad \text{if } T \leq 5^{\circ}\text{C} \\ &g(T) = -1 + 0.2T, \quad \text{if } 5^{\circ}\text{C} < T \leq 20^{\circ}\text{C} \\ &g(T) = -25 + 1.4T, \quad \text{if } T > 20^{\circ}\text{C} \end{aligned} \tag{5.11}$$

$$\begin{aligned} f(RN) = 10 \cdot \left[1 - \text{e}^{-0.001(RN-250)}\right] \cdot \left(1 - \frac{N_{\text{dm}}}{6}\right), \\ \text{if } RN > 250 \text{ mm and } 0 \leq N_{\text{dm}} \leq 6 \\ \text{Otherwise } f(RN) = 0 \end{aligned} \tag{5.12}$$

where T is the average annual temperature (°C), RN is the annual rainfall (mm/yr), and N_{dm} is the number of months per year where the rainfall is less than 5 mm (Wang and Wang, 2012). Given the influence of climatic conditions on timber decay, it is clear that if the climate changes in the future, as the IPCC predict it will, there may be implications for the reliability and management of timber-based infrastructure assets worldwide.

Possible changes in wind speed are also an important consideration for power distribution infrastructure as increases or decreases in annual maximum wind speed will reduce or increase the reliability performance of power pole infrastructure, respectively. The location and scale parameters developed by Wang et al. (2013) discussed in the previous section are based on the data recorded from 1939 to 2007. Thus, this wind field does not accounted for the possible future reductions or increases in wind field magnitude due to climate change. A modification to Eq. (5.7) has been suggested by Stewart (2015) to allow climate change-related effects to be incorporated into the Gumbel distribution under the assumption that the relationship between mean wind speed (averaged over 10 min), and peak gust wind speed is constant (i.e., the proportionality constant between gust wind and mean wind speeds is one). Eq. (5.7) thus becomes.

$$F(v) = e^{-A} \text{ where } A = e^{-\frac{\left(\frac{v}{1 + \frac{\gamma_{\text{mean}}(t)}{100}}\right) - v_g}{\sigma_g}} \tag{5.13}$$

where $\gamma_{\text{mean}}(t)$ represents the time-dependent percentage change in gust wind speed for a given Monte Carlo simulation. The Wang et al. (2013)

statistical parameters for the Gumbel distribution are assumed to reflect pre-climate change wind speeds.

5.3.1.2 Climate and Hazard Modelling

The case study presented in this section examines the climate change impacts on power distribution poles across five different Australian cities; Brisbane, Sydney, Canberra, Melbourne, and Perth. Regionally specific climate change predictions for each location, developed by the Australian Commonwealth Scientific and Industrial Research Organisation (Webb and Hennessy, 2015), were used in the analysis. Three climate change scenarios were examined for each city; (a) no climate change, whereby climatic conditions remain at 1990 levels, (b) the IPCC RCP4.5 (median) emissions scenario, and (c) the IPCC RCP8.5 (worst) emission scenario. Table 5.1 presents the median, 10th percentile, and 90th percentile predicted changes in temperature, rainfall, and wind speed for the RCP4.5 and RCP8.5 scenarios for the five Australian cities. The values represent the total change to the year 2090, relative to 1990 levels. In line with the framework set out by Stewart (2015) and Stewart and Deng (2015) truncated normal distributions were used to represent the uncertainty associated with climatic predictions, whereby the 50th, 10th, and 90th percentiles provided by the CSIRO (Webb and Hennessy, 2015), allow the standard deviation of two truncated normal distributions, each with a cumulative probability of 50%, to be calculated. In the absence of yearly data resolution for regional climate change predictions (the CSIRO provides total change predictions at year 2090), a time-dependent linear change in climatic conditions for the RCP4.5 and RCP8.5 scenarios was assumed (Stewart, 2015; Stewart and Deng, 2015). A Monte Carlo simulation approach was used to incorporate the climatic uncertainty in the probabilistic model, with one million Monte Carlo iterations carried out for each of the cities considered.

Examination of the prediction patterns in Table 5.1 reveals two important points. First, the predicted climate change effects vary from location to location. For instance, Brisbane has a predicted median wind speed increase of 2.2% under the RCP8.5 (most severe) scenario, while three of the four other cities have a predicted median reduction in wind speed under this emission scenario. A reduction in wind speed means a reduction in structural loading, which could result in positive climate change impacts for some regions. Second, from Table 5.1, it is noted that temperature is expected to increase and rainfall is expected to decrease (by varying amounts) for all five locations. A rise in temperature will increase deterioration rates,

Table 5.1 Statistical Parameters for Climate Change Predictions to 2090 for Five Australian Locations

Parameter	RCP 4.5 10th P	RCP 4.5 50th P	RCP 4.5 90th P	RCP 8.5 10th P	RCP 8.5 50th P	RCP 8.5 90th P
Brisbane						
Temperature (°C)	+1.2	+1.8	+2.6	+2.5	+3.7	+4.7
Rainfall (%)	−21	−9	+7	−32	−16	+17
Wind speed (%)	−2.5	+0.5	+3.6	−1.2	+2.2	+6.5
Sydney						
Temperature (°C)	+1.3	+1.8	+2.5	+2.9	+3.7	+4.6
Rainfall (%)	−16	−2	+9	−20	−3	+16
Wind speed (%)	−4.2	−1.0	+0.2	−6.9	−1.1	+4.2
Canberra						
Temperature (°C)	+1.3	+1.8	+2.4	+2.7	+3.8	+4.5
Rainfall (%)	−16	−6	+4	−27	−5	+ 9
Wind speed (%)	−4.6	−1.3	+0.8	−5.0	−0.6	+2.6
Melbourne						
Temperature (°C)	+1.1	+1.5	+1.9	+2.4	+3.0	+3.8
Rainfall (%)	−15	−7	+3	−27	−9	+4
Wind speed (%)	−5.2	−1.1	+0.9	−5.9	−1.7	+1.7
Perth						
Temperature (°C)	+1.1	+1.7	+2.1	+2.6	+3.5	+4.2
Rainfall (%)	−22	−12	−1	−37	−18	−5
Wind speed (%)	−2.6	−0.3	+1.4	−2.7	+0.3	+3.3

Note: P, percentile.

increasing network wind vulnerability, while a decrease in rainfall will decrease deterioration rates, decreasing network vulnerability. This illustrates how climate change impacts are not intuitive for many infrastructure networks, with the possibility of positive or negative net impacts dependent on the interactions of opposing influencing climate change effects.

5.3.2 Case Study Findings

The aim of this case study was to determine the likely climate change impacts on a notional network of one million newly installed CCA-treated timber power distribution poles, for the period 2015–2090 (76 years), across five Australian cities. To facilitate fair comparison between locations, both the new 2015 poles, and replacement poles, were designed for each location in accordance with existing standards [AS/NZS 1170.2:2011 (Standards Australia/New Zealand, 2011) and AS/NZS 7000:2010 (Standards Australia/New Zealand, 2010)], for the typical Australian power distribution pole layout presented in Fig. 5.2. Spotted gum (*Corymbia citriodora*, *Corymbia henryi*, and *Corymbia maculate*) is the most common timber species used for treated poles in Brisbane, Sydney, Canberra, and Melbourne (Spencer and Elder, 2009, Francis and Norton, 2006), and thus was utilised in the design and modelling for these locations. The majority of newly installed poles used in Western Australia, however, are CCA-treated radiata pine, and thus this pole type was used in the design and modelling for Perth. The statistical parameters for model inputs can be found in Appendix 5A.

The impact of the three climate change scenarios on the average annual pole condemning rates for the five Australian cities is shown in Fig. 5.7. Fig. 5.8 presents the percentage increase in pole condemning rates for the three scenarios, for each of the locations, that is, increase relative to "no change," or baseline condemning rates for each location. As can be seen from Fig. 5.7, the predicted climate change impact can be significant with Brisbane experiencing the largest impact. The RCP8.5 climate scenario for Brisbane results in an increase in average annual condemning rates over the monitoring period of approximately 0.11%, representing a 11.2% increase in pole condemning rates when compared to the Brisbane "no change" scenario. Sydney also experiences a notable proportional increase in pole condemning rates due to climate change under the RCP8.5 scenario with a 10.7% increase when compared with the no change scenario. In contrast to Brisbane and Sydney, Melbourne actually experiences a slight reduction in annual pole condemning rates for both the RCP4.5 and RCP8.5 emission scenarios, while the climate change impacts for Canberra were relatively small.

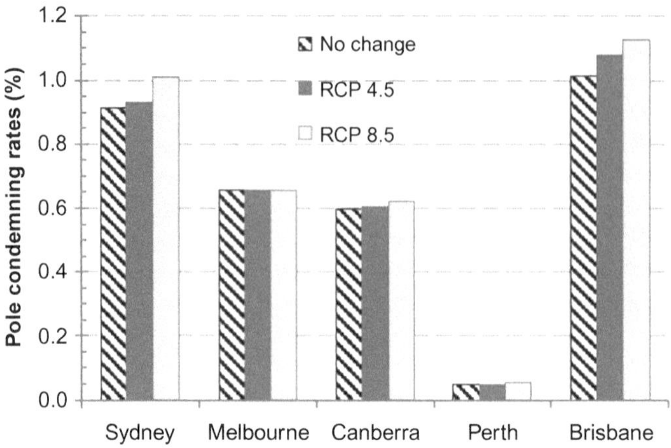

Fig. 5.7 Annual pole condemning rates for three climate change scenarios for the five locations.

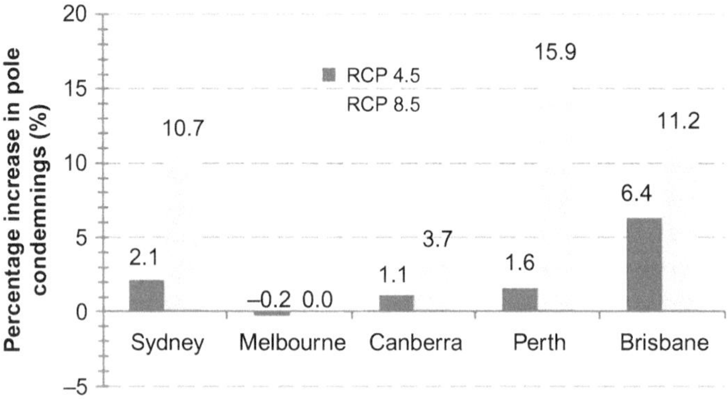

Fig. 5.8 Percentage change in annual pole condemning rates due to climate change for five locations.

Insight into the nature of the regional variability of climate change impacts can be obtained by comparing the findings for Sydney and Canberra, which exhibit very different results for the RCP8.5 emission scenario, despite having similar predicted changes in temperature and rainfall for this scenario (see Table 5.1). This is due to the influence of the cities' baseline 1990 temperature and rainfall levels on probabilistically modelled timber deterioration in accordance with the Wang et al. (2008b) timber decay model. First, considering temperature, it can be noted from Eqs. (5.10) and (5.11) that the relationship between temperature and decay rate in the model of Wang et al. is bimodal, with a greater increase in decay rate for unit change in temperature after the

20°C point. Thus, 1°C increase in temperature in a location with a baseline annual average temperature of 22°C will have a greater influence on decay than a 1°C change in temperature at a location with an annual average temperature of 16°C. Sydney annual average temperature is much closer to the critical 20°C mark than Canberra, with baseline temperatures of 18.2°C and 13.6°C in the two cities, respectively. Thus, with Sydney and Canberra both having a median climate change-related temperature increase of approximately 3.7°C, the temperature change has a higher influence on decay rates for Sydney than for Canberra. This is also a key contributing factor in Brisbane's high predicted climate change impacts, as Brisbane has a baseline temperature of 21°C.

When considering rainfall, the nature of the $f(RN)$ expression in Eq. (5.12) means that once annual rainfall drops below approximately 600 mm, the decay rate begins to drastically reduce. Canberra has a baseline annual rainfall of 607 mm, while Sydney's baseline annual rainfall is 1302 mm. This results in a greater reduction in decay rates due to annual rainfall change for Canberra than for Sydney, despite both locations experiencing similar predicted climate change related percentage reductions in rainfall.

Finally, in the context of Figs. 5.7 and 5.8, it is noted that the Perth pole network experiences a low annual condemning rate for the three emission scenarios. This was due to the influence of pole timber species utilised, with the radiata pine poles resulting in far less poles condemned than the spotted gum poles over the 76-year monitoring period for a given location. As shown in Fig. 5.8, however, the small increases in pole condemning rates due to the RCP8.5 emission scenario represent a notable proportional increase in condemning rates, with a 15.9% increase observed when compared to the no change scenario.

Fig. 5.9 shows the average annual wind failure rates for the three emission scenarios for each of the five Australian locations, with Fig. 5.10 showing the percentage change in annual wind failure rates for the RCP4.5 and RCP8.5 emission scenarios for each location. Brisbane can be seen to experience the most severe increase in annual wind failures, which is partially due to the fact that Brisbane is the only one of the five cities examined to experience a notable median increase in wind speed (Table 5.1). Sydney also experiences a significant increase in average annual wind failure rates for the RCP8.5 scenario, with a rise of 0.047%, representing a 17.5% increase in the average annual wind failure rates when compared to "no climate change" scenario. This is primarily related to increased decay and increased

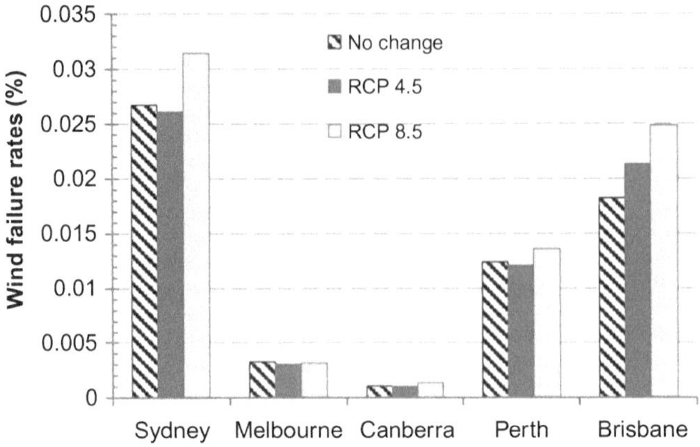

Fig. 5.9 Annual pole wind failure rates for three climate change scenarios for the five locations.

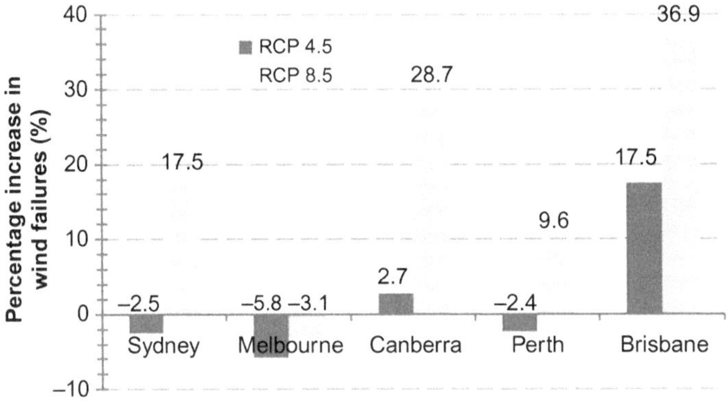

Fig. 5.10 Percentage change in annual pole wind failure rates due to climate change for five cities.

variability in wind speed caused by climate change uncertainty. Canberra experiences a small increase in average annual wind failure rates, but a significant proportional increase in wind failure rates. This is due to the fact that Canberra city experiences low wind failures for the "no change" scenario as median wind speed for this location is approximately 20 m/s, 9 m/s less than Sydney, despite both locations having the same wind speed classification from a design perspective (Standards Australia/New Zealand, 2011). It is also interesting to observe that Melbourne experiences a decrease in wind failures for both emission scenarios.

It is noted that there is considerable uncertainty about climate change predictions relating to extreme wind speed. Thus, by way of a sensitivity study, the authors examined how predicted climate change impacts for Brisbane would be effected if the predicted change in wind speed was omitted from the analysis, that is, what are the climate change impacts if only temperature and rainfall are considered. The results are shown in Table 5.2 for both climate change scenarios. As expected the condemning rates, which are influenced by decay rates, are not affected by the omission of wind speed change to the year 2090. However, it is more interesting to note that the impact of climate change on wind failures in Brisbane only reduces by 1.6% for RCP4.5 (mean predicted wind speed change is +0.5%) and 5.4% for RCP8.5 (mean predicted wind speed change is +2.2%). This verifies that deterioration drives the magnitude of impacts for both failure rates and condemning rates. Consequently, the lack of confidence in predicted changes in extreme wind speed does not greatly impact upon the findings presented in Figs. 5.7–5.10.

Overall, the climate change case study indicates that impacts on power pole networks are likely to be substantial for some locations in Australia, with Brisbane experiencing a projected 11% increase in pole condemning rates, and an approximate 37% increase in wind failure rates over the period 2015–90

Table 5.2 Effect of Removing Wind Speed Changes from Climate Change Impact Analysis

	Wind Failures		Pole Condemnings	
Scenario	**Total Failures**	**% Impact of Climate Change**	**Total Condemnings**	**% Impact of Climate Change**
No Climate Change	13,640	–	760,420	–
RCP 4.5	16,030	17.5	808,820	6.4
RCP 8.5	18,670	36.9	845,620	11.2
Neglecting wind speed change				
RCP 4.5 with no wind speed change	15,802	15.9	807,836	6.2
RCP 4.5 with no wind speed change	17,936	31.5	845,410	11.2

under RCP8.5. Such increases could have substantial societal impacts when one considers the consequences of power pole failures, which range from interruptions of power for businesses and homes, to loss of life due to wildfires. However, the predicted impacts of climate change were found to be highly regionally variable. In addition to the interaction between three climatic change parameters (temperature, rainfall, and wind speed), predicted climate change impacts are also influenced by a regions baseline climatic conditions, type of timber used in a region, and the uncertainty associated with climate change predictions, something that can only be captured in a probabilistic assessment. This complex regionally variable nature of impact emphasises the need for detailed regional-specific probabilistic analysis before effective climate adaptation strategies can be put in place, with neither impacts, nor regional effects intuitive, that is, Sydney experiences an increase in predicted wind failures despite a median reduction in wind speed, and Canberra and Sydney experience very different condemning rate impacts, despite having very similar temperature and rainfall climate change predictions. Thus, Sydney and Canberra are likely to require different adaptation strategies, while the analysis indicates the implementation of an adaptation strategy for Melbourne based on perhaps a national assessment would have been a waste of finances. This reinforces the point made in Chapter 1 that engineering adaptation plans and measures need to be tailored for local conditions. Such an exercise is described in the following section, which presents a detailed, regionally specific, climate adaptation analysis investigating a range of adaptation measures for a single location.

5.4 Cost-Benefit Analysis of Adaptation Measures

The ability to implement cost-effective climate change adaptation strategies will be a key factor in determining how well our critical infrastructure copes in a changing climate. However, as shown in the previous section, climate change impacts are highly complex and not always intuitive, especially for infrastructure networks. This is also true for adaptation feasibility, which adds another layer of complexity to the analysis, that is, quantifying costs and benefits of adaptation. Consequently, detailed and robust analysis is normally required to ascertain if a given infrastructure climate adaptation strategy is effective for a given country, or indeed region. Failure to conduct such analysis could lead to significant opportunity costs, with actions typifying cost neglect taken after extreme events due to public pressure or other nonscientific forcers.

This section presents a numerical assessment of the appropriateness of different climate change adaptation strategies for power distribution infrastructure. Probability-based cost-benefit analysis is used to assess the impact of each adaptation strategy on monetary risk. Presentation of the results in this form facilitates a robust and quantitative universal metric which infrastructure asset managers, owners, and researchers across disciplines can relate to, that is, do mean net present values (NPVs) and benefit-to-cost ratios (BCRs) indicate that annual monetary risk would be reduced or increased by implementation of a given climate adaptation strategy. In this section, the nature of the case study will be presented, followed by the cost-benefit methodology which was used, together with the methodologies presented in Sections 5.2.1 and 5.3.1, to examine the appropriateness of climate adaptation strategies incorporating uncertainty at the various levels. Subsequently, the cost-benefit analysis findings are discussed. The reader is referred to Ryan and Stewart (2017) for more detailed discussion of the cost-benefit case study presented in this section.

5.4.1 Case Study Details and Methodology

Based on the findings from the multiregional climate impact case study presented in the previous section, Brisbane was selected for this climate change adaptation analysis. Climate adaptation is thus examined in the context of a notional network of one million CCA-treated timber power distribution poles in Brisbane region. This case study uses the IPCC Assessment Report (AR5) climate change projections for Australia (Webb and Hennessy, 2015). Again, in line with the large uncertainty and variability associated with long-term climate projections, climatic changes (wind speed, temperature, and rainfall) were modelled probabilistically for the no change scenario, the RCP4.5 (medium) scenario and the RCP8.5 (severe case) scenario. The simulation period for the notional network of one million new poles was 2015–90.

The methodology utilised in this section builds on the frameworks developed in Sections 5.2.1 and 5.3.1 through addition of a life-cycle cost-benefit analysis approach, enabling feasibility assessment of the climate adaptation strategies. The cost-benefit analysis herein examines both the NPV and BCR of each adaptation strategy over a life cycle (i.e., 2015–90 for this case study). The NPV is calculated as:

$$NPV(t) = LCC_{BAU}(t) - LCC_{adapt}(t) \tag{5.14}$$

where $LCC_{BAU}(t)$ and $LCC_{adapt}(t)$ are the life-cycle costs for 'business as usual' (BAU, i.e., existing practice) and the adaptation measure, respectively, discounted to a present value. The LCC for a given pole in the network is:

$$LCC(T) = C_C + C_{IN}(T) + E_{damage}(T) \tag{5.15}$$

where C_C is the construction materials cost, $C_{IN}(T)$ is the cost of inspections during the service life T, and $E_{damage}(T)$ is the expected cost of repair or loss during service life T. The expected cost of repair and loss can be described as a present value:

$$E_{damage}(T) = \sum_{j=1}^{DS} \sum_{i=1}^{T} P_{f,i} \frac{C_{damage}}{(1+r)^i} \tag{5.16}$$

where $P_{f,i}$ is the probability of damage in year i, C_{damage} is the cost of repair and loss, r is the discount rate, and DS is the number of different damage states. In the present case, there are two damage states, namely power pole condemnings and power pole wind failures. The probabilities used in the estimation of E_{damage} are calculated using the Monte Carlo simulation techniques. The benefit-to-cost ratio of adaptation over the life cycle is determined by:

$$BCR = \frac{Benefit_{adapt}}{C_{adapt}} \tag{5.17}$$

Thus, the economic performance is determined by cost of pole construction materials, inspection costs, and the expected damage costs, all of which are impacted by the cost of adaptation strategy implemented, and the effectiveness of the adaptation strategy. The benefit ($Benefit_{adapt}$) is defined as the reduced losses due to reductions in pole condemnings, inspections, or pole wind failures, when compared to the BAU case. Similarly, the costs (C_{adapt}) associated with implementation of an adaptation strategy can range from additional pole material cost, to additional wind failure and condemning costs, to additional inspection costs, depending on the nature of the adaptation measure employed. A NPV greater than zero, and a BCR value greater than one, indicates that an adaptation measure is cost-effective and monetary risk is reduced by adaptation implementation. The monetary risk reduction (ΔR) value was also calculated for each adaptation strategy. This represents the percentage reduction in life-cycle costs (monetary risks) brought about by the adaptation strategy in question, when compared to the BAU. The life-cycle costs considered in the analysis are in accordance with Eq. (5.15).

A discount rate of 4% was used for the cost-benefit analysis (Stewart, 2015). The case study considered both direct and indirect costs and benefits. These costs were obtained from the power industry and from Australian and International literature and reports. All cost values are presented in Table 5.3, together with the sources of costs provided. The selection of these values is discussed in detail in Ryan and Stewart (2017). The parameters for lost load (LL) in Table 5.3 allow quantification of the energy which is not supplied in the event of a power outage due to power pole wind failure. The LL is calculated using Eq. (5.18) whereby, N_{res} and $N_{business}$ are the number of residential and business customers affected by the outage, respectively, and R_{res} and $R_{business}$ are the power consumption rates for residential and business customers, respectively. The direct cost to the energy company, and the user cost can then be calculated by multiplication of the relevant LL figure and cost of energy (CE) and value of lost load (VOLL), respectively. The influence of power outage duration (OD) on VOLL for residential customers was incorporated in accordance with Praktiknjo et al. (2011). It is also noted that the safety cost, which arises from high consequence, low probability events, such as bushfires or wildfires, death or injury to people or livestock due to downed power lines, etc. was not considered herein due to a lack of cost data for such events.

$$LL = OD\left[N_{res}R_{res} + N_{business}R_{business}\right] \quad (5.18)$$

5.4.2 Climate Adaptation Strategies

The five adaptation strategies and the BAU strategy investigated in this study are as follows:

- BAU: pole diameter design in accordance with AS/NZS standards, inspection intervals set at 5 years, and pole condemning criteria set at 50% of original pole capacity.
- Adaptation strategy 1: same as the BAU with exception that pole condemning criteria is increased from 50% to 55% of original pole capacity. This means that poles are replaced at an earlier stage of deterioration, reducing the vulnerability of the network to power pole wind failures.
- Adaptation strategy 2: same as the BAU with exception that pole inspection intervals are reduced from five to four years. Poles are thus inspected more frequently, meaning severely deteriorated poles are more likely to

Table 5.3 Direct and Indirect Cost Parameters for Power Pole Networks

Parameter	Symbol	Unit	Distribution	Mean	COV	Source
Pole and inspection costs						
Standard spotted gum pole cost	C_c	$	Deterministic	561	–	Industry supplier
Larger spotted gum pole cost	C_c+	$	Deterministic	715	–	Industry supplier
Inspection cost	C_{IN}	$	Uniform (25 to 125)	75	–	(AER, 2015)
Pole replacement cost	–	$	Uniform (3,500 to 15,000)	9,250	–	(AER, 2015)
Lost load (LL) for power outage						
Outage Duration	OD	mins	Deterministic	380	–	(AER, 2015)
Number of residential customers effected	N_{res}	customers	Deterministic	83.6	–	(AER, 2015)
Number of business customers effected	$N_{business}$	customers	Deterministic	11.4	–	(AER, 2015)
Average residential power consumption rate	R_{res}	kWh/h/customer	Deterministic	0.67	–	(ESSA, 2015)
Average business power consumption rate	$R_{business}$	kWh/h/customer	Deterministic	13.21	–	(ESSA, 2015)
Cost of lost load						
Cost of energy unsupplied	–	$/kWh	Deterministic	0.3	–	(AEMO, 2015)
Value of lost load residential customers	–	$/kWh	Log-normal	23.57	1.22	(Praktiknjo et al., 2011)
Value of lost load business customers	–	$/kWh	Log-normal	54.12	1.22	(Praktiknjo et al., 2011, AEMO, 2014)

be removed before wind failure occurs, increasing the overall resilience of the network.

- Adaptation strategy 3: same as the BAU with exception that both original poles and replacement poles used are one size grade larger than required under existing Australian design procedures. Practical implementation for this strategy would mean that power companies only install larger poles than standards require in their network from this point in time onwards.
- Adaptation strategy 4: as with strategy 3, both original poles and replacement poles are one size grade larger than required under current Australian design procedures, however, the pole condemning criteria is lowered from 50% to 45%. Thus, this adaptation strategy balances measures to increase and decrease vulnerability, respectively, in an attempt to optimise the cost-benefit result.
- Adaptation strategy 5: this strategy combines poles one size grade larger than required under AS/NZS design procedure, with an increase in inspection intervals from 5 to 6 years. Again this dual approach seeks to increase network resilience to climate change through increased pole size, while reducing network operating costs by increasing inspection intervals.

5.4.3 Cost-Benefit Outcomes

This section provides the risk-based decision support for power infrastructure asset managers through assessment of five different climate adaptation measures. Both the effectiveness and the cost-benefit outcomes for the five climate adaptation strategies under the RCP4.5 and RCP8.5 scenarios for Brisbane are presented in Table 5.4, for the period 2015–90. The effectiveness of each adaptation strategy is presented in terms of ability to reduce pole condemnings and pole wind failures to pre-climate change levels. These metrics are independent of cost and are expressed as a percentage of condemning rates, and pole wind failures rates, occurring under the adaptation strategy and RCP scenario, as a percentage of those occurring for the BAU approach under pre-climate change conditions. For instance, the BAU under RCP8.5 reads as 137% in Table 5.4, reflecting the fact that if no action is taken the AR5 RCP8.5 scenario is projected to increases wind failures by 37%. Looking at adaptation strategy 1 in Table 5.4, this strategy reduces wind failures under RCP8.5 to 105% of those experienced under pre-climate change conditions (a 32% reduction compared to the BAU). The

Table 5.4 Climate Adaptation Strategy Results for Brisbane Under RCP4.5 and RCP8.5 Scenarios

	RCP 4.5					RCP 8.5				
	Effectiveness		Cost-Benefit			Effectiveness		Cost-Benefit		
Adaptation	**% of Condemning Failures**	**% of Wind Failures**	**Mean NPV ($/pole)**	**Mean BCR**	**ΔR (%)**	**% of Condemning Failures**	**% of Wind Failures**	**Mean NPV ($/pole)**	**Mean BCR**	**ΔR (%)**
BAU (i.e., Climate Change impacts)	106%	118%	–	–	–	111%	137%	–	–	–
Strat. 1–CC 55%	116%	88%	−102	0.47	−4.5	121%	105%	−99	0.66	−11.1
Strat. 2–Insp 4 yrs	109%	83%	−30	0.81	−1.3	114%	98%	−25	0.87	−1.1
Strat. 3–L.P.	93%	40%	163	1.13	7.2	98%	46%	184	1.18	7.8
Strat. 4–L.P. and CC 45%	83%	69%	241	1.79	10.7	87%	82%	272	1.87	11.5
Strat. 5–L.P. and Insp 6 yr	92%	69%	164	1.39	7.3	97%	82%	193	1.44	8.2

Strat. = adaptation strategy, CC = condemning criteria, L.P. = larger poles, Insp = inspection interval.

cost-benefit results for each adaptation strategy are expressed in terms of mean NPV, mean ΔR, and mean BCR, indicating if an adaptation strategy makes sense from a monetary risk or cost perspective. The effectiveness values thus provide information for a risk adverse asset manager, while the cost-benefit outcomes provide insight for an asset manager who wishes to utilise a risk-based decision framework grounded in detailed probabilistic modelling. A negative NPV or ΔR and a BCR <1.0 are all indicators that the investment required to implement the strategy outweighs its financial benefit, when costs are discounted to present value. The BAU case is also shown in the table for comparative purposes.

Examination of the results for adaptation strategies 1–3 provides insight into the individual effect of changes in condemning criteria, inspection intervals or pole design procedure, respectively. As can be seen from Table 5.4, increasing condemning criteria to 55% is effective in reducing the number of pole failures to almost no climate change levels (from 137% to 105% for RCP8.5), however, this occurs at the expense of a significant increase in the numbers of poles condemned and replaced. Due to the relatively rare nature of power pole wind failures, replacement due to pole condemning is in the region of 50 times more likely to occur over the modelling period. Consequently, while the increase in condemning criteria to 55% was found to reduce pole wind failures, the significant increase in pole replacements due to condemning meant the adaptation strategy was highly cost ineffective, with the lowest mean NPV and ΔR values, and the smallest mean BCR, of the five climate adaptation strategies considered.

Reduction in inspection intervals from 5 to 4 years was found to be a more effective measure. This strategy reduced pole wind failures below pre-climate change rates for both RCP scenarios. This came at a cost of a 9% and 14% increase in the annual pole condemning rates for RCP4.5 and RCP8.5, respectively, and the cost associated with a greater number of inspections over the 76-year network monitoring period. These costs resulted in a negative cost-benefit result. By contrast, increasing the pole dimensions by one size grade was found to have a highly positive cost-benefit outcome, while also being highly effective in reducing climate change impacts. Failure rates are less than half those experienced under the no climate change condition for both RCP scenarios, while the mean NPV is \$184 per pole for RCP8.5, equating to a saving of \$2.4 million per year over the BAU approach for the notional network of one million Brisbane poles.

Strategy 4 illustrates how combined adjustments to design procedure and pole condemning criteria can be used to optimise climate change adaptation. This adaptation strategy, which combines the use of larger poles with less stringent pole condemning criteria, is the best performing adaptation strategy. The BCR metric under RCP8.5 indicates that for each $1 of investment $1.87 of benefit is obtained, while the NPV indicates a mean saving of $272 million dollars to 2090, for a network of one million Brisbane poles. Finally, adaptation strategy 5, which combines larger poles with longer inspection intervals, can be seen from Table 5.4 to be the second best adaptation strategy, with a BCR of approximately 1.4. It is noted, however, that this adaptation strategy may be easier to implement from a practical perspective, and thus may be more favourable for power companies. The safety cost is assumed to be zero in the above analysis given the uncertainty associated with this cost. Inclusion of a safety cost in the assessment increases the cost-benefit performance of the climate adaptation strategies (Ryan and Stewart, 2017). It can be seen from Table 5.4 that cost-benefit results were not hugely sensitive to the emission scenarios considered. In fact, climate adaptation strategy 4 was found to exhibit a strongly positive cost-benefit outcome for the 'no climate change' scenario, with a mean of BCR of 1.73 and a mean NPV of $223 per pole, representing a win-win adaptation strategy.

Overall the results show that it is not difficult to develop adaptation strategies which are effective in reducing climate change impacts on power infrastructure, with projected impacts of climate change substantially reduced in some aspect for all adaptation options considered herein. However, it is far more challenging to develop cost-effective climate adaptation strategies. The region-specific risk-based decision support framework developed in this chapter allows the distinction to be made between adaptation choices which are financially feasible, and those that are not. It is noted that this modelling does not attempt to replicate actual networks. It should rather be seen as an asset management tool, which allows infrastructure owners and operators to assess risk implications of network management decisions made. This is an important facility given that, maintenance cycles, the scale of networks and the slow nature of deterioration mean it can take decades before the impact of asset management decisions can be quantified in a real network through monitoring. Importantly, it can also take decades to reverse the implications of poor asset management or adaptation strategies.

5.4.4 Adaptation Payback Period

While selection of the most appropriate climate adaptation strategy is of key importance, the timing of adaptation implementation is also an important consideration for asset managers. The results presented in the previous section consider a 76-year monitoring period which runs from the year 2015 to 2090. However, what are the cost-benefit outcomes if this monitoring period is shortened, that is, what is the payback period for the adaptation strategy for the notional network of one million poles. This is examined for adaptation strategy 4 under RCP8.5 in Fig. 5.11, which shows the mean NPV for various policy timeframes up to 2090. As can be seen from the figure the adaptation strategy, which involves an initial investment of $154 per pole, has a strongly negative cost-benefit result in the year 2030. This is because the network is just 16 years old at this point, meaning few poles have failed and no poles have been replaced for either the BAU case or the adaptation strategy 4 case, that is, little chance for the adaptation strategy to perform better than the BAU case as neither case experiences pole failures in early years. However, the payback period for the adaptation strategy is reached at approximately 2041 (26 years). Following this point, the mean NPV of the adaptation per pole continues to grow with time as the poles in the network continue to age, and are thus more likely to fail and require replacement, giving the adaptation strategy more of a chance to outperform the BAU strategy. It is noted, however, that benefits realised later in the monitoring period will have diminished effect on the NPV as a result of future costs and benefits being discounted.

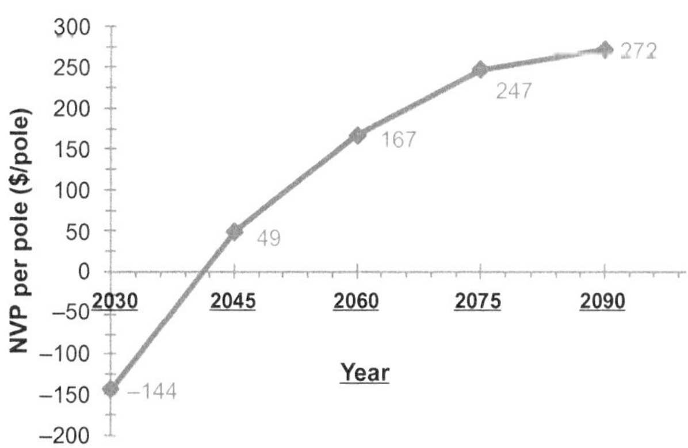

Fig. 5.11 Sensitivity of NPV to end of monitoring period selected for adaptation Strategy 4 under the RCP 8.5 scenario.

5.5 Summary

A probabilistic framework was developed with the aim of assessing climate change impacts and investigating the feasibility of subsequent climate adaptation strategies for timber power pole networks. Probabilistic event-based time-dependent modelling was used to represent the existing vulnerability of a notional network of one million power poles. The model was then adjusted to allow for assessment of the impact of climate change on power pole infrastructure performance. The case study results indicated that climate change impacts on timber power poles can be substantial, however, the magnitude of the impacts are highly regionally variable. The complex regionally variable nature of impact emphasises the need for detailed regional-specific probabilistic analysis before effective climate adaptation strategies can be put in place, with neither impacts, nor regional effects intuitive, that is, Sydney experienced an increase in predicted power pole wind failures despite a median reduction in wind speed, and Canberra and Sydney experienced very different condemning rate impacts, despite having very similar climate change predictions. This reinforces the point made in Chapter 1 that climate adaptation engineering measures need to be tailored for local conditions.

Information on climate change risks and impacts for a given infrastructure network provides the first line of decision support for climate change adaptation planning, that is, is any adaptation necessary? Having established the need for adaptation in the Brisbane region of Australia in this chapter, the feasibility of five adaptation strategies were examined using probabilistic cost-benefit analysis. The strategies incorporated a range of adjustments to current practice in design and network maintenance. For the Brisbane region, an adaptation measure that allows for the installation of larger poles, but less stringent pole replacement criteria, was found to have the highest net benefit. Overall the case study results showed that it is not difficult to develop adaptation strategies which are effective in reducing climate change impacts on power infrastructure, with projected impacts of climate change substantially reduced in some way by all adaptation strategies considered. However, it is far more challenging to develop cost-effective climate adaptation strategies. The region-specific risk-based decision support framework developed in this chapter allows distinction to be made between adaptation choices which are likely to be financially viable, and those which are not. This is a valuable asset management tool in the face of climate change challenges for critical infrastructure owners. Particularly when it is considered that it can take decades (if at all) before asset managers can quantify the impact of an asset management decision through monitoring. Importantly, it can also take decades to reverse the implications of poor asset management or adaptation decisions, resulting in substantial opportunity costs.

5.6 Potential Design and Practice Evolutions

The work presented in this chapter highlights the scale of the challenge faced by design committees and asset managers when attempting to cope with climate change impacts on critical infrastructure. The case study illustrates that climate change impacts can range from significant, to negligible, from one region to another within Australia. It was also shown that, while it is relatively straightforward to reduce the impact of climate change through evolutions in design and maintenance practices, it is far more difficult to do this in a cost-effective manner. It can thus be concluded that detailed, regionally specific analysis will be required to inform: (i) where changes to design codes and industry practices are required for power distribution systems, and (ii) what exactly these measures should be to ensure cost-effective outcomes.

5.7 Open Research Questions

The authors are currently working on a number of open research questions relating to the work discussed in this chapter. These include:

- Will similar findings be identified when the framework employed here in an Australian context, is utilised to investigate climate change impacts and adaptation in other countries?
- How can the probabilistic model be further developed, with greater depth and complexity, so that it becomes more and more representative of timber power pole networks in service?
- Climate change projecting is constantly evolving as climate science improves. How will these changes in climate projections affect the outcomes of this study?
- How can we best work with infrastructure owners to improve the quality of data collected for pole failures, asset deterioration, maintenance and network parameters, etc.?

Acknowledgments

The authors gratefully acknowledge the support of the Irish Environmental Protection Agency (EPA) (Grant No. 2016-CCRP-MS.41), and the Commonwealth Scientific and Industrial Research Organisation (CSIRO) Flagship Cluster Fund through the project Climate Adaption Engineering for Extreme Events, in collaboration with the Sustainable Cities and Coasts Theme of the CSIRO Climate Adaption Flagship.

Appendix 5.A

Table 5.A.1 Statistical Parameters for Probabilistic Model

Property	Units	Distribution	Mean	COV	Source
f_b Spotted Gum	MPa	Normal	104.1	0.14	(Spencer, 2013)
f_b Radiata Pine	MPa	Normal	43.0	0.10	(Engineering, 2014)
Radiata Pine corewood depth	Growth rings	Uniform (3–5)	4	–	(Harris and Cown, 1991)
Spotted Gum corewood depth	Growth rings	Uniform (5–10)	7.5	–	(Ryan et al., 2014)
Spotted Gum sapwood depth	mm	Uniform (25–50)	37.5	–	(Bamber, 1987)
Radiata Pine sapwood depth	mm	Deterministic	100	0	(Bamber, 1987)
ME_{decay}[a] Spotted Gum (D.C. 2)	–	Log-normal	1.00	0.92	(Wang et al., 2008a)
ME_{decay}[a] Radiata Pine (D.C. 4)	–	Log-normal	1.00	1.20	(Wang et al., 2008a)
HV and LV wire diameter	mm	Normal	13.5	0.06	Fig. 5.2 and (Bjarnadottir et al., 2013)
Street lighting wire	mm	Normal	9.0	0.06	Fig. 5.2 and (Bjarnadottir et al., 2013)
Pole height	m	Normal	10.7	0.03	(Crews and Horrigan, 2000, Bjarnadottir et al., 2013)
High voltage wire height	m	Normal	10.95	0.03	(Crews and Horrigan, 2000, Bjarnadottir et al., 2013)
Low voltage cables height	m	Normal	9.5	0.03	
Communication wire height	m	Normal	8.1	0.03	

Table 5.A.1 Statistical Parameters for Probabilistic Model—cont'd

Property	Units	Distribution	Mean	COV	Source
ME_w	–	Log-normal	1.00	0.05	(Henderson and Ginger, 2007)
ρ_{air}	kg/m^3	Log-normal	1.2	0.02	(Henderson and Ginger, 2007)
$M_{z,cat}$	–	Log-normal	0.95	0.10	(Henderson and Ginger, 2007, Standards Australia/ New Zealand, 2010)
$C_{d\ pole}$	–	Log-normal	1.2	0.05	(Standards Australia/ New Zealand, 2010, Ryan et al., 2014)
$C_{d\ wires}$	–	Log-normal	1.0	0.05	(Standards Australia/ New Zealand, 2010, Ryan et al., 2014)
G (poles)	–	Normal	0.96	0.11	(Bjarnadottir et al., 2013)
G (wires)	–	Normal	0.81	0.11	(Bjarnadottir et al., 2013)
Brisbane wind field	m/s	Gumbel	24.276	3.208	(Wang et al., 2013)
Sydney wind field	m/s	Gumbel	29.289	2.572	(Wang et al., 2013)
Canberra wind field	m/s	Gumbel	20.076	2.104	(Wang et al., 2013)
Melbourne wind field	m/s	Gumbel	25.803	1.825	(Wang et al., 2013)
Perth wind field	m/s	Gumbel	29.043	2.780	(Wang et al., 2013)

[a]ME_{decay} distribution truncated at 10.0.

References

AEMO, 2014. Value of customer reliability review. Australian Energy Market Operator (AEMO).

AEMO, 2015. AEMO Average Pricing Tables 2015. (Australian Energy Market Operator).

AER 2015. AER Regulatory Information Notices (RIN) 2015. https://www.aer.gov.au/taxonomy/term/1495:Australian Energy Regulator (AER).

Auld, H., MacIver, D. & Klaassen, J. Adaptation options for infrastructure under changing climate conditions. 2006 IEEE EIC Climate Change Technology Conference, EICCCC 2006, May 10, 2006–May 12, 2006, Ottawa, ON, Canada. Inst. of Elec. and Elec. Eng. Computer Society.

Bamber, R.K., 1987. Technical Publication Number 2—Sapwood and Heartwood. (Forestry Commission of New South Wales).

Barone, G., Frangopol, D.M., 2014. Reliability, risk and lifetime distributions as performance indicators for life-cycle maintenance of deteriorating structures. Reliab. Eng. Syst. Safety 123, 21–37.

Billinton, R., Wang, P., 1999. Teaching distribution system reliability evaluation using Monte Carlo simulation. IEEE Trans. Power Syst. 14, 397–403.
Bjarnadottir, S., Li, Y., Stewart, M.G., 2013. Hurricane risk assessment of power distribution poles considering impacts of a changing climate. ASCE J. Infrastruct. Syst. 19, 12–24.
Bjarnadottir, S., Li, Y., Stewart, M.G., 2014. Risk-based economic assessment of mitigation strategies for power distribution poles subjected to hurricanes. Struct. Infrastruct. Eng. 10, 740–752.
Bolin, C.A., Smith, S.T., 2011. Life cycle assessment of pentachlorophenol-treated wooden utility poles with comparisons to steel and concrete utility poles. Renew. Sustain. Energy Rev. 15, 2475–2486.
Crews, K.I., Horrigan, A., 2000. Strength assessment of timber utility poles in Australia. New Zealand Timber Des. J. 9.
Electricity Supply Association of Australia, 1991. Guidelines for the Design and Maintenance of Overhead Distribution and Transmission Lines. C(b)-I.
Engineering, U., 2014. Strength Testing of SA and WA Kiln Dried Pine Poles. Newcastle, Australia.
ESSA 2015. Data and Statistics—Energy in Australia. http://www.esaa.com.au/policy/data_and_statistics-_energy_in_australia: Energy Supply Association of Australia.
Francis, L., Norton, J., 2006. Australian Timber Pole Resources for Energy Networks: A Review. Department of Primary Industries & Fisheries, Queensland.
Goldenberg, S., 2014. Pentagon: Global warming will change how US military trains and goes to war. The Guardian. 14 October 2014.
Gustavsen, B., Rolfseng, L., 2004. Asset management of wood pole utility structures. In: International Conference on Probabilistic Methods Applied to Power Systems. Elsevier Ltd, pp. 641–646.
Harris, J.M., Cown, D.J., 1991. Basic Wood Properties. In: Kininmonth, J., Whitehouse, L. (Eds.), Properties and Uses of New Zealand Radiata Pine. Ministry of Forestry, New Zealand.
Henderson, D.J., Ginger, J.D., 2007. Vulnerability model of an Australian high-set house subjected to cyclonic wind loading. Wind Struct. 10, 269–285.
IPCC, 2014. Summary for Policymakers. In: Editors, F. (Ed.), Climate Change 2014: Impacts, Adaptation and Vulnerability. Contribution of Working Group II to the Fifth Assessment Report of the Intergovernmental Panel on Climate Change Cambridge, United Kingdom and New York, NY, USA.
Irish Independent, 2017. Ophelia recovery operation: gale force winds may disrupt recovery efforts as tens of thousands still without electricity and water. Irish Independent Newspaper.
Johansson, J., Hassel, H., Zio, E., 2013. Reliability and vulnerability analyses of critical infrastructures: Comparing two approaches in the context of power systems. Reliab. Eng. Syst. Saf. 120, 27–38.
Li, G.-Q., Li, J.-J., 2002. A semi-analytical simulation method for reliability assessments of structural systems. Reliab. Eng. Syst. Safety 78, 275–281.
Pfautsch, S., Macfarlane, C., Ebdon, N., Meder, R., 2012. Assessing sapwood depth and wood properties in Eucalyptus and Corymbia spp. using visual methods and near infrared spectroscopy (NIR). Trees Struct. Funct. 26, 963–974.
Praktiknjo, A.J., Hähnel, A., Erdmann, G., 2011. Assessing energy supply security: Outage costs in private households. Energy Policy 39, 7825–7833.
Ryan, P.C., Stewart, M.G., 2017. Cost-benefit analysis of climate change adaptation for power pole networks. Clim. Change 143, 519–533.
Ryan, P.C., Stewart, M.G., Spencer, N., LI, Y., 2014. Reliability assessment of power pole infrastructure incorporating deterioration and network maintenance. Reliab. Eng. Syst. Saf. 132, 261–273.

Ryan, P.C., Stewart, M.G., Spencer, N., LI, Y., 2016. Probabilistic analysis of climate change impacts on timber power pole networks. Int. J. Electr. Power Energy Syst. 78, 513–523.

Salman AM, Li Y, Bastidas-Arteaga E. Maintenance optimization for power distribution systems subjected to hurricane hazard, timber decay and climate change. Reliab. Eng. Syst. Safety 2017; 168: 136–149.

Spencer, N., 2013. Western Power East-Coast Hardwood Pole Testing Report. APD Power Engineeing, Perth.

Spencer, N., Elder, L., 2009. Pole Service Life—An Analysis of the Country Energy Data. In: Energy 21C. Australia, Melbourne.

Standards Australia/New Zealand, 2010. AS/NZS 7000-2010: Overhead Line Design—Detailed Procedures. In: Sydney.

Standards Australia/New Zealand 2011. AS/NZS 1170–2:2011 Structural Design Actions, Part 2: Wind Actions. Sydney

Standards Australia/New Zealand, 2013. AS/NZS 5334-2013 Climate Change Adaptation for Settlements and Infrastructure—a Risk Based Approach. In: Sydney.

Stewart, M., 2015. Risk and economic viability of housing climate adaptation strategies for wind hazards in Southeast Australia. Mitig. Adapt. Strat. Glob. Chang. 20, 601–622.

Stewart, M., Deng, X., 2015. Climate impact risks and climate adaptation engineering for built infrastructure. ASME J. Risk Uncert. Eng. Syst. A: Civ. Eng. 1.

Stewart, M., Val, D., 1999. Role of load history in reliability-based decision analysis of aging bridges. J. Struct. Eng. 125, 776–783.

Walker, J.C.F., 2006. Primary Wood Processing—Principals and Practice. Springer.

Wang, C.-H., Wang, X., 2012. Vulnerability of timber in ground contact to fungal decay under climate change. Clim. Change 115, 777–794.

Wang, C.-H., Leicester, R.H., Nguyen, M., 2008a. Probabilistic procedure for design of untreated timber poles in-ground under attack of decay fungi. Reliab. Eng. Syst. Safety 93, 476–481.

Wang, C.-H., Leicester, R. H. & Nguyen, M. 2008b. Timber durability technical report. Manual No. 3—Decay in ground contact. CRISO and FWPRDC.

Wang, C.-H., Wang, X., Khoo, Y.B., 2013. Extreme wind gust hazard in Australia and it's sensitivity to climate change. Nat. Hazards 67, 549–567.

Webb, L.B., Hennessy, K., 2015. Climate Change in Australia—Projections for Selected Australian Cities. Commonwealth Scientific and Industrial Research Organisation (CSIRO) and Bureau of Meteorology, Australia.

Winkler, J., Duenas-Osorio, L., Stein, R., Subramanian, D., 2010. Performance assessment of topologically diverse power systems subjected to hurricane events. Reliab. Eng. Syst. Saf. 95, 323–336.

Yates, A., 2014. A framework for studying mortality arising from critical infrastructure loss. Int. J. Crit. Infrastruct. Prot. 7, 100–111.

Further Reading

Stewart, M.G., 2013. Break-Even Economic Analysis of Residential Housing Climate Adaption Stratagies for Wind Hazard in South-East Australia. In: CAEx Report 1/2013, Climate Adaption Engineeirng for Extreme Events Cluster, CSIRO Climate Adaption Flagship, Australia.

CHAPTER SIX

Climate Change Impact for Bridges Subjected to Scour and Corrosion

Boulent Imam
Department of Civil and Environmental Engineering, University of Surrey, Guildford, United Kingdom

6.1 Introduction

The built infrastructure, including buildings, transport systems, and bridges can be adversely affected by climate change (Kirshen et al., 2008; Meyer and Weigel, 2011; Posey, 2012; Kumar and Imam, 2013). Natural disasters such as hurricanes, extreme precipitation, and flooding can have major socioeconomic impacts, including among others loss of life and damage to the built infrastructure (e.g., loss of service in transport and other infrastructure systems, damage to buildings/bridges, etc.) leading to potentially significant macroeconomic effects (Lehner et al., 2006; Swiss Re, 2013).

Consideration of environmental conditions for the design and maintenance of bridge structures is vital for the safety of bridge infrastructure. Changing environmental conditions, brought about through climate change, may lead to detrimental impacts on bridge structures and the materials that they have been constructed by. Such impacts may be short lived, acute, and intensive when derived by extreme weather conditions (e.g., more frequent heat waves and extreme rainfall) and long lived and slow when derived by changing climatic conditions (e.g., increase in the average annual temperature, experiencing overall drier summers or wetter winters). Both the changes in long-term average climatic conditions and short-term extreme events carry a major potential to affect the sustainability of built infrastructure.

Bridges are most vulnerable to natural hazards such as flooding, storms, hurricanes, and winds. Statistics on bridge collapses worldwide reveal that natural hazards are the predominant cause of failure (Imhof, 2004). This demonstrates, bearing in mind the adverse climate change impacts, the high

Climate Adaptation Engineering
https://doi.org/10.1016/B978-0-12-816782-3.00006-1

risk present in the future for bridge structures and transport networks with respect to weather-related extreme events. One of the key effects of climate change on the bridge population will be the increased risk of scour of bridge piers and abutments (TRB, 2008). This will arise from more frequent and more intense river flooding due to the expected increases in precipitation in the future. Scour is caused by the erosive action of flowing water, removing sediment from around bridge foundations. This has been one of the most common causes of bridge collapses in the past as per the failure statistics reported by Wardhana and Hadipriono (2003), Imhof (2004), JBA (2004), and Imam and Chryssanthopoulos (2012). A recent risk assessment has shown that bridge scour may increase by between 5% and 50% over the present value by the 2080s in the United Kingdom, depending on the local bridge site conditions (DEFRA, 2012).

In terms of long-term changes in environmental and/or climatic conditions, these may affect the long-term deterioration behaviour of bridges and other infrastructure. For example, corrosion is a commonly observed deterioration mechanism in metallic bridge structures, resulting in loss of material with time, which may lead to impaired performance and premature in-service failures due to loss of resistance. A number of factors are known to influence the initiation and subsequent rate of corrosion in metallic bridges, including climatic parameters such as relative humidity (RH) and temperature, atmospheric pollutants (e.g., SO_2), and high airborne salinity. These factors and their interactions may have a detrimental influence on the rate of corrosion. Alterations of exposure conditions, for instance, due to climate change or/and adoption of government policies related to the levels of atmospheric pollutants, are likely to influence the long-term corrosion rates in metallic bridges.

Both the interaction of the bridge structures with atmospheric pollutants as well as the changing environmental conditions are equally important for the safety and economy of *transport infrastructure*. The vast value of transport infrastructure assets shows the risk for large economic losses due to the effects of both climate change and extreme weather conditions. For example, the highway and railway networks in the United Kingdom alone have asset values in excess of £87 and £54 billion, respectively (Highways Agency, 2009; Network Rail, 2015). Deterioration caused by atmospheric pollution is likely to be realised over a long-term by increase in the deterioration rate of construction materials such as steel, cast and wrought iron, concrete, masonry, and timber. Conversely, extreme environmental conditions can have a significant impact in the short-term by disrupting road and rail networks (Booij, 2005; Nicholls, 2004; UNEP, 2007) that can lead to noticeable economic losses (Larsen et al., 2008). In many cases, elements of the transportation infrastructure (e.g., bridge structures) also form part of electricity, telephone, water, and gas networks. Therefore, the economic cost of transport asset and network failures may extend far wider than the boundaries of transportation systems to other forms of critical infrastructure (ICE, 2009).

The aim of this chapter is to present risk-based frameworks that have been developed as part of a number of case studies to assess the potential impacts of climate change on rail bridge structures. These risk frameworks can be used to identify adaptation strategies in the future. In order to capture both long-term changes in average environmental conditions as well as extreme weather events, risk frameworks are presented for the case of rail bridges subjected to scour due to extreme river flooding and for the case of metallic bridges subjected to long-term corrosion deterioration. This chapter concludes by providing insight into assessing the importance of risks arising from climate change compared with another source of risk which is asset stock data uncertainty.

6.2 Climate Change Impacts on Bridge Deterioration

6.2.1 Background

Corrosion, which is associated with time-dependent material loss, is a widespread cause of deterioration in metallic bridges. If left untreated, increasing corrosion damage may lead to reduced safety, premature in-service failures as well as altered structural behaviour and failure modes. The exposure conditions of the structure, which are determined by environmental and atmospheric parameters, that is, RH, temperature (T), atmospheric pollution, etc., have a significant influence on the deterioration process due to corrosion. Hence, changes on the exposure conditions are likely to be responsible for changes on the deterioration process, including the rate of corrosion.

As a large part of resources is allocated for managing the effects of corrosion in metallic bridges, it is important to assess the potential impacts of changing environmental conditions on the structural performance of deteriorating structures. Understanding the process of corrosion and its influence on the long-term structural performance of bridges can assist their rational assessment, optimise their maintenance, and prioritise their repair when needed. To this end, a number of models are available in the literature for the estimation of the rate of corrosion of steel. These models, which consider, indirectly or directly, the influence of exposure conditions, can be used to estimate the long-term material loss due to corrosion. To quantify the effect of changing exposure conditions, such as increasing temperature and variations in concentrations of atmospheric pollutants, the corrosion model should be able to capture the individual contribution of each of these factors on the rate of corrosion. Such models, which relate the rate of corrosion to several climatic and atmospheric factors, are known as dose-response functions (DRFs). In general, a high degree of uncertainty is associated with the predictions of the different corrosion models due to the variability of the local exposure conditions with time.

The quantification of the residual capacity in deteriorating metallic bridges involves a high degree of uncertainty due to the variability in the material and geometrical properties of the structure as well as the deterioration process itself (Nowak and Szerszen, 1998; Sharifi and Paik, 2011; Deco and Frangopol, 2011). The rational assessment of deteriorating steel bridges should account for these uncertainties in the analysis. To this end, reliability analysis has been widely proposed for the assessment of deteriorating structures. A number of methods are available (e.g., the First-Order Reliability Method (FORM), the Monte Carlo simulation (MCS), etc.), which can consider the variability of the corrosion process and resistance variables, for the estimation of the probability of bridge failure, p_f, with respect to time. Risk assessment, which considers the consequences of bridge failure, is an important tool in the long-term decision-making process and planning. A number of studies examined the long-term reliability of corroding metallic bridges (e.g., Czarnecki and Nowak, 2008; Sharifi and Paik, 2011; Deco and Frangopol, 2011). In these studies, which focused on the assessment of highway bridges, the influence of exposure conditions on corrosion rate was modelled indirectly and hence the influence of changing environmental conditions was not quantified.

This section presents a methodology, which is based on the MCS, for the estimation of time-dependent reliability and, subsequently, risk of corroding metallic railway bridges, considering the impact of long-term changes in climatic and atmospheric pollutant variables. The evolution of the variables related to the deterioration process (e.g., climatic parameters and atmospheric pollutants) and bridge resistance are treated as random using suitable distributions. The procedure is demonstrated through a case study using a steel railway bridge in the United Kingdom. Results are presented for the moment-capacity limit state, considering several exposure scenarios based on the UK emission scenarios and changes in the concentrations of atmospheric pollutants.

6.2.2 Framework for Risk Assessment

Reliability analysis allows the estimation of the failure probability, p_f, of a deteriorating structure for different limit states, for example, moment capacity of corroding bridge members. The performance function for a limit state of a corroding metallic bridge is given by:

$$G(t) = R(t) - S(t) \tag{6.1}$$

where $G(t)$ is the time-dependent safety margin and $R(t)$ and $S(t)$ are the time-dependent resistance and load effects, respectively. $G(t) \leq 0$ indicates the failure realisation of the limit state. Assuming statistical independence between $R(t)$ and $S(t)$, the annual probability of failure, $p_f(t)$, is given by:

$$p_f(t) = \Pr[G(t) \leq 0] = \Pr\{[R(t) - S(t)] \leq 0\} \tag{6.2}$$

Assuming that the failures in individual years are independent, the cumulative time-dependent probability of failure, at any point within a time period, is given by:

$$p_f(0, t_i) = 1 - \prod_{i=1}^{k} \{1 - \Pr[G(t) \leq 0]\} \tag{6.3}$$

where k is the number of years within the analysis period.

As discussed in Chapter 1, the time-dependent risks $E(L,t)$ is the product of the failure (or damage) probability multiplied with the failure consequences:

$$E(Lt) = p_f(t) \times L(t) \tag{6.4}$$

where $L(t)$ are the direct and indirect consequences of bridge failure, usually expressed in monetary terms (GBP, £).

6.2.3 Changing Environmental Conditions

Changing environmental conditions, within the context of long-term bridge infrastructure performance, refers to the changes in the exposure conditions of a structure, such as alterations of the average air temperature, RH, precipitation patterns, etc., due to climate change and/or changing concentrations of atmospheric pollutants (e.g., SO_2). This implies that the assessment of the potential impacts of man-made climate change and/or evolution in the levels of atmospheric pollutants is largely determined by the location of the structure.

In the United Kingdom, the UK Climate Projections (UKCP) were established to provide a framework for the quantification of the potential effects of different assumed emission levels on several climatic parameters for the entire United Kingdom area. The probabilistic projections are provided for alternative 'low,' 'medium', and 'high' emissions pathways, corresponding to the IPCC SRES B1, A1B, and A1FI scenarios, respectively. These predate the representative concentration pathway (RCPs) discussed in Chapter 1, and were used extensively in the IPCC AR4; the new UKCP18 projections will align with the latest IPCC RCPs. The UKCP09, which can be considered as an important tool in developing adaptation strategies for tackling the effects of climate change, can be accessed through an online interface (UKCP09, 2009). In the UKCP09, the United Kingdom is divided in 25 km^2 grids in which probabilistic projections for a number of climatic variables are available up to year 2099 in 30-year overlapping periods. For example, Fig. 6.1 shows the evolution of the mean annual temperature (values are plotted in the centre of each 30-year time period) for the three emission scenarios in the UKCP09 for the London Borough of Hillingdon, in the Greater London area including the Heathrow International Airport. This is an area with relatively high levels of atmospheric pollution.

The future predictions of the UKCP09 indicate that the mean annual air temperature in this particular location is set to rise for all three emission scenarios. The differences in the projected mean temperatures become noticeable beyond 2040, with a maximum difference between the low and high emission scenarios being approximately 1.6°C within time period

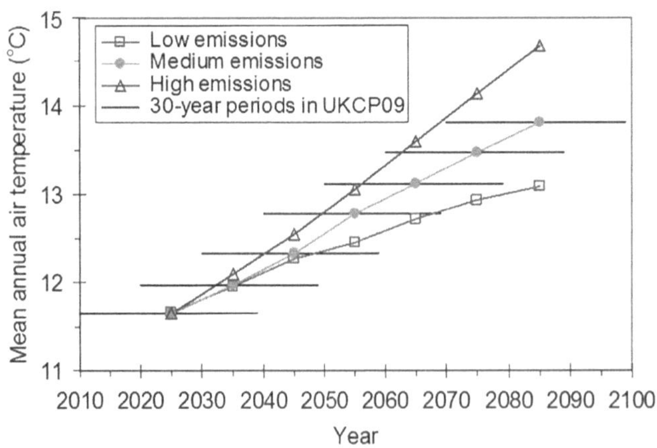

Fig. 6.1 Evolution of mean annual temperature for the Borough of Hillingdon obtained from the UKCP09.

2070–99. The predicted increases of temperature in period 2070–99 relative to the period 2010–39 are 1.41°C, 2.16°C, and 3.02°C for the low, medium, and high emission scenarios, respectively.

6.2.4 Corrosion Damage Modelling

In this study, corrosion damage is modelled as a uniform reduction of thickness in the flanges and web of the affected structural steel member. Experimental evidence suggests that corrosion may affect the mechanical properties of steel in the affected components due to pit formation (Appuhamy et al., 2011). However, in this study, the mechanical properties are assumed to be unaffected by the increasing corrosion damage, that is, the statistical properties for the yield strengths of the web, flanges, and doublers remain constant.

Several models are available in literature for the estimation of long-term corrosion loss. A well-known model for the long-term estimation of corrosion loss is given by Eq. (6.5) (Feliu et al., 1993a,b; Haagenrud and Henriksen, 2006; Sharifi and Paik, 2011):

$$C(t) = At^{B} \tag{6.5}$$

where $C(t)$ is the average loss of thickness after t years of corrosion and parameters A and B are the experimentally determined coefficients obtained through regression analysis on test results in different environmental exposure conditions. Hence, the influence of different exposure conditions, including climatic factors and levels of atmospheric pollutants, can be considered by adjusting the values of A and B. An obvious drawback of using Eq. (6.5) is its inability to consider separately the impact and future evolution of each of the climatic and atmospheric variables, which in-turn determine the local exposure conditions. To overcome this problem, models (also known as DRFs) have been developed to directly relate the climatic and atmospheric variables to the rate of corrosion (Feliu et al. 1993a,b; Haagenrud and Henriksen, 2006; Klinesmith et al., 2007; BS EN ISO 2012a,b). However, not all DRF are suitable for the long-term corrosion prediction in structures exposed to the changing environmental conditions. For example, BS EN ISO (2012a) provides a useful DRF (Eq. 6.6) which can be used to calculate the coefficient A in Eq. (6.5) as a function of several environmental and atmospheric variables:

$$A(t) = 1.77\mathrm{SO}^{0.52}\exp(0.02\mathrm{RH} + f_{St}) + 0.102\mathrm{Cl}^{0.62}\exp(0.033\mathrm{RH} + 0.04T) \tag{6.6}$$

where SO is the average annual deposition of SO_2 ($mg/m^2/day$), RH is the average annual RH (%), Cl is the average annual deposition of Cl^- ($mg/m^2/day$), and T is the annual (average) temperature (°C). Furthermore, BS EN ISO (2012b) provides different values for the mean and standard deviation of coefficient B in Eq. (6.5), depending on the exposed material (i.e.,for steel: $B_{mean}=0.523$ and $B_\sigma=0.026$). This model is very useful in estimating the first-year corrosion (i.e., coefficient A) based on the actual exposure conditions (SO_2, T, etc.). However, the influence of long-term changes of exposure conditions (due to climate change and/or increase/decrease of the concentration of atmospheric pollutants) on the rate of corrosion cannot be considered in the analysis (i.e., coefficient B is a constant). The model of Klinesmith et al. (2007), given by Eq. (6.7), allows the calculation of thickness loss (μm) due to corrosion considering the evolution of the different climatic and atmospheric variables:

$$C(t)=At^B\left(\frac{TOW}{C_1}\right)^D\left(1+\frac{SO_2}{E}\right)^F\left(1+\frac{Cl}{G}\right)^H e^{J(T+T_0)} \tag{6.7}$$

where TOW is the time of wetness (h/year), SO_2 is the sulphur dioxide concentration ($\mu g/m^3$), Cl is the chloride deposition rate ($mg/m^2/day$), $C_1=3800$ h/year (mean value), E=mean of measured values of SO_2 ($\mu g/m^3$), T is the air temperature (°C), and $T_0=20$°C. For flat carbon steel specimens (Klinesmith et al., 2007): $A=13.4$, $B=0.98$, $D=0.46$, $F=0.62$, $H=0.34$, and $J=0.016$.

Fig. 6.2 shows the effects of different climatic and atmospheric variables on the long-term predictions of corrosion thickness losses calculated using Eq. (6.7). The values of the variables used for the predictions are typical values suggested in BS EN ISO (2012a). By comparing the figures, it is evident that airborne salinity and SO_2 concentration have the greatest impact on the long-term rate of corrosion.

6.2.5 Evaluation of Structural Performance

In this study, the limiting moment of resistance, M_R (Eq. 6.8), of the main girders, which considers U-frame action in plate girder railway bridges, is calculated based on the provisions of BS 5400 (BSI, 2000):

$$\frac{M_R}{M_{ult}}=0.5\left[\left\{1+(1+\eta)\frac{5700}{\beta^2}\right\}-\sqrt{\left\{1+(1+\eta)\frac{5700}{\beta^2}\right\}^2-\frac{22800}{\beta^2}}\right] \tag{6.8}$$

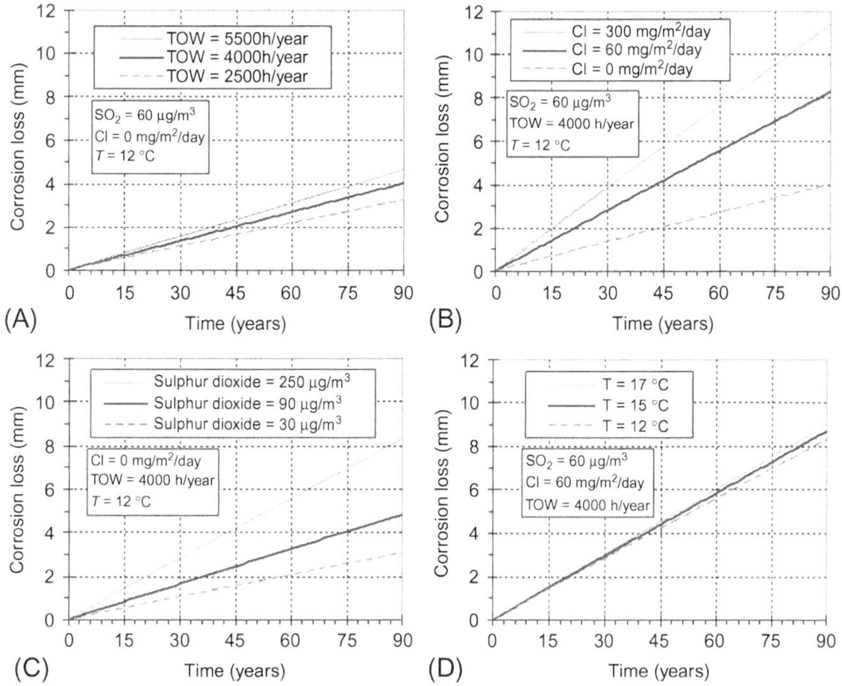

Fig. 6.2 Effect of varying climatic and atmospheric parameters on the corrosion losses with time, predicted using Eq. (6.7): (A) time of wetness (TOW), (B) Cl deposition rate, (C) SO_2 concentration, and (D) temperature.

with

$$\eta = 0.008(l_w/l_e)(\beta - 30) \tag{6.9}$$

and

$$\beta = \lambda_{LT}\sqrt{\left(\frac{\sigma_{yc}}{355}\right)\left(\frac{M_{ult}}{M_{pe}}\right)} \tag{6.10}$$

where l_w is the half-wavelength of buckling, l_e is the effective length, λ_{LT} is the slenderness parameter for lateral torsional buckling, σ_{yc} is the yield stress for the compression flange material (in MPa), M_{ult} is the moment of resistance if lateral torsional buckling is prevented, and M_{pe} is the plastic moment of resistance.

6.2.6 Consequences of Rail Bridge Failure

Estimation of the consequences of failure plays an essential role in both qualitative and quantitative risk-based design and assessment of bridges. Failure consequences can come about in many forms ranging from human and economic consequences and may extend into environmental as well as social consequences, depending on the system boundaries. A classification framework for the consequences of bridges failures has been developed by Imam and Chryssanthopoulos (2012). For the purposes of this study, only human and economic consequences have been considered for the development of risk profiles. For the former, it is here arbitrarily assumed that the bridge failure will result in five casualties. By considering the recommended cost for a railway fatality (£3.46 million) (Railway Safety, 2003), this can be converted into a total monetary value for human consequences. It is also assumed that the unavailability of the bridge will result in train detour and delay costs with economic consequences of £20 million in total. Furthermore, the cost of bridge replacement is assumed to be £20 million. The total value of the failure consequences is then multiplied by the instantaneous and the time-dependent probabilities of failure to determine the corresponding risk profiles for the different exposure scenarios examined. Although the costs of the failure consequences are assumed to remain constant in this study, these are likely to change overtime depending on the use of the bridge and the railway traffic demands.

6.2.7 Computational Procedure

In this study, the annual and cumulative probabilities failure for the moment capacity limit state and the corresponding risk of the corroding bridge are computed using the MCS with $n = 10^6$ samples/year. Corrosion damage, which is computed using Eq. (6.7), is assumed to occur in the form of a uniform thickness loss in the flanges, doublers, and web of the main girders. The applied load is treated deterministically, since this chapter focuses on the quantification of the effect of changing environmental conditions on the evolution of bridge reliability and risk. The procedure can be used for a new structure where no corrosion is present or for an existing structure where some thickness loss has already taken place in the past and, hence, the thickness at the beginning of the analysis is lower than the original thickness. The following sequential steps summarise the computation procedure used:

1. Future projections of climatic and atmospheric pollutants variables are obtained considering the location of the bridge, using databases such as the UKCP09.
2. In each year, n values of the resistance, climatic, and atmospheric variables are randomly generated using suitable distributions. In this study, the examined time period starts in 2012 and ends in 2099.
3. The randomly generated climatic and atmospheric variables are used in Eq. (6.7) to calculate the thickness loss due to corrosion.
4. The moment capacity, M_R, of the bridge is computed for each of the n runs in every year and compared with a specified externally applied moment due to railway loading.
5. The number of failures, n_f, is counted in each year and the annual probability of failure is calculated using n_f/n.
6. Cumulative probabilities of failure are calculated for each time period using Eq. (6.3).
7. Failure consequences are used to calculate the risk associated with the probability of bridge failure.

6.2.8 Case Study of a Steel Railway Bridge

6.2.8.1 Bridge Description and Location

The methodology is demonstrated through a case study using a 36 m single-span simple supported (half-through) steel bridge located in the United Kingdom and taken from Iles (2004). Fig. 6.3 shows the geometrical characteristics of the bridge. The mean thicknesses of both the bottom and top flanges and their doublers are 60 mm. The web of the main girders has a mean thickness of 20 mm.

Uniform corrosion is assumed to occur along the entire length of the main girders of the bridge. This assumption is supported by the study of Tamakoshi et al. (2006), where differences are reported in the location of corrosion occurrence in composite and noncomposite plate girder bridges. In their study, outer noncomposite girders were found to be even more susceptible to corrosion damage, compared to composite inner girders. Finally, in this study, the stiffeners as well as the crossbeams and slab are assumed to remain uncorroded.

The bridge is assumed to be located in the London Borough of Hillingdon in the United Kingdom, which includes the Heathrow International Airport. In this area, the levels of atmospheric pollutants, including SO_2, for 2003 and 2010 appear to be among the highest in the Greater London

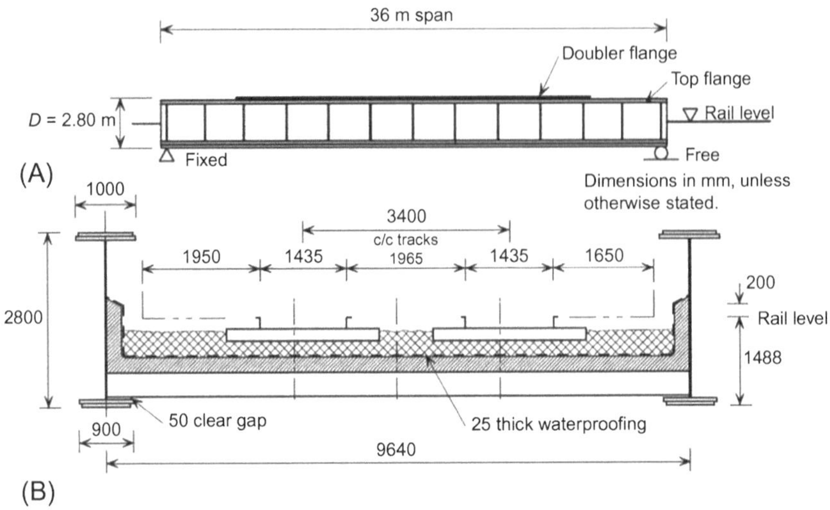

Fig. 6.3 Geometrical details of the half-through railway bridge used in the case study. (A) Side view; (B) cross-section (Iles, 2004).

area (Mattai and Hutchinson, 2006). In this location, the airborne salinity is assumed to be negligible (hence Cl in Eq. 6.7 is set to zero) since this is mainly critical in areas which are in close proximity to the sea.

6.2.8.2 Random Variables

The resistance random variables include the thicknesses of the flanges/doublers and web and their corresponding yield stresses (the values of f_y for the flanges and doublers are assumed to be fully correlated). Other geometrical variables taken as random include the length, height, and spacing of the main girders. The variables related to the corrosion process include the annual concentration levels of climatic factors and atmospheric pollutants. Table 6.1 summarises the statistical properties for the variables used in this study.

6.2.8.3 Analysis Scenarios

A number of exposure scenarios are considered which investigate the influence of long-term changes in the environmental conditions on the evolution of reliability and risk with time. Exposure scenarios S1 and S2 are used to examine the influence of different levels of SO_2 concentrations, assuming no future change of temperature or TOW. In S1 and S2, the values of mean and standard deviation for the temperature are taken as the average values obtained from the three emission scenarios in the UKCP09 for the period

Table 6.1 Statistical Properties of Random Variables for Corrosion Analysis

Variable	Mean	COV	Distribution
Flange/doubler yield stress, f_y (MPa)[a]	335	0.05	Log-normal
Web yield stress, f_{yw} (MPa)[a]	345	0.05	Log-normal
Flange/doubler thickness (mm)	60	0.05	Log-normal
Web thickness (mm)	20	0.1	Log-normal
Girder height, D (mm)	2,800	0.05	Log-normal
Girder length (mm)	36,000	0.02	Log-normal
Spacing of main girders (mm)	9,640	0.02	Log-normal
Time of wetness, TOW (h/year)	1,375/4,000[b]	Ranges: 250–2,500 and 2,500–5,500	Triangular
C (Eq. 6.7)	3,800	Range: 3,300–4,300	Triangular
SO_2 (μg/m^3)	17.5/60/170[c]	Ranges: 5–30, 30–90 and 90–250	Triangular
Temperature (°C)	In each emission scenario μ and σ evolve with time (Fig. 1)		Normal

[a]COV based on Melchers (1999) and distribution based on Deco and Frangopol (2011).
[b]Mean taken as the central values of the TOW ranges defined in Table B.1 of BS EN ISO (2012a).
[c]Mean taken as the central values of the SO_2 ranges defined in Table B.3 of BS EN ISO (2012a).

2010–39. The effect of increasing temperature, due to climate change, on the bridge reliability is examined using exposure scenarios S3–S5, based on the emission scenarios of the UKCP09 for the studied location. In these analyses cases, μ and σ for the temperature, obtained by fitting distributions to the UKCP09 data, are assumed to evolve linearly with time between periods 2010–39 and 2070–99. The influence of the atmospheric SO_2 concentration on the evolution the probability of bridge failure and the corresponding risk is investigated using cases S1 and S5–S7. Exposure scenarios S1, S5, and S6 assume constant atmospheric concentrations of SO_2 throughout the time period examined, while in S7 the effect of reducing SO_2 concentration with time (e.g., due to enforcement of government policy) is examined. In S7 scenario, it is assumed that the initial highly polluted industrial concentration of SO_2 reduces to urban levels within a 50-year period (i.e., period 2012–62); thereafter, the SO_2 concentration remains constant. Table 6.2 summarises the analysis cases examined.

Table 6.2 Summary of Exposure Scenarios Examined

Exposure Scenario	Variables		
	Temperature[a] (°C)	Sulphur Dioxide[b] (μg/m³)	TOW[c] (h/year)
S1	$\mu=11.66$, $\sigma=0.51$	5–30	2,500–5,500
S2	$\mu=11.66$, $\sigma=0.51$	30–90	2,500–5,500
S3	Low emissions	30–90	2,500–5,500
S4	High emissions	30–90	2,500–5,500
S5	Medium emissions	30–90	2,500–5,500
S6	Medium emissions	90–250	2,500–5,500
S7	Medium emissions	Evolution from 90–250 to 5–30	2,500–5,500
S8	Medium emissions	30–90	250–2,500

[a]Based on emission scenarios of UKCP09.
[b]Ranges of outdoor concentrations according to different environments of BS EN ISO (2012a); urban: 5–30 μg/m³, industrial: 30–90 μg/m³, highly polluted industrial: 90–250 μg/m³.
[c]Ranges for outdoor environments based on Table B.1 of BS EN ISO (2012a).

6.2.9 Results and Discussion

Fig. 6.4 shows the evolution of cumulative probability of failure, p_f, with time for exposure scenarios S1 and S2, where it can be seen, that there are small differences in the cumulative p_f up to around year 2035. Beyond this year, the results clearly show that the cumulative probability of bridge failure progressively becomes much higher for exposure scenario S2. Scenarios S1 and S2 assume no future increase of the mean annual air temperature. Exposure scenarios S3–S5 are used to investigate the potential effect of increasing temperature based on the UKCP09 projections for the assumed bridge location (see Fig. 6.5).

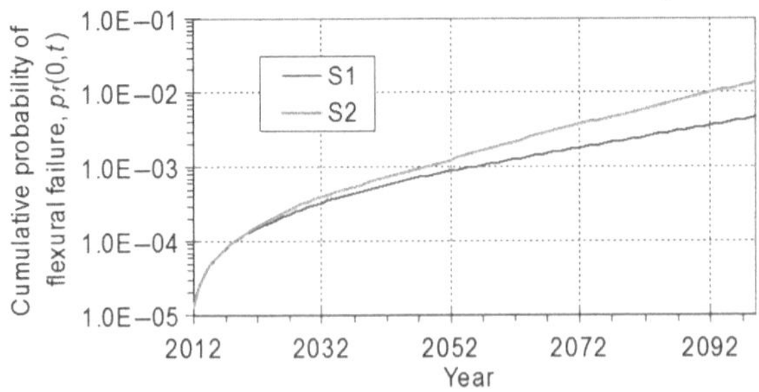

Fig. 6.4 Effect of urban and industrial environments in terms of SO_2 concentrations on the cumulative p_f.

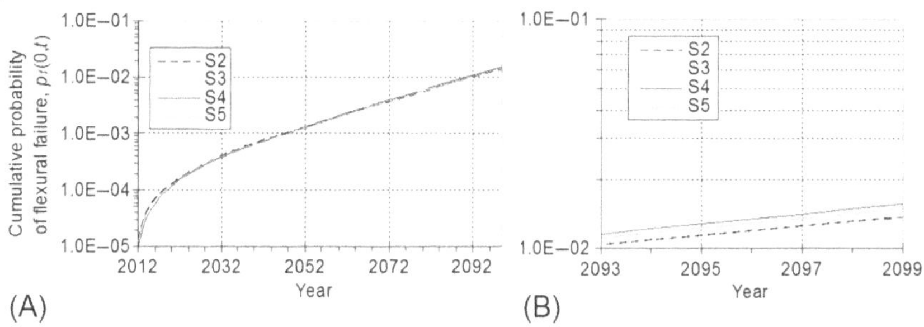

Fig. 6.5 Effect of increasing annual air temperature on the cumulative probabilities of failure: (A) entire time period 2012–99 and (B) close-up period 2093–99.

Fig. 6.5A and B shows the effect of increasing temperature, based on the UKCP09 emission scenarios, on the cumulative p_f with time. The results indicate that among the exposure scenarios used to investigate the effect of increasing air temperature, the high emissions UKCP09 scenario (scenario S4 in Table 6.2) resulted in the higher values of cumulative p_f (Fig. 6.5B). The results in Fig. 6.5 indicate that the overall direct effect of increasing average (annual) air temperature on bridge reliability is relatively small for all three emission scenarios examined. However, the potential interactions of increasing temperature with other environmental variables (e.g., RH and precipitation patterns) may affect bridge reliability and thus they require further investigation. A similar finding was observed for corrosion of reinforced concrete structures (Bastidas-Arteaga and Stewart, 2015, 2016).

Fig. 6.6 shows the results for scenarios S5–S7, which examine the influence of SO_2 concentration on the long-term bridge reliability. The results reveal small differences between scenarios S5, S6, and S7 up to year 2032;

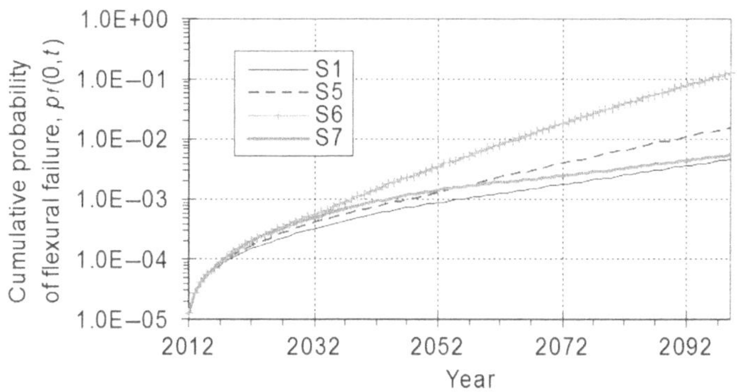

Fig. 6.6 Effect of SO_2 concentration on the cumulative probabilities of failure.

beyond this year, the predicted cumulative p_f for these exposure scenarios begins to diverge. As expected, the most unfavourable cumulative p_f were predicted for scenario S6 (i.e., scenario with the highest SO_2 concentration). The cumulative p_f for scenario S7, which assumes a linear reduction of SO_2 concentrations with time, are similar with scenario S6, up to year 2032, despite a 35% reduction of the mean SO_2 concentration. Beyond 2072, the predicted cumulative p_f for scenario S7 begins to converge to the predictions for scenario S1.

Fig. 6.7 shows the results for exposure scenarios S5 and S8, which examine the influence of TOW on the cumulative p_f. Similar cumulative p_f are predicted for both scenarios up to year 2025. Beyond this year, the lower values of annual TOW (exposure scenario S8) have a beneficial effect on the long-term reliability of the bridge. This is acceptable from a physical point of view, since the presence of water is a fundamental requirement for corrosion to occur (Klinesmith et al., 2007). Future changes in the annual TOW due to climate change for this particular location are likely to have an effect on the bridge reliability. In the case of gradually reducing future values of TOW, the cumulative p_f is likely to be placed between exposure scenarios S5 and S8.

Fig. 6.8A and B show the time-dependent annual and cumulative economic risk $E(L,t)$ associated with bridge failure, respectively, for exposure scenarios S1 and S5–S7. The results in Fig. 6.8 indicate that both the annual and cumulative risks are similar for the different scenarios examined up to around year 2017. Beyond this year, the annual and cumulative risks of scenarios S1 and S5 are lower compared to S6 due to the their less polluted, in terms of SO_2, exposure conditions (see Table 6.2). Fig. 6.8A indicates that

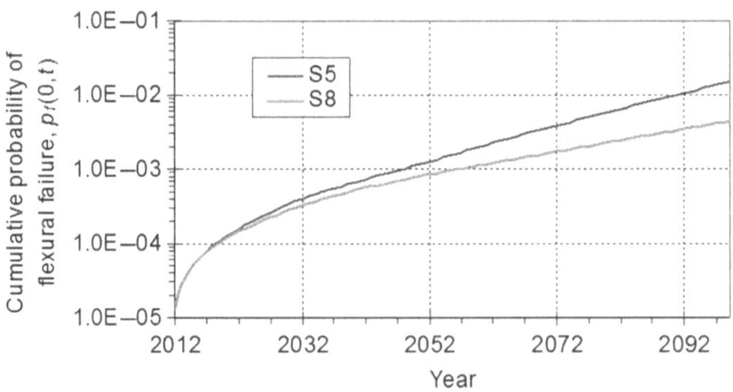

Fig. 6.7 Effect of time of wetness (TOW) on the cumulative probabilities of failure.

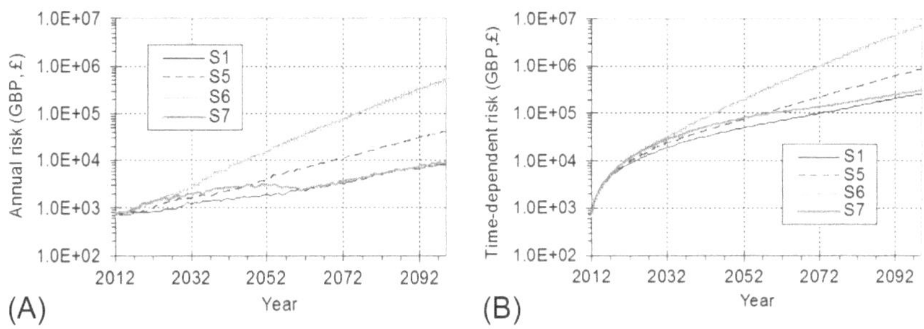

Fig. 6.8 Effect of SO_2 concentration on risk (GBP, £) associated with bridge failure: (A) annual risk and (B) time-dependent risk.

the annual risk for scenario S7 (gradual reduction of SO_2 concentration) increases at a rate similar to scenario S6 (highly polluted environment) up to around year 2025. Beyond this year, the annual risk of scenario S6 steadily increases up to 2099. In contrary, the annual risk of scenario S7 stabilises between years 2035 and 2060 (a small increase is observed for S7 during this time period) and it eventually converges with scenario S1 at around year 2060. The results in Fig. 6.8B, however, indicate that throughout the examined time period, the cumulative risk for scenario S7 remains higher than the corresponding cumulative risk of S1.

6.3 Climate Change Impacts on Bridge Scour

6.3.1 Background

Scour is identified as one of the most widespread causes of structural failure in bridges spanning over rivers (Imam and Chryssanthopoulos, 2012; Wardhana and Hadipriono, 2003). Scour is characterised by the erosion/removal of (underwater) river bed material in the vicinity of the piers and abutments (Melville and Coleman, 2000), leading to the development of scour holes. Scour hole depths exceeding the foundation depth (FD) of the piers/abutments can lead to structural instability and sudden bridge failure (Maddison, 2012; Van Leeuwen and Lamb, 2014). A number of factors are associated with the scour depths which may potentially develop at the pier and/or abutments of a bridge including, among others, the geometrical characteristics of the pier/abutment, the river characteristics including bed material and angle of attack as well as the flow magnitude at the bridge location. The scour phenomenon has been extensively investigated and a number of—mainly empirical—models are available which allow the quantification of scour depths considering the most significant influencing factors such as pier geometry, river, and flow characteristics (Breusers et al., 1977; Melville, 1997; Melville and Coleman, 2000; Richardson and Davis, 2001; Sheppard et al., 2014).

The prediction of scour depths in practice involves significant uncertainty caused by the variability associated with the different influencing factors, including bridge, river, and flow characteristics as well as the scour prediction models themselves, which have been developed empirically through small-scale laboratory experiments. A source of uncertainty which is becoming increasingly relevant when predicting future scour performance of bridges is the potential influence of climate change. The increased risk of scour of bridges due to climate change has been recognised worldwide (DEFRA, 2012, 2016; TRB, 2008). The potential consequences of climate change on bridge scour performance are currently not well known, on a quantitative basis, and need to be investigated to assist the development and implementation of adaptation strategies (Meyer and Weigel, 2011). Hence, capturing the aforementioned uncertainties in the analysis would allow a more reliable performance assessment of scour prone bridges and assist towards more efficient decision-making under conditions of uncertainty.

Some previous studies examined the probability of scour failure, without, however, considering the potential influence of climate change in performance assessment (Briaud et al., 2007, 2014; Deco and Frangopol, 2011; Johnson, 1992; Muzzammil and Siddiqui, 2009; NCHRP, 2003; Stein et al., 1999). On the other hand, risk-based frameworks considering the impact of climate risks on the long-term performance of civil engineering structures are becoming increasingly relevant within the context of infrastructure management (Stewart and Deng, 2015).

An important parameter which directly affects the depth of the developing scour hole is the magnitude of the flow which may potentially be encountered in a river at the bridge location. The flow magnitude in a river is governed by a several factors such as catchment characteristics, precipitation patterns, etc. Climate change is predicted to cause changes in the river flow characteristics due to changes in the precipitation patterns and catchment characteristics (IPCC, 2012; Robson, 2002; UKCP09, 2009).

In this section, a methodology is presented for the reliability assessment of scour prone bridges considering the potential effects of climate change. Statistical analysis of the expected maximum annual flow is combined with the MCS to compute the probability of scour failure, through scenario-based modelling. The Flood Estimation Handbook (FEH) and the statistical software WINFAP-FEH 3 are used to estimate the statistical properties of the expected maximum annual flow of the river. The potential influence of climate change on the flow characteristics is considered through gradual changes in the mean and variability (i.e., standard deviation) of the expected maximum annual flow distribution to assess the sensitivity of the probability of failure of the bridge to such changes. The uncertainties associated with the different factors which influence scour performance are considered through suitable distributions. Results are presented from a case study using a bridge in the United Kingdom, in which scour reliability profiles are presented for a number of scenarios which assume time-dependent changes in the distribution of the expected maximum annual flow. Furthermore, the influence of FD and local scour model parameters is also investigated in relation to the changing flow characteristics.

6.3.2 Climate Change and River Flows

Alterations in the climatic and weather conditions due to climate change can potentially increase the uncertainty associated with the magnitude as well as the prediction of extreme weather events including extreme precipitation and river flows (IPCC, 2012). The available evidence suggests that these changes may prevail as temporal changes in the statistical properties and distribution of key climatic parameters such as temperature and precipitation (IPCC, 2012; Katz, 1993). To enable both the quantification of the potential effects of climate change and the development of future adaptation strategies, a number of potential emission scenarios have been devised by the IPCC which cover a period up to the end of the 21st century (IPCC, 2013). Increasing flood frequency and magnitude due to increasing precipitation and/or changes to the catchment characteristics can have a significant effect on the scour performance of bridges.

In the United Kingdom, during the years 1961–95, an increase was observed in the extreme precipitation events during winter (Osborn et al., 2000). A decreasing annual trend of rainfall due to climate change in a particular area can include significant increases in precipitation associated with seasonal trends (e.g., during winter). In general, the climatic projections for the United Kingdom (UKCP09, 2009) indicate this situation. The impacts of climate change, however, are catchment specific and their magnitude can differ remarkably even for catchments located close to each other (Prudhomme et al., 2013). Recently, increasing trends have been observed in relation to flood frequency and magnitude in the United

Kingdom (Prudhomme et al., 2003); however, the findings of Robson (2002) suggest that these increasing trends are associated with natural climatic variability rather than climate change. Despite these observations, Robson (2002) suggested that climate change expressed as increases in rainfall should not be ignored as a potential source of future increase of flooding in the United Kingdom.

The preceding discussion indicates that at present it is difficult to precisely quantify the effect of climate change in terms of precipitation and temperature changes on fluvial flood frequency and magnitude. However, the results of several studies suggest that in some areas flood frequency and magnitude will increase in the future (i.e., occurrence of extreme events will become more frequent). In view of these limitations, the methodology developed here aims to quantify the effect of potential changes of flood characteristics (i.e., changes in the frequency and magnitude of the expected maximum annual flow due to climate change) on the probability of local scour failure. Results are presented from a case study using a bridge in the United Kingdom to investigate the potential effects of changing mean and variability of the expected maximum annual flow on the probability of local scour failure in bridge piers.

6.3.3 Local Scour Prediction Model

A number of models have been proposed in literature for the estimation of local scour in bridge piers (Sheppard et al., 2014). Here, local scour is estimated using the HEC-18 design equation (Arneson et al., 2012), given by Eq. (6.11), which considers scour as a time-independent process, that is, temporal effects of local scour development are not modelled (Melville and Chiew, 1999):

$$y_{max} = 2y_0 K_1 K_2 K_3 K_4 \left(\frac{D}{y_0}\right)^{0.65} F_0^{\,0.43} \tag{6.11}$$

where $y_{\max}$ is the maximum scour depth (m), y_0 is the depth (m) of the flow upstream of the bridge pier, K_1, K_2, K_3, and K_4 are the coefficients which allow for pier shape, angle of attack, streambed conditions, and the river bed material size, D is the pier diameter, and F_0 is the Froude number given by:

$$F_0 = \frac{V}{(gy_0)^{0.5}} \tag{6.12}$$

where g is the gravity acceleration and V is the flow velocity given by:

$$V = \frac{Q}{By_0} \tag{6.13}$$

and the flow depth y_0 is given by (BD 97/12, 2012):

$$y_0 = \left(\frac{nQ}{Bs^{1/2}}\right)^{3/5} \tag{6.14}$$

where Q is the flow (m^3/s), B is the river width (m), n is the Manning's coefficient, and s is the longitudinal slope of the channel.

K_1 depends on the pier nose shape and it can take the following values: 1.1 for a square nose, 1.0 for round nose or circular cylinder, and 0.9 for a sharp nose. Coefficient K_2 is a function of the angle of attack of the river flow with respect to the pier (Arneson et al., 2012). K_3 depends on whether there is clearwater scour or the river bed is plane ($K_3 = 1.1$) as opposed to the case of having dune bed configurations with different dune heights (K_3 ranging between 1.1 and 1.3 depending on the dune height). K_4 depends on the diameter of the river bed material and can range between 0.4 (fine material) and 1.0 (coarse material).

Eq. (6.14) can be used for wide rivers with ratios B/y_0 exceeding about 10, giving conservative predictions for cases where this is less than about 10 (BD 97/12, 2012). In this chapter, the flow Q is estimated using the statistical analysis procedures implemented in the software WINFAP-FEH 3 (Robson and Reed, 1999).

6.3.4 Probabilistic Assessment of Local Scour

Reliability analysis allows the estimation of the failure probability, p_f, of a structure for different limit states. The performance function for the limit state of a bridge is given by:

$$G(t) = R - S(t) \tag{6.15}$$

where $G(t)$ is the time-dependent safety margin, R and $S(t)$ are the resistance (i.e., FD) and time-dependent load effects (i.e., maximum scour depth given as a function of flow magnitude, river and pier characteristics, etc.), respectively. The performance function for local scour in bridge piers is given by:

$$G(t) = D_F - y_{max}(t) = D_F - \left[2y_0(t)K_1K_2K_3K_4\left(\frac{D}{y_0(t)}\right)^{0.65} F_0(t)^{0.43}\right] \quad (6.16)$$

where D_F is the FD. Using the statistical properties, including distribution type, for the random variables of Eq. (6.16) and assuming that the resistance D_F and the load effects $y_{\max}(t)$, are statistically independent, the instantaneous (annual) probability of failure, $p_f(t)$, is given by:

$$p_f(t) = \Pr[G(t) \leq 0] = \Pr\left\{\left[D_F - \left[2y_0(t)K_1K_2K_3K_4\left(\frac{D}{y_0(t)}\right)^{0.65} F_0(t)^{0.43}\right]\right] \leq 0\right\} \quad (6.17)$$

Here, Eq. (6.17) is evaluated using the MCS implemented in MATLAB software. Alternatively, the availability of a closed-form expression for the limit state function enables the use of other reliability analysis methods (e.g., FORM). The use of the MCS, however, allows to estimate the failure probability as well as to obtain the distribution of the scour depth (Eq. 6.11)—which is not known a priori—for the analysis cases examined. The cumulative probability of failure, at any point within a time period, is given by Eq. (6.3), provided that the failures are statistically independent.

The uncertainties associated with the random variables of Eq. (6.17), can be separated as aleatory and epistemic uncertainties (Frangopol and Liu, 2007; Kiureghian and Ditlevsen, 2009; Merz and Thieken, 2005). Within the context of scour assessment under changing flow magnitude (i.e., expected maximum annual flow) due to climate change, the variability of flow magnitude is associated with both natural variability and epistemic uncertainty. Aleatory uncertainty in relation to flow modelling is due to the lack of a specific emission scenario for future climate change since climate change itself is a function of several variables which are not always possible to be objectively quantified such as future global population, government policy, technological breakthroughs, etc. On the other hand, epistemic uncertainty is caused by the lack of understanding of how a specific climate change scenario (e.g., increased precipitation, etc.) will affect the flow conditions of a particular catchment. Furthermore, the available models for scour are associated with epistemic uncertainty, since the inherently complex nature of scour is modelled through approximations developed from laboratory experiments of small-scale pier models (e.g., Eq. 6.11). This uncertainty type can be reduced by developing more accurate models of the phenomenon through, for instance, additional experimentation (this is discussed later in this chapter).

The Monte Carlo-based procedure for the probabilistic assessment of local scour considering the potential effects of climate change is implemented in MATLAB using a sample size of $N=2\times10^6$ per year. A number of methods have been devised to predict the potential flow in a particular river, for instance, through statistical analysis of historic flow records or alternatively using the rainfall-runoff method which requires knowledge of the precipitation patterns and the catchment descriptors (Robson and Reed, 1999). The statistical analysis of historic flow records through the widely used WINFAP-FEH 3 software, which has been used for this case study allows the estimation of the statistical properties of the expected maximum annual flow for any river in the United Kingdom. Flow events in different years are assumed to be independent. The expected maximum annual flow, which is denoted here as QMED, is modelled by fitting a suitable distribution to flood data using the statistical procedures implemented in WINFAP-FEH 3 (Robson and Reed, 1999). This approach is based on the creation and analysis of a pooling group of several catchments of similar hydrological characteristics, with the available years of flow records for each station (in the different catchments) contributing to the total number of station years.

Eqs. (6.12)–(6.14) are used to calculate the Froude number, flow velocity, and flow depth, respectively, while Eq. (6.11) is used to compute the depth of local scour, for each set of the N randomly generated values for the variables (e.g., expected maximum annual flow, pier width, river slope, etc., see Table 6.3) in each year, Eqs. (6.15)–(6.18) are used to calculate the annual (instantaneous) and time-dependent (cumulative) failure probability for a specific value of FD. The FD and the load effects $y(t)$, that is, the scour depth, are assumed to be statistically independent. In actual bridges, the FD is governed by several factors, one of which is the maximum expected scour depth. However, this correlation is not considered in the present analysis due to the lack of sufficient information. The procedure is demonstrated through a case study; the different analysis cases examined are discussed in the subsequent sections of this chapter.

6.3.5 Bridge Case Study

The bridge considered in this case study is assumed to be located on the river Earn in Scotland, United Kingdom, assuming alluvial riverbed conditions

Table 6.3 Statistical Properties of Variables for Scour Analysis

Variables		Mean	COV	Distribution	References
River	Width (m)	65	0.05	Normal	Assumed
	Streambed conditions (K_3)	1.1	0.05	Uniform	NCHRP (2003)
	Bed material size (K_4)	1.0	–	–	–
	Slope	0.0032	0.05	Log-normal	Assumed
	Manning's coefficient	0.035	0.28	Log-normal	NCHRP (2003)
Bridge piers	Foundation depth (m)	4.5	–	–	Assumed
	Pier nose shape (K_1)	1.0	–	–	–
	Angle of attack (K_2)	1.0	–	–	–
	Pier width, D (m)	2	0.05	Normal	Assumed

(Gilvear and Black, 1999). Initially, the location of the examined bridge is established on the maps and the available nearby stations are identified. This station is then used in WINFAP-FEH 3 software to create the pooling group using stations from similar catchments with a total of 1000 station years [annual maxima (AM) series]. WINFAP-FEH 3 provides a number of options for estimating QMED, which is the maximum annual flow with a return period $T=2$ years, for instance, using the catchment descriptors or AM series (for more details, see Robson and Reed (1999)). In this chapter, a QMED of $250.2\,m^3/s$ is estimated from AM series of the station.

Analysis of pooling group flood data (i.e., AM series) using the WINFAP-FEH 3 software indicates that the generalised extreme value (GEV) distribution, given by Eq. (6.18) (Kottegoda and Rosso, 1997), is the most suitable distribution for modelling the magnitude of the expected maximum annual flow; the cumulative distribution function is expressed as follows:

$$F_{X_{max}} = \exp\left\{-\left[1-\frac{k(x-\varepsilon)}{\alpha}\right]^{1/k}\right\} \tag{6.18}$$

where k is the shape parameter, ε is the location parameter, and α is the scale parameter. The GEV distribution parameters obtained from the statistical analysis of stations in similar catchments (i.e., pooling group) in WINFAP are $\alpha=0.222$, $\varepsilon=0.919$, and $k=0.002$. The scale and location parameters of the GEV distribution are given by Eqs. (6.19), (6.20), respectively (Kottegoda and Rosso, 1997):

$$\alpha = \sqrt{\frac{k^2\sigma^2}{\Gamma(1+2k) - \Gamma^2(1+k)}} \tag{6.19}$$

$$\varepsilon = \mu - \frac{\alpha}{k}[1 - \Gamma(1+k)] \tag{6.20}$$

where μ is the sample mean, σ is the sample standard deviation and k, ε, and α are the shape, location, and scale parameters of the GEV distribution, respectively (Kottegoda and Rosso, 1997), and Γ () is the gamma function.

The potential effects of climate change on scour are examined through a parametric study in which the scale and location parameters are gradually changed for increasing values in the mean μ and variability (i.e., standard deviation σ) of the flood magnitude in Eqs. (6.19) and (6.20).

Within the probabilistic framework developed, several variables associated with the bridge and river characteristics are treated as random. Table 6.3 summarises the statistical properties of the random variables considered in this case study. In this study, the bed material is assumed to be deterministic and time invariant; however, overtime, changing flow properties may modify the size of bed material. The river and pier dimensions (i.e., widths of the river and pier, respectively) are also treated as random. Examination of historic drawings from existing bridges has assisted the selection of the mean values for these variables, with the selected bridge being considered as a representative example for this area. Actual measurements taken during a site visit could allow for deterministic values to be used for these variables. To this end, actual measurements on several (similar) piers within the same bridge may be associated with some degree of variability. Similarly, the river width is not constant and its value depends on the location that the measurement is obtained.

Climate change is likely to impact the precipitation patterns and catchment characteristics of a specific area which in turn may affect the magnitude and frequency of expected maximum annual flow. In this study, it is assumed that climate change will cause a certain amount of temporal changes in the statistical properties of the flood predicted using the FEH (e.g., 20% increase of sample mean). The different analysis cases examined in relation to changing flow conditions are summarised in Table 6.4. These changes are assumed to evolve linearly with time over a 60-year period. In this way, the effects of climate change on the precipitation patterns and catchment descriptors and hence on the flood frequency and magnitude are implicitly considered in the analysis of the different scenarios examined. A FD of 4.5 m is assumed to

Table 6.4 Analysis Cases for Assessing the Influence of Assumed Changes in the Expected Maximum Annual Flow

Analysis Case	Foundation Depth (m)	Mean, μ	Standard Deviation, σ
45M0000	4.5	No change	
45M2000		20% increase	No change
45M4000		40% increase	No change
45M6000		60% increase	No change
45M0020		No change	20% increase
45M0040		No change	40% increase
45M0060		No change	60% increase
45M2020		20% increase	20% increase
45M4040		40% increase	40% increase
45M6060		60% increase	60% increase

facilitate the estimation of the scour failure probability; the effect of this variable is investigated later.

6.3.6 Results and discussion

6.3.6.1 Effect of Changing Flow Characteristics on Pier Scour Failure Probability

Figs. 6.9–6.11 show the annual and cumulative probabilities of pier scour failure for the scenarios examined with increasing mean, increasing variability, and simultaneous increase of mean and variability in the expected maximum annual flow, respectively, for a given FD equal to 4.5 m. In all the figures, (A) and (C) refer to annual and cumulative failure probability, respectively.

The results in Fig. 6.9 show the effect of increasing mean (Eq. 6.19) in the expected maximum annual flow on the annual and cumulative probabilities of failure for the scenarios with mean increases of 20%, 40%, and 60% over a 60-year period. The results in this figure indicate that the effect of increasing mean is relatively small for the initial 10 years and it gradually becomes noticeable between 10 and 20 years and significant beyond the initial 20-year period. At the end of the examined period, the cumulative probabilities of failure for the scenarios with 20%, 40%, and 60% increase in mean (i.e., see Fig. 6.9 for cases 45M2000, 45M4000, and 45M6000; in this nomenclature, 45 stands for FD of 4.5 m, M stands for model, and 2000 stands for 20% increase in mean and 0% increase in variability) are predicted to be 59%, 176%, and 337%, respectively, higher than the base-line scenario which assumes no changes in the statistical properties of the expected maximum annual flow overtime.

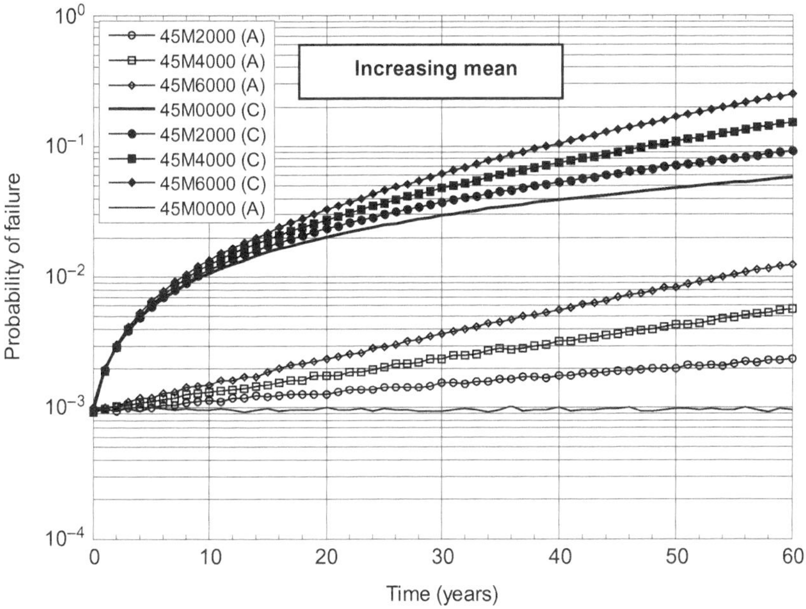

Fig. 6.9 Effect of increasing mean in the expected maximum annual flow on the annual (A) and cumulative time-dependent (C) probabilities of scour failure.

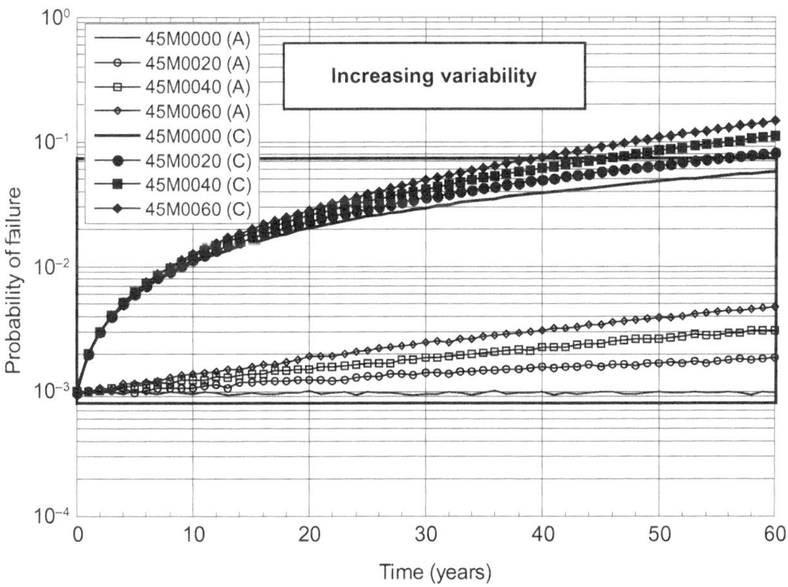

Fig. 6.10 Effect of increasing variability (standard deviation, σ) in the expected maximum annual flow on the annual (A) and cumulative time-dependent (C) probabilities of scour failure.

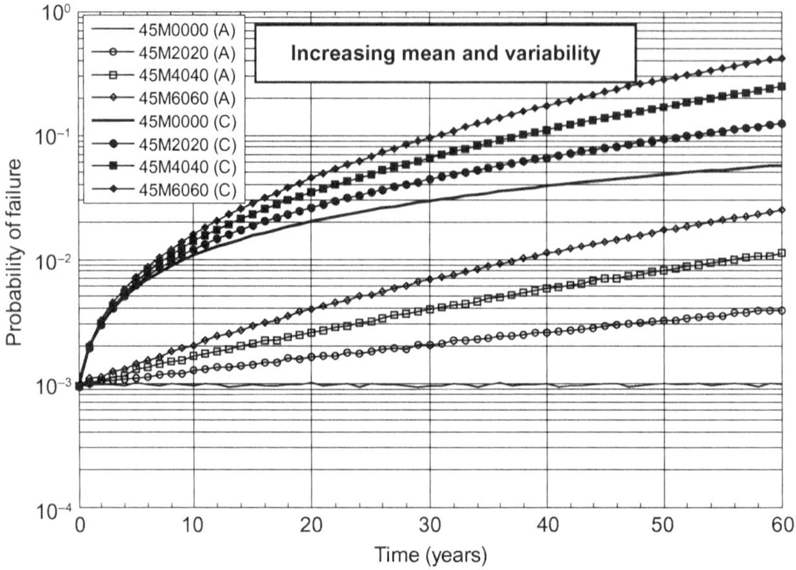

Fig. 6.11 Effect of simultaneous increase of mean and variability in the expected maximum annual flow on the annual (A) and cumulative time-dependent (C) probabilities of scour failure.

The results in Fig. 6.10 show the effect of increasing variability (Eq. 6.20) in the expected maximum annual flow on the annual and cumulative probabilities of failure for the scenarios with variability increases of 20%, 40%, and 60% over a 60-year period. The results in this figure indicate that the effect of increasing variability is relatively small for the initial 15 years and it gradually becomes significant beyond the initial 20-year period. For the analysis cases examined, the effect of increasing variability in the expected maximum annual flow is predicted to have a relatively smaller effect on the failure probabilities compared to the previous case of assuming a gradually increasing mean. In general, the results in Fig. 6.10 follow a similar trend to the results obtained for increasing mean (Fig. 6.9); that is as the variability of the expected maximum annual flow increases, the probability of failure also increases (cumulative failure probabilities at the end of the 60-year period are predicted to increase by approximately 41%, 92%, and 153% for cases 45M0020, 45M4040, and 45M0060, respectively, relative to the base-line case 45M0000).

The results in Fig. 6.11 show the effect of simultaneously increasing mean and standard deviation up to 60% (see Eqs. 6.19 and 6.20) of the expected maximum annual flow on the annual and cumulative probabilities of failure for scenarios 45M2020, 45M4040, and 45M6060 (see Table 6.4). For the cases examined, the effect of simultaneous increase of mean and variability has the highest relative effect on the predicted probabilities of local scour failure. More specifically, the results for cases 45M2020, 45M4040, and 45M6060 show that at the end of the 60-year period, an increase in the cumulative failure probabilities by approximately 117%, 332%, and 633%, respectively, compared to the base-line case 45M0000. It is interesting to note that the combined effects of increasing mean and variability associated with the maximum expected annual flow (i.e., analysis cases 45M2020, 45M4040, and 45M6060) is greater than the sum of the individual effects (i.e., cases where the increasing mean and variability are considered separately), see Figs. 6.9–6.11. Based on these observations, it can be concluded that the combined effects of increasing mean and variability have the greatest effect on the probability of local scour failure. When considering the impact of increasing mean or variability individually, increasing mean in the maximum expected annual flow is predicted to have a more significant (adverse) effect on the predictions compared to the increasing variability.

The cumulative failure probability profiles have been developed assuming that the flow characteristics occur gradually (linearly) over a 60-year time period. It would be interesting to investigate the effect of potentially sudden changes in the statistical properties of the expected maximum annual flow-due to climate change-on the scour probability of failure. Furthermore, the preceding analyses assume that the distribution type (i.e., GEV) of the expected maximum annual flow remains the same throughout the examined period. At present, however, it is not possible to confidently predict the timing and magnitude of potentially sudden changes in the flow characteristics that may occur in the future (IPCC, 2012; Robson, 2002). One of the key challenges that still remain is to establish a link between the altered precipitation patterns and/or catchment properties (due to climate change) with the flow (i.e., expected maximum annual flow) characteristics. To this end, significant uncertainty still exists as to which climate change scenario will be realised.

6.3.6.2 Effect of FD on Scour Probability of Failure

A common challenge with existing bridges is that the FD is often unknown even if the original drawings of the superstructure are available (RSSB, 2005). The results presented in the previous sections have been obtained by assuming, conservatively, a FD of 4.5 m. A number of additional analysis cases are considered to investigate the influence of assuming FD, due to having unknown foundations, on the cumulative probability of scour failure. The FDs examined are 4 and 5 m. These foundations depths are examined for a number of cases which assume increasing mean or/and variability of the expected maximum annual flow distribution; these are: (a) a case of 40% increase in mean, (b) 40% increase of variability, and (c) simultaneous increase of mean and variability by 40%.

Fig. 6.12 shows the cumulative probabilities of scour failure for the analysis cases with varying assumed FDs. As expected, the results in this figure clearly show the influence of FD on the predicted cumulative p_f, with smaller FDs having higher probabilities of failure. For the case of FD = 4 m, the effect of changes in the statistical properties of the flow (i.e., increasing mean or/and variability) has a very small effect on the cumulative probabilities of scour failure. The results in Fig. 6.12 indicate that, for the FD = 4 m case, scour failure becomes almost certain during the last 5 years (of the 60-year period) for the cases with increasing mean and simultaneous increase of mean and variability of the expected maximum annual flow distribution.

As the FD increases, the probabilities of scour failure decrease as shown in Fig. 6.12. The results in this figure further indicate that the effect of increasing mean and/or variability is not constant when different foundations

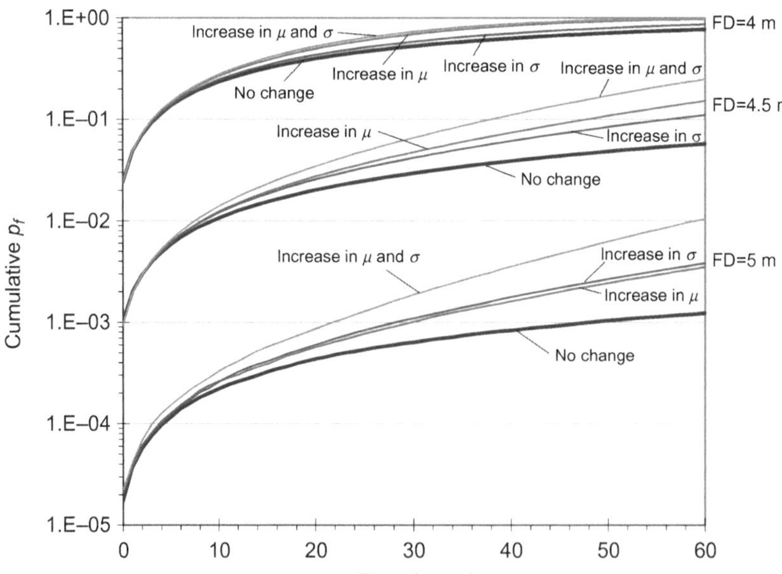

Fig. 6.12 Effect of foundation depth on time-dependent probability of scour failure under changing expected maximum annual flow.

depths are considered. For example, for the case of FD = 5 m, the increasing variability of the expected maximum annual flow has a greater effect than the increasing mean. Conversely, the increasing mean of the expected maximum annual flow distribution has a greater effect compared to the effect of increasing variability for decreasing FDs. As shown in Fig. 6.12, the simultaneous increase of mean and variability of the flow distribution produces the highest probabilities of failure in all cases examined. The influence of the simultaneous increase of mean and variability of the expected maximum annual flow reduces for decreasing FDs.

These results indicate that FD has a significant effect on the predictions. In practice, this variable is deemed with high uncertainty while in many cases no data are available on the actual foundation type and depth of a particular bridge (JBA, 2004). In such cases, conservative values of FD are recommended in assessing scour performance (JBA, 2004). To this end, the systematic collection of actual FD measurements of piers in scour prone bridges would reduce the uncertainty and hence improve the accuracy of the scour failure predictions during assessments.

6.3.6.3 Effect of Model Parameter λ_{sc} on Predicted Scour Depths and Failure Probabilities

It has been shown that the scour equation (Eq. 6.11) of HEC-18 leads to conservative predictions of local scour (Landers and Mueller, 1996; NCHRP, 2003). This is due to the fact that the scour prediction models used in codes of practice have been developed through small-scale laboratory experiments. Comparisons of these prediction models with field measurements of scour depths on real bridges have shown that there is a discrepancy between them. It has been suggested that a model parameter λ_{sc} can be introduced in Eq. (6.11) to reduce its conservatism (NCHRP, 2003) inherently taking into account epistemic uncertainty. The local scour model is now given by:

$$y_{max}(t) = 2\lambda_{sc} y_0(t) K_1 K_2 K_3 K_4 \left(\frac{D}{y_0(t)}\right)^{0.65} F_0(t)^{0.43} \quad (6.21)$$

Several values have been proposed for the statistical properties of λ_{sc}; for a summary see NCHRP (2003). To investigate the influence of this parameter, a number of scenarios are considered using values for the statistical properties of λ_{sc} suggested in NCHRP (2003). The analysis cases in this section assume a 40% gradual increase in mean, standard deviation, and the simultaneous increase of both for the expected maximum annual flow considering two FDs of 4.5 and 5 m. The model parameter λ_{sc} is modelled using a normal distribution with mean = 0.55 and COV = 52% (NCHRP, 2003). In this way, comparisons can be made with results presented in the previous sections.

Fig. 6.13A and B shows the predicted time-dependent failure probabilities for the analysis cases with FDs 4.5 and 5 m, respectively. The results in these figures indicate that although the mean value of λ_{sc} is lower than 1 (i.e., mean value of $\lambda_{sc} = 0.55$), the introduction of this model parameter causes an increase of the predicted time-dependent failure probabilities. This observation can be explained by considering the influence of λ_{sc} on the predicted scour depths in Figs. 6.14 and 6.15. More specifically, the results in Figs. 6.14 and 6.15 indicate that the high variability associated with model parameter λ_{sc} results in an increase of the scour depth values exceeding 4.5 and 5 m (generated using the MCS in MATLAB). As expected, the mean value of the predicted scour depths is lower than the predicted scour depths when $\lambda_{sc} = 1$. The results in Fig. 6.13A and B also indicate that the very high variability of the model parameter λ_{sc} (COV = 0.52) overshadows the effect of the assumed increasing variability in the expected maximum annual flow on the predicted time-dependent probabilities of failure (i.e., in Fig. 6.15 the increasing variability of flow is predicted to have an insignificant effect on cumulative p_f).

Figs. 6.14 and 6.15 indicate that negative scour depths are predicted in the cases where λ_{sc} is modelled using normal distribution, which is not acceptable from a physical point of view. In contrary, λ_{sc} remains positive

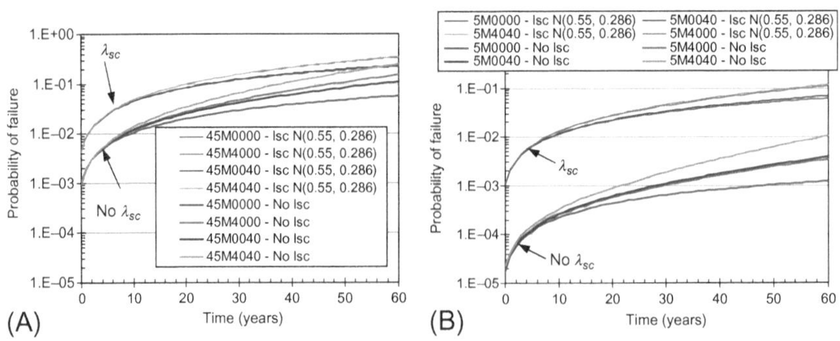

Fig. 6.13 Effect of model parameter λ_{sc} on the time-dependent probabilities of local scour failure for up to 40% increase in the mean or/and variability of the expected maximum annual flow distribution: (A) foundation depth = 4.5 m and (B) foundation depth = 5 m.

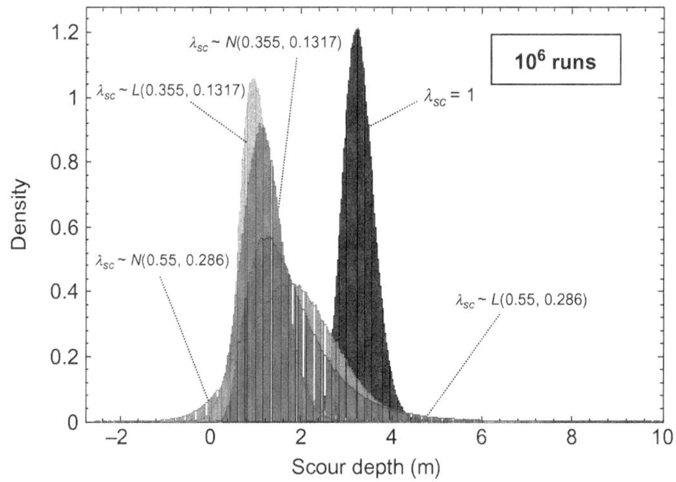

Fig. 6.14 Effect of model parameter λ_{sc} (Eq. 6.21) on the predicted scour depths for cases with no changes in the distribution of the expected maximum annual flow. *N* and *L* mean normal and log-normal distributions, respectively.

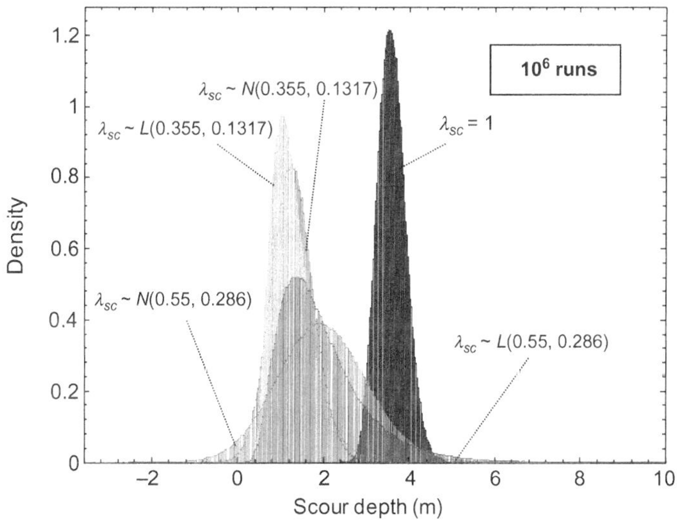

Fig. 6.15 Effect of model parameter λ_{sc} (Eq. 6.21) on the predicted scour depths for cases with 40% increase in the mean of the expected maximum annual flow distribution. *N* and *L* mean normal and log-normal distributions, respectively.

when a log-normal distribution is used for its modelling; in this case, however, the magnitudes of the predicted maximum scour depths (near the upper tail) are significantly larger than the cases where a normal distribution is used.

Based on the results presented in this section, it can be concluded that the use of λ_{sc}, which is a source of epistemic uncertainty, has a significant effect on the predictions. However, at present, the available statistical properties for this variable are not consistent (e.g., negative or very large scour depths) with actual observations. To this end, further research is needed to obtain accurate statistical properties including the distribution type for this variable through additional field measurements and comparison with the code predictions.

6.4 Climate Change Versus Asset Stock Data Uncertainties

Many infrastructure operators worldwide manage large numbers of assets. The number of rail bridges in the European Union is approximately 217,000 (Bell, 2004), while the total number of highway bridges in the United States is over 600,000 (ASCE, 2013). Infrastructure operators often manage bridge stocks with limited budgets through prioritisation techniques based on risk assessments, using network- and asset-specific data. However, when managing a large asset base, maintaining reliable and up-to-date data records can be a challenging task. The challenge is exacerbated when the infrastructure is old and robust data records may have been lost or nonexistent; for example, most of the 28,000 bridges on the British railway network were built in the 19th century (Bell, 2004). Thus, asset managers often need to make decisions on the basis of incomplete and uncertain information, prompting the need for a robust risk management framework (Piyatrapoomi et al., 2004). Numerous additional uncertainties from various sources affect the decision-making process in bridge management. Notably, climate change is notoriously uncertain and its effects on infrastructure are still not well understood by engineers.

This section explores uncertainty associated with data used in bridge scour risk assessments at a bridge stock level. This uncertainty is explored in the context of a changing climate, to explore the propagation of climate change uncertainty through the scour risk assessment and its interaction with other sources of uncertainty. To understand the relative importance of different parameters on the scour risk of bridges, sensitivity analyses have been carried using a number of United Kingdom railway bridges as case studies and considering the United Kingdom (Network Rail) railway scour assessment code. Scour assessments in the United Kingdom take place at two stages; stage 1 is a preliminary assessment, which all bridges over water are subjected to. Any bridge highlighted as being at risk is assessed at stage 2, which includes a detailed hydrological and hydraulic assessment and a more robust data collection process. Local scour is estimated through an empirical model, similar to the one presented in Section 6.3. Scour risk is estimated in terms of a final priority rating (FPR) which is a dimensionless value that is used to compare the relative risks at different structures, so that they can be prioritised with respect to scour vulnerability and criticality. A more detailed description of the procedure and the models used for the sensitivity analysis can be found in Dikanski et al. (2018).

Fig. 6.16 shows a summary of the results from the sensitivity analysis for the different parameters that have been investigated for the case study bridges analysed. The diagram shows the overall change of risk score FPR resulting from variations across the entire studied parameter ranges, which have been

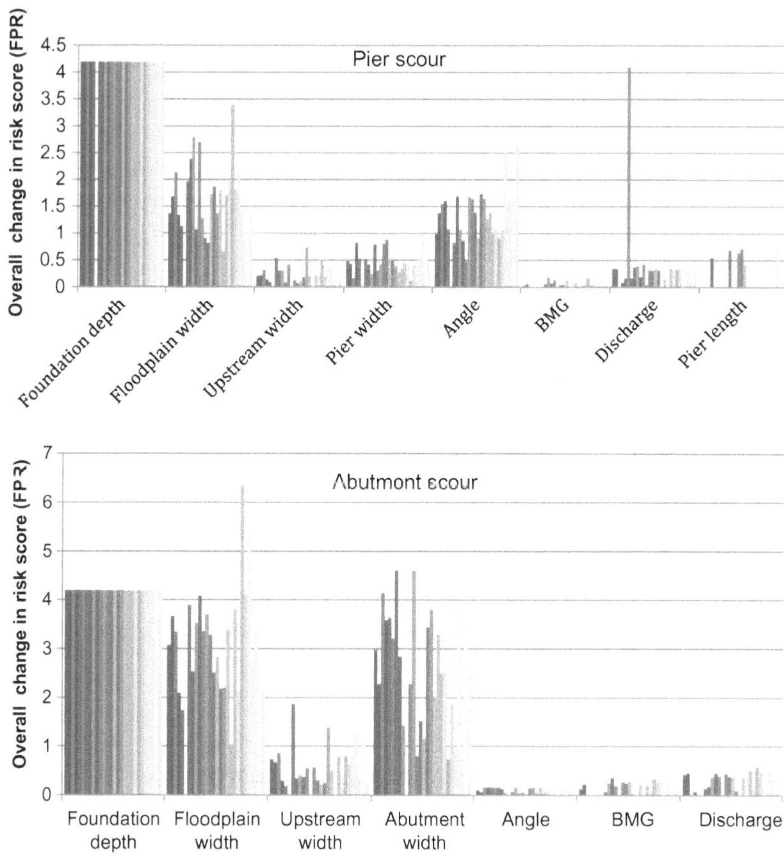

Fig. 6.16 Sensitivity analysis for piers (*top*) and abutments (*bottom*).

identified by comparing data between stage 1 and stage 2 assessments, as explained in Dikanski et al. (2018). The most influential parameters are shown to be FD, pier or abutment width and floodplain width. Angle of attack is also significant for pier scour, but not for abutments.

It was also found that, in most cases, the errors in parameters would lead to an underprediction of actual risk, meaning that exposure to scour risk is likely to be higher than current assessments suggest. Errors in floodplain width records are shown to have the largest effect on the estimation of risk both for piers and abutments.

Overall, parameters that are often assumed to be deterministic in the literature were found to be subject to large and variable uncertainties due to the data used for assessments. In the case of unknown FD, the high resulting uncertainty affecting scour risk estimates is a well-known issue; the results presented above re-emphasise that it is one of the largest sources of uncertainty. Furthermore, some additional parameters, for example, abutment width and floodplain width, are also found to have a major influence on the estimation of scour risk. In practice, this reduces the confidence in scour assessments, which determine investment priorities for scour interventions. By understanding the sensitivities of a scour risk model to its inputs, infrastructure operators can target their efforts in reducing the uncertainty in the most significant parameters, thereby increasing confidence in scour assessments. Thus, scour intervention schemes can truly target the highest priority sites and the value of infrastructure investments can be maximised, while risks are effectively reduced. As a result of research presented in this chapter, a review of bridge data used for scour assessments globally is recommended. This would be an essential first step towards improving bridge resilience both in the context of current and future climate.

Climate change is associated with significant uncertainty, which is often quoted as a major barrier to increasing climate resilience. However, results show that resulting changes in river flows have a relatively low impact on the estimation of scour risk, as compared to other parameters. This is due to several reasons, including the use of high initial assessment flows (200-year return period), exclusion of flow considerations from contraction scour estimation and exclusion of cumulative effects of scour. In practice, the increase of flood frequency and severity expected to be brought about by climate change is likely to increase scour risk, but this is not being detected by the selected case study model, as shown by a range of previous studies (Khelifa et al., 2013; Wright et al., 2012; Nemry and Demirel, 2012; Dikanski et al., 2016; HR Wallingford, 2014). In order to begin to understand the true impacts of climate change on bridge scour risk, asset managers need to reduce epistemic uncertainties, associated with bridge data, which currently dwarf the effects of climate change uncertainty on scour risk assessments. This would be a valuable climate change adaptation measure and an essential first step towards increasing bridge resilience in the context of current and future climate.

Although for bridges with unknown foundations, bridge data uncertainties have been found to be even more influential, compared to climate change uncertainties, this may not be the case with bridges with known foundations or better defined asset data. For the latter, the climate change uncertainties may become relatively more significant; an extension of the current work can provide insight in these relative differences.

6.5 Summary

This chapter presented two frameworks that have been developed recently to quantify, within a risk-based assessment, the potential impact of climate change on bridge infrastructure. Two distinct case studies, the first one capturing long-term gradual changes in environmental and atmospheric pollutant parameters and the second one capturing changes on the magnitude and/or frequency of extreme events, have been used to demonstrate the applicability of the framework. The results obtained from these case studies, which cover long-term environmental deterioration of metallic bridges and scour of bridge piers under river flooding, have provided insight into some of the key impacts that may be experienced by bridge structures, both in the short- and long-term. The frameworks that quantify the time-dependent reliability and risk of bridge failure can be integrated within wider climate change adaptation and mitigation planning frameworks for decision-making in the long-term.

6.6 Potential Design and Practice Evolutions

The availability of advanced deterioration models for metallic bridges, termed DRFs, which directly link the rate of corrosion with atmospheric pollution and climate variables provide a pathway towards capturing climate change effects on the long-term deterioration of metallic bridges. Hence, the importance of accessing and/or collecting atmospheric pollution and environmental/climate data is expected to increase overtime. The analyses carried out on the effects of climate change on long-term corrosion of metallic bridges showed that mitigation of the atmospheric pollution can have a positive influence but only after a period of several decades has evolved. Hence, any mitigation actions on improving air quality in urban areas may only be beneficial in the long-term. However, although the long-term reduction of SO_2 concentration in an area can enhance the reliability and hence reduce the risk of the exposed structure, it was shown that the time-dependent probability of failure and associated risk remain higher compared to a bridge exposed to less-polluted environment throughout its service life.

In terms of bridge scour, it was found that the risk of bridges with larger FDs is expected to be affected to a higher extent by climate change as compared to bridges with unknown or shallow foundations. Even if mean climate might not change in a location, an increase in the variability of the climate may also have a significant effect of the scour risk of bridges. It was also shown that our ability to foresee climate change impacts on bridge scour risk is not only limited by the aleatory uncertainty of climate change projections but also is greatly influenced by avoidable uncertainty in basic asset data, which can outweigh aleatory uncertainty by an order of magnitude. Results presented in this chapter further indicate that in order to be able to understand and account for this increase, scour management processes must effectively address data uncertainty. Active measures to control data quality would be an effective step towards understanding and managing bridge resilience in the context of current and future climatic conditions. Efforts of asset owners towards data collection, that is, FDs, bridge/river-related geometric data, can be justifiable considering the potential gains received towards better asset stock management.

6.7 Open Research Questions

- How can both long-term and short-lived, time-dependent, processes/hazards be integrated within a universal multihazard risk assessment of bridge infrastructure?
- What types of data would be most beneficial to be collected/monitored towards carrying out more efficient risk assessment of bridge structures?
- How can the effectiveness of potential adaptation and mitigation strategies be captured within a risk assessment framework?
- How can a time-dependent risk assessment methodology be integrated within a wider climate change adaptation framework?

Acknowledgement

The financial support (Grants EP/I00744X/1 and EP/G037612) provided by the Engineering and Physical Sciences Research Council (EPSRC) for these projects in the United Kingdom is gratefully acknowledged.

References

Appuhamy, J.M.R.S., Ohga, M., Kaita, T., Fujii, K., Dissanayake, P.B.R., 2011. Development of analytical method for predicting residual mechanical properties of corroded steel plates. Int. J. Corros. https://doi.org/10.1155/2011/385083.

Arneson, L.A., Zevenbergen, L.W., Lagasse, P.F., Clopper, P.E., 2012. Evaluating scour at bridges. In: Hydraulic Engineering Circular No. 18 (HEC-18), FHWA-HIF-12-003. fifth ed. U.S. Department of Transportation, Federal Highway Administration.

ASCE, 2013. Report Card for America's Infrastructure. ASCE.

Bastidas-Arteaga, E., Stewart, M.G., 2015. Damage risks and economic assessment of climate adaptation strategies for design of new concrete structures subject to chloride-induced corrosion. Struct. Saf. 52 (Part A), 40–53.

Bastidas-Arteaga, E., Stewart, M.G., 2016. Economic assessment of climate adaptation strategies for existing RC structures subjected to chloride-induced corrosion. Struct. Infrastruct. Eng. 12 (4), 432–449. Special Issue: Decision-Making in a Changing Climate, https://doi.org/10.1080/15732479.2015.1020499.

BD 97/12, 2012. The assessment of scour and other hydraulic actions at highway structure. In: Design Manual for Roads and Bridges, Highway Structure: Inspection and Maintenance Assessment.vol. 3. Section 4, Part 21, UK.

Bell, B., 2004. European railway bridge demography. Technical Report. Deliverable D1.2 of the Sustainable Bridges Project.

Booij, M.J., 2005. Impact of climate change on river flooding assessed with different spatial model resolutions. J. Hydrol. 303, 176–198.

Breusers, H.N.C., Nicollet, G., Shen, H.W., 1977. Local scour around cylindrical piers. J. Hydraul. Res. 15 (3), 211–252.

Briaud, J.L., Brandimarte, L., Wang, J., D'Odorico, P., 2007. Probability of scour depth exceedance owing to hydrologic uncertainty. Georisk Assess. Manag. Risk Eng. Syst. Geohazards 1 (2), 77–88.

Briaud, J.L., Gardoni, P., Yao, C., 2014. Statistical, risk and reliability analyses of bridge scour. J. Geotech. Geoenviron. 140(2). 04013011.

BS EN ISO, 2012a. BS EN ISO 9223:2012. In: Corrosion of Metals and Alloys—Corrosivity of Atmospheres—Guiding Values for the Corrosivity Categories. BSI.

BS EN ISO, 2012b. BS EN ISO 9224:2012. In: Corrosion of Metals and Alloys—Corrosivity of Atmospheres—Classification, Determination and Estimation. BSI.

BSI, 2000. BS 5400-3. In: Steel, Concrete and Composite Bridges—Part 3: Code of Practice for Design of Steel Bridges. BSI.

Czarnecki, A.A., Nowak, A.S., 2008. Time-variant reliability profiles for steel girder bridges. Struct. Saf. 30, 49–64.

Deco, A., Frangopol, D.M., 2011. Risk assessment of highway bridges under multiple hazards. J. Risk Res. 14 (9), 1057–1089.

DEFRA, 2012. The UK Climate Change Risk Assessment 2012 Evidence Report. Committee on Climate Change, London.

DEFRA, 2016. The UK Climate Change Risk Assessment 2017. Synthesis report, Committee on Climate Change, London.

Dikanski, H., Hagen-Zanker, A., Imam, B., Avery, K., 2016. Climate change impacts on railway structures: bridge scour. Proc. Inst. Civ. Eng.—Eng. Sustain. 170 (5), 237–248.

Dikanski, H., Imam, B., Hagen-Zanker, A., 2018. Effects of uncertain asset stock data on the assessment of climate change risks: a case study of bridge scour in the UK. Struct. Saf. 71, 1–12.
Feliu, S., Morcillo, M., Feliu Jr., S., 1993a. The prediction of atmospheric corrosion from meteorological and pollution parameters—I. Annual corrosion. Corros. Sci. 34 (3), 403–414.
Feliu, S., Morcillo, M., Feliu Jr., S., 1993b. The prediction of atmospheric corrosion from meteorological and pollution parameters—II. Long-term forecasts. Corros. Sci. 34 (3), 415–422.
Frangopol, D.M., Liu, M., 2007. Maintenance and management of civil infrastructure based on condition, safety, optimization, and life-cycle cost. Struct. Infrastruct. Eng. 3 (1), 29–41.
Gilvear, D.J., Black, A.R., 1999. Flood-induced embankment failures on the river Tay: implications of climatically induced hydrological change in Scotland. Hydrol. Sci. J. 44 (3), 345–362.
Haagenrud, S.E., Henriksen, J.F., 2006. Degradation of built environment—review of cost assessment model and dose-response functions. In: Sjostrom, C. (Ed.), Durability of Building Materials and Components, vol. 7. Routledge, London, p. 1.
Highways Agency, 2009. Climate Change Adaptation Strategy and Framework. Department for Transport.
HR Wallingford, 2014. Indicators to Assess the Exposure of Critical Infrastructure in England to Current and Projected Climate Hazards. Committee on Climate Change, London.
ICE, 2009. The State of the Nation: Defending Critical Infrastructure. Institution of Civil Engineers, London.
Iles, D.C., 2004. Design Guide for Steel Railway Bridges. Publication Number: SCI P318. The Steel Construction Institute. ISBN 1-85942-150-4.
Imam, B.M., Chryssanthopoulos, M.K., 2012. Causes and consequences of metallic bridge failures. Struct. Eng. Int. 22 (1), 93–98.
Imhof, D., 2004. Risk Assessment of Existing Bridge Structures. PhD Thesis, University of Cambridge.
IPCC, 2012. Managing the risks of extreme events and disasters to advance climate change adaptation. In: Field, C.B., Barros, V., Stocker, T.F., Qin, D., Dokken, D.J., Ebi, K.L., Mastrandrea, M.D., Mach, K.J., Plattner, G.K., Allen, S.K., Tignor, M., Midgley, P.M. (Eds.), A Special Report of Working Groups I and II of the Intergovernmental Panel on Climate Change. Cambridge University Press, Cambridge and New York, NY.
IPCC, 2013. Summary for policymakers. In: Stocker, T.F., Qin, D., Plattner, G.K., Tignor, M., Allen, S.K., Boschung, J., Nauels, A., Xia, Y., Bex, V., Midgley, P.M. (Eds.), Climate Change 2013: The Physical Basis, Contribution of Working Group I to the Fifth Assessment Report of the Intergovernmental Panel on Climate Change. Cambridge University Press, Cambridge and New York, NY.
JBA, 2004. Scour & flood risk at railway structures. In: Infrastructure Integrity (4) Research Theme: Project Number T112. JBA Consulting, Rail Safety & Standards Board Final report.
Johnson, P.A., 1992. Reliability-based pier scour engineering. J. Hydraul. Eng. 118 (10), 1344–1358.
Katz, R.W., 1993. Towards a statistical paradigm for climate change. Clim. Res. 2, 167–175.
Khelifa, A., Garrow, L.A., Higgins, M.J., Meyer, M.D., 2013. Impacts of climate change on scour-vulnerable bridges: assessment based on HYRISK. J. Infrastruct. Syst. 19, 138–146.
Kirshen, P., Ruth, M., Anderson, W., 2008. Interdependencies of urban climate change impacts and adaptation strategies: a case study of metropolitan Boston USA. Climate Change 86, 105–122.

Kiureghian, A.D., Ditlevsen, O., 2009. Aleatory or epistemic? Does it matter? Struct. Saf. 31, 105–112.
Klinesmith, D.E., McCuen, R.H., Albrecht, P., 2007. Effect of environmental conditions on corrosion rates. J. Mater. Civ. Eng. 19 (2), 121–129.
Kottegoda, N.T., Rosso, R., 1997. Statistics, Probability and Reliability for Civil and Environmental Engineers. McGraw-Hill, New York.
Kumar, P., Imam, B., 2013. Footprints of air pollution and changing environment on the sustainability of built infrastructure. Sci. Total Environ. 444, 85–101.
Landers, M.N., Mueller, D.S., 1996. Channel Scour at Bridges in the United States. Report No. FHWA-RD-95-184U.S. Department of Transportation, Federal Highway Administration.
Larsen, P.H., Goldsmith, S., Smith, O., Wilson, M.L., Strzepek, K., Chinowsky, P., Saylor, B., 2008. Estimating future costs for Alaska public infrastructure at risk from climate change. Glob. Environ. Chang. 18, 442–457.
Lehner, B., Doll, P., Alcamo, J., Henrichs, T., Kaspar, F., 2006. Estimating the impact of global change on flood and drought risks in Europe: a continental integrated analysis. Climate Change 75, 273–299.
Maddison, B., 2012. Scour failure of bridges. Proc. Inst. Civ. Eng.—Forensic Eng. 165 (1), 39–52.
Mattai, J., Hutchinson, D., 2006. London Atmospheric Emissions Inventory 2003, Second annual report, Mayor of London, Greater London Authority, ISBN 10: 1-85261-977-5.
Melchers, R.E., 1999. Structural Reliability Analysis and Prediction, Second ed. John Wiley & Sons.
Melville, B.W., 1997. Pier and abutment scour: integrated approach. J. Hydraul. Eng. ASCE 123 (2), 125–136.
Melville, B.W., Chiew, Y.M., 1999. Time scale for local scour at bridge piers. J. Hydraul. Eng. 125 (1), 59–65.
Melville, B.W., Coleman, S.E., 2000. Bridge Scour. Water Resources Publications, LLC, Highland Ranch, Colorado.
Merz, B., Thieken, A.H., 2005. Separating natural and epistemic uncertainty in flood frequency analysis. J. Hydrol. 209, 114–132.
Meyer, M.D., Weigel, B., 2011. Climate change and transport engineering: preparing for a sustainable future. J. Transp. Eng. 137 (6), 393–403.
Muzzammil, M., Siddiqui, N.A., 2009. A reliability-based assessment of bridge pier scour in non-uniform sediments. J. Hydraul. Res. 47 (3), 372–380.
NCHRP, 2003. Design of Highway Bridges for Extreme Events. Report 489, National Cooperative Highway Research Program, Transportation Research Board, New York.
Nemry, F., Demirel, H., 2012. Impacts of Climate Change on Transport: A Focus on Road and Rail Transport Infrastructures. European Commission Joint Research Centre, Luxembourg.
Network Rail, 2015. A Short Guide to Network Rail. National Audit Office.
Nicholls, R.J., 2004. Coastal flooding and wetland loss in the 21st century: changes under the SRES climate and socio-economic scenarios. Glob. Environ. Chang. 14, 69–86.
Nowak, A.S., Szerszen, M.M., 1998. Bridge load and resistance models. Eng. Struct. 20, 985–990.
Osborn, T.J., Hulme, M., Jones, P.D., Basnett, T.A., 2000. Observed trends in the daily intensity of United Kingdom precipitation. Int. J. Climatol. 20, 347–364.
Piyatrapoomi, N., Kumar, A., Setunge, S., 2004. Framework for investment decision-making under risk and uncertainty for infrastructure asset management. Res. Transp. Econ. 8, 199–214.
Posey, J., 2012. Climate change impacts on transportation in the Midwest. U.S. National Climate Assessment Midwest Technical Input Report, Winkler J, Andresen J, Hatfield J, Bidwell D, Brown D. Coordinators.

Prudhomme, C., Jakob, D., Svensson, C., 2003. Uncertainty and climate change impact on the flood regime of small UK catchments. J. Hydrol. 277, 1–23.
Prudhomme, C., Kay, A.L., Crooks, S., Reynard, N., 2013. Climate change and river flooding: part 2 sensitivity characterisation for British catchments and example vulnerability assessments. Climate Change 119, 949–964.
Rail Safety & Standards Board (RSSB), 2005. Safe Management of Railway Structures—Flooding and Scour Risk. In: Final Report. JBA Consulting, London.
Railway Safety, 2003. Railway Group Safety Plan 2002/2003. Railway Safety, London.
Richardson, E.V., Davis, S.R., 2001. Evaluating scour at bridges. In: United States Department of Transportation, fourth ed., Federal Highway Administration, Washington.
Robson, A.J., 2002. Evidence for trends in UK flooding. Philos. Trans. R. Soc. Lond. A 360, 1327–1343.
Robson, A., Reed, D., 1999. Statistical procedures for flood frequency estimation. In: Flood Estimation Handbook. Centre for Ecology & Hydrology, vol. 3. Natural Environment Research Council, Wallingford, Oxfordshire.
Sharifi, Y., Paik, J.K., 2011. Ultimate strength reliability analysis of corroded steel-box girder bridges. Thin-Walled Struct. 49, 157–166.
Sheppard, D.M., Melville, B., Demir, H., 2014. Evaluation of existing equations for local scour at bridge piers. J. Hydraul. Eng. 140 (1), 14–23.
Stein, S.M., Young, G.K., Trent, R.E., Pearson, D.R., 1999. Prioritizing scour vulnerable bridges using risk. J. Infrastruct. Syst. 5 (3), 95–101.
Stewart, M.G., Deng, X., 2015. Climate impact risks and climate adaptation engineering for built infrastructure. J. Risk Uncertain. Eng. Syst. 1(1). 04014001.
Swiss Re, 2013. Mind the Risk—A Global Ranking of Cities Under Treat From Natural Disasters. Swiss Reinsurance Company Ltd., Switzerland
Tamakoshi, T., Yoshida, Y., Sakai, Y., Fukunaga, S., 2006. Analysis of damage occurring in steel plate girder bridges on national roads in Japan. In: 22nd US-JAPAN Bridge Engineering Workshop, Seattle, WA.
Transportation Research Board (TRB), 2008. Potential impacts of climate change on U.S. transportation. In: Transportation research board special report 290. Committee on Climate Change and U.S. Transportation, National Research Council, Washington D.C.
UKCP09, 2009. UK Climatic Projections. Accessible online at: http://ukclimateprojections.metoffice.gov.uk//. Department for Environment, Food and Rural Affairs (Defra), Crown Copyright.
UNEP, 2007. Buildings and Climate Change: Status, Challenges and Opportunities. United Nations Environment Programme. ISBN 978-92-807-2795-1.
Van Leeuwen, Z., Lamb, R., 2014. Flood and Scour Related Failure Incidents at Railway Assets between 1846 and 2013. Project W13-4224, JBA Trust Limited.
Wardhana, K., Hadipriono, F.C., 2003. Analysis of recent bridge failures in the United States. J. Perform. Constr. Facil. 17 (3), 144–150.
Wright, L., Chinowsky, P., Strzepek, K., Jones, R., Streeter, R., Smith, J.B., Mayotte, J.M., Powell, A., Jantarasami, L., Perkins, W., 2012. Estimated effects of climate change on flood vulnerability of U.S. bridges. Mitig. Adapt. Strateg. Glob. Chang. 17, 939–955.

CHAPTER SEVEN

Climate Change-Related Risks and Adaptation of Interdependent Infrastructure Systems

Dimitri V. Val*, **Daniil Yurchenko**[†], **Maria Nogal**[‡], **Alan O'Connor**[‡]
*Institute for Infrastructure & Environment, EGIS, Heriot-Watt University, Edinburgh, United Kingdom
[†]Institute of Mechanical, Process & Energy Engineering, EPS, Heriot-Watt University, Edinburgh, United Kingdom
[‡]School of Engineering, Trinity College Dublin, Dublin, Ireland

7.1 Introduction

Critical infrastructure systems such as energy, water/wastewater, transport, and information and communication technology (ICT) are subject to ever increasing loads due to population growth and suffer from deterioration associated with ageing that makes them more vulnerable. They also face new challenges due to climate change and become more complex, centralised, and interdependent. All these further increase their vulnerability. To ensure their continuous and reliable functioning in the future, their resilience needs to be improved and their adaptation to climate change is an essential part of this process.

In this chapter, changes in global climate that may have adverse impacts on the performance of the critical infrastructure systems will be considered, with the main emphasis on extreme weather events and how the latter are affected by a changing climate. Past examples of infrastructure failures caused by extreme weather events will be described, including cases when interdependencies between various infrastructure systems played a major role. The examples will highlight the need in infrastructure adaptation to a changing climate. Potential adverse impacts of climate change on individual critical infrastructure systems then will be summarised and presented.

Simulation of the performance of infrastructure systems, including the effects of interdependencies, may provide major assistance in developing cost-efficient adaptation strategies. Thus, different types of the interdependencies will be considered in more detail and their classification, which can be useful for their modelling, will be discussed. After that, an overview of

Climate Adaptation Engineering
https://doi.org/10.1016/B978-0-12-816782-3.00007-3

various approaches/methods that have been proposed and applied for the analysis of interdependent infrastructure systems (IIS) under normal and extreme conditions will be presented. As discussed in Chapter 1, there are major uncertainties associated with modelling/predicting the performance of infrastructure systems, especially in relation to weather-related hazards, so that results of such analyses are usually expressed in terms of risk. A risk-based framework for analysis of IIS under single and multiple hazards will briefly be described. Special attention will be paid to modelling the cascading failures within such systems.

Finally, examples of climate adaptation measures, which can be implemented for various critical infrastructure systems, will be considered. More detailed case studies will also be provided.

7.2 Risk to Infrastructure Systems Due to Climate Change

7.2.1 Effect of Climate Change on Extreme Weather Events and Weather-Related Hazards

The performance of infrastructure systems depends on their climatic environment, which currently undergoes transformations caused by climate change. These transformations include gradual changes in climatic variables (e.g., temperature, precipitation, and wind) and associated environmental parameters (e.g., sea level and CO_2 atmospheric concentration) as well as variations in frequency and intensity of extreme weather events (e.g., heat waves, floods, droughts, and storms). Since the latter can be very harmful for infrastructure systems the trends, which have been detected, that the frequency (or intensity, or both) of such events has increased over the last few decades and will continue to increase in the future due to a changing climate are especially worrying. For instance, there is strong evidence of an increase in the intensity of tropical cyclones in the North Atlantic and North Western Pacific Oceans, although there is no evidence that their frequency increases as well (Mendelsohn et al., 2012; Walsh et al., 2014). This information is important since damage due to tropical cyclones is usually the highest among all extreme weather events, for example, recent estimates put the average global damage from tropical cyclones as high as \$26 billion (USD) per year (Mendelsohn et al., 2012).

Another type of cyclones is extratropical cyclones, which represent any synoptic-scale low-pressure system that develops in the middle latitudes, that is, between 30° and 60°; most of cyclones affecting North America and

Europe are of this type. Strong extratropical cyclones, storms, blizzards in winter, bring high winds and heavy precipitation and also can be very damaging to infrastructure (Chapters 5 and 9). Currently available evidence indicates an increase in both frequency and intensity of this type of storms during the winter season in the Northern Hemisphere since 1950; however, future projections regarding changes in their intensity are less conclusive (Vose et al., 2014).

Another extreme weather event, whose frequency and intensity are on increase in most places in the world as a result of climate change, is heavy precipitation (Trenberth, 2011). This has been observed even in places, in which mean precipitation decreases. For example, available data confirm this trend across most of the United States, although in some areas the average amount of precipitation has decreased (Kunkel et al., 2013; Walsh et al., 2014). At the basic level, this trend can be explained by global warming and the Clausius-Clapeyron equation, according to which the water-holding capacity of air increases by about 7% per 1°C increase in its temperature. Thus, in a warmer climate, more water vapour accumulates in the atmosphere that leads to more intense precipitation events (Trenberth, 2011). The latter are directly linked to flooding, another weather-related hazard. In fact, global warming increases the risks of both floods and droughts. For instance, there is an evidence of this in relation to flooding in England and Wales in 2000 (Pall et al., 2011), as well as in a projected risk of widespread drought across large parts of central and southern United States (Walsh et al., 2014).

Global warming also leads to an increase in the frequency and intensity of heat waves (Chapter 11). There is strong evidence that probability of extreme heat waves with record-breaking temperatures has increased and will continue to increase in the future as the global climate becomes warmer (Walsh et al., 2014). For example, a recent study (Dosio et al., 2018) has shown that under 1.5°C global warming, a significant increase in the frequency and intensity of heat waves is expected across most of the world but especially in Africa, Southeast Asia, and South America. In this case, 13.8% of the world population will be exposed to severe heat waves on average every 5 years. However, if the average global temperature grows by 2°C, the proportion of the world population under such exposure will be 36.9%, that is, nearly three times larger.

An environmental phenomenon directly linked to global warming and having a major influence on infrastructure is sea-level rise (Chapter 8). It occurs due to water expansion as the latter warms up and also due to melting of ice sheets and glaciers. Global sea level has risen by about 0.2 m since reliable records began in 1880 and, according to the current projections, it is expected to rise another 0.3–1.2 m by 2100 depending on an emission scenario (e.g. Walsh et al., 2014). Major coastal cities all over the world, for example, Alexandria (Egypt), Bangkok (Thailand), Dhaka (Bangladesh), Jakarta (Indonesia), Lagos (Nigeria), Miami (United States), New York (United States), Shanghai (China), and Tokyo (Japan), are currently in danger of coastal flooding, the risk of which will significantly increase by the end of this century due to sea-level rise (Hanson et al., 2011). This risk may be even higher for many critical infrastructure assets in these cities, which historically were placed at low-lying areas in order to have easier access to waterways for transporting supplies and products, for example, in New York (Zimmerman et al., 2017).

Other weather-related hazards affected by climate change that pose major risks to infrastructure assets and systems include landslides, coastal erosion, subsidence, scour (Chapter 6), wildfires, etc. Examples of failures of various infrastructure systems due to extreme weather events and related to them hazards will be discussed in the following section.

7.2.2 Impacts of Extreme Weather Events and Weather-Related Hazards on Infrastructure Systems

As noted above, tropical cyclones are the most damaging among extreme weather events. In part, this is due to hurricanes, that is, tropical cyclones that form in the North Atlantic Ocean and impact the Gulf and East Coasts of the United States. Direct damage due to hurricanes is usually inflicted by the combination of strong winds and flooding caused by heavy precipitation and storm surge. The US Department of Energy provides data on damages to energy infrastructure by several hurricanes that hit the United States between 2005 and 2012 (DoE, 2009; DoE, 2013)—see Table 7.1.

Of course, hurricane-induced damage is not limited to energy infrastructure, it affects other infrastructure systems as well. One of the two most costliest hurricanes in history, Katrina (the other one is Harvey), with the estimated cost of damage of $125 billion (USD), caused such widespread devastation along the US Gulf Coast that it was compared to a weapon of mass destruction attack (Miller, 2006). All critical infrastructure systems in the region were heavily damaged: impacts on energy infrastructure are shown in Table 7.1; many water pumping stations, water and wastewater treatment plants were flooded or/and damaged by wind; many roads were also flooded, a few bridges collapsed while many others were damaged. ICT

Table 7.1 Energy Impacts of Hurricanes, 2005–2012, United States

	Hurricane						
Energy Type	**Katrina 08/05**	**Rita 09/05**	**Wilma 10/05**	**Gustav 09/08**	**Ike 10/08**	**Irene 08/11**	**Sandy 10/12**
Electricity customer outages (10^6)	2.7	1.5	3.5	1.1	3.9	6.7	8.7
Lost natural gas production ($10^6 m^3$/day)	250	230	0	200	210	0	0
Lost oil/petroleum production (10^6 barrels/day)	1.4	1.5	0	1.3	1.3	0.24	0.31

was one of the most severely affected infrastructure systems—many of central switching centres of the wire-line communication network were flooded, while countless miles of its lines were destroyed or damaged, more than 3 million customer phone lines were out of services, about 2000 of cell phone towers were knocked out (Miller, 2006).

This catastrophic event also very clearly demonstrated interdependencies between different infrastructure systems, when failure in one system had cascading effects on others. For instance, many water and wastewater facilities, which were not flooded, were out of service because of power loss—many of them did not have backup generators, while where generators were available they soon run out of fuel so that the facilities became also inoperable, for example, sewer lift stations stopped working that resulted in raw sewage overflowing into streets and homes (Leavitt and Kiefer, 2006). A few lessons have been learned and when hurricane Isaac hit Louisiana in August 2012, water and wastewater systems performed much better because practically all water pumping stations and many sewer lift stations were equipped with backup generators. However, there was shortage of petrol because almost all petrol stations were without electricity and backup generators so that their pumps could not be operated (Miles et al., 2015). The same cascading effects of power outage on wastewater system and petrol supply system were observed after hurricane Sandy hit the US East Coast in late October 2012, in particular because of a lack of backup generators. For example, in the New York metro area, dozens of wastewater treatment plants along the coast stopped working due to power outage so that about a million of cubic metres of raw sewage were discharged into waterways; numerous petrol stations were left without power that led to petrol shortage (New York State 2100 Commission, 2013).

Tropical storms, although less powerful, also affect the West Coast of the United States. For instance, in December 2014, a tropical storm hit the northern California coast and led to flooding that knocked out a major electricity substation leaving 85,000 customers without power. It also had a cascading effect on San Francisco's transport system by shutting down its electrical bus service (Gibson, 2017).

Extratropical storms are usually less powerful than hurricanes but can inflict severe damages on infrastructure systems, especially in winter period. The North American Ice Storm of 1998 (also known as Great Ice Storm of 1998) that affected the Canadian provinces of Quebec, Ontario, and New Brunswick and the US states in New England was one of the most devastating winter storms in history. In fact, it was a series of smaller storms in quick succession, one after another, that brought large amounts of precipitation, frozen rain, to the affected areas. In the context of infrastructure, the electricity system suffered the most. Overhead power lines collapsed under the weight of ice, for example, Hydro Quebec, the Canada's largest electric utility, lost 1500 steel pylons, 4500 transformers, and 26,000 wooden poles; at the peak of the disaster more than 2.2 million people in Canada and the United States were without power. Overhead power lines were damaged not only directly by ice but also by falling tree branches. The latter blocked many roads causing major disruptions to the transportation system (Dupigny-Giroux, 2000). A series of storms in winter 2013/14 led to severe flooding, both coastal and river, in England and Wales (United Kingdom) and inflicted major damages to infrastructure, especially to the transportation system. Many roads were closed due to flooding and/or damage caused directly by it or by associated with it erosion and landslips, and also because they were clogged by trees falling during high wind; several main roads remained closed for several weeks. The rail network was also severely affected, mainly the lines close to the coast that were damaged due to coastal erosion; a part of the main line connecting Devon to Cornwall was washed away after a sea wall collapsed. There were major disruptions to services of Gatwick Airport in London because of flooding of electricity substations serving the airfield; in particular, more than 140 flights (both departing and arriving) were cancelled. Total economic damages due to this hazard were around £1.3 billion (Environment Agency, 2016).

Flooding due to heavy raining in Queensland, Australia, at the end of 2010–beginning 2011, inflicted massive damage, in excess of $4 billion

(AUD), on the state transportation system: 9170 km of state-controlled roads, 29% of the state rail network, and three major ports were impacted (Weerakoon et al., 2012). As a result of that (and also due to flooding of mines), Queensland's $20 billion coal export industry was nearly brought to a standstill. Intense, far exceeding existing records, rainfall in Colorado, United States, in September 2013 caused devastating and widespread flash flooding, which resulted in extensive infrastructure damage, especially to road and rail networks: 30 state highway and several railroad bridges were destroyed, another 20 highway bridges were severely damaged along with 485 miles (780 km) of roads and 150 miles (240 km) of rail track. Total direct and indirect losses due to this flooding were estimated to be around $3 billion (USD) (DoC, 2014).

Drought, another weather-related hazard, may affect different infrastructure systems, in particular water and energy. Between 2011 and 2014, the state of California, United States, experienced one of the worst droughts in its history. As a result of that, the state water resources were significantly depleted—in 2014, the water level in key fresh water reservoirs dropped to 50% below average (Gibson, 2017). Moreover, the drought had a major impact on electricity generation using hydropower; for the 3 years of the drought, the amount of electricity which came from hydropower was about 34,000 GWh less compared with the long-term average and the share of hydropower in the state electricity mix dropped from 18% to 12% (Gleick, 2015).

Heat waves impact energy infrastructure as well, in particular electricity generation. The heat wave in Europe in 2003 significantly affected electricity generation in France due to warming of river waters that were used for cooling of nuclear power plants. This had an adverse effect on the nuclear power generation by reducing its capacity by 4000 MW and the amount of generated electricity by 5.3 TWh (Añel et al., 2017). Conditions created by a heat wave can also initiate failure in a power network. For instance, a heat wave in Melbourne, Australia, in January 2009 caused failure of an electricity substation, which cascaded through the power network. This left 500,000 customers without power, shutdown train and tram services, and caused nationwide disruptions to internet services (Climate Institute, 2012).

High temperatures and drought produce ideal conditions for wildfires, which can be very damaging to infrastructure systems. For example, wildfires in California in 2007 and 2015 severely impacted the state power network by destroying poles and damaging overhead electric wires, leaving 80,000 and 12,700 customers, respectively, without power (Gibson,

2017). Wildfires during a heat wave in the state Victoria, Australia, in 2007 damaged two key power transmission lines that led to cascading failure and disruptions in the national grid; 690,000 customers were affected by power outage (Añel et al., 2017).

Generic examples of climate change impacts on the critical infrastructure systems can be found in Table 7.2.

7.3 Modelling Interdependent Infrastructure Systems and Associated Risks

7.3.1 Types of Infrastructure Interdependencies

To model interaction between IIS, it is important to understand what type of dependencies the infrastructure systems may have. First of all, there are unidirectional and bidirectional dependencies, where the former implies that one infrastructure consumes products or services from another. For example, ATM machines use electricity to provide cash to customers upon demand. Bidirectional interdependence infers mutual dependences of two or more infrastructure systems like the power-transport or power-ICT pairs. Mutual integration of different infrastructure systems seems to be very natural and is meant to improve their efficiency. An excellent example of such integration is wide proliferation of information technologies in the last two decades that has significantly changed the way other infrastructure systems operate today. At the same time, this integration has made other infrastructure systems highly dependent on ICT increasing their vulnerability to various ICT malfunctioning.

According to Rinalidi et al. (2001), the interdependencies may be divided into four types:

- physical,
- cyber,
- geographical, and
- logical.

The physical interdependencies are based on the physical flow of commodities or raw materials that are transferred from one place to another to produce other goods or provide services. The cyber or ICT interdependencies cover a wide range of dependencies from simple communication and data management to automated or remote control of various subsystems. Geographical interdependencies are those, which arise from the geographical closeness of different infrastructure systems so that the same local hazard can simultaneously affect all (or several) of them. The rest of interdependencies have been combined under the last category, where the dependencies are

Table 7.2 Examples of Climate Change Impacts on Infrastructure Systems

	Infrastructure Systems			
Climate Change Impacts	**Energy**	**Water/ Wastewater**	**Transport**	**ICT**
Increased storm intensity	• Damage to facilities • Damage to power lines • Damage to fuel transportation • Disruption to operation • Change in capacity for wind and solar power	• Disruption to operation • Damage to water/ wastewater treatment facilities • Pipeline ruptures • Increased risk of water contamination	• Closure of roads due to debris • Disruptions to ports and airports • Inundation of roads and rail tracks • Disruption to traffic due to high winds	• Increased damage to above-ground transmission infrastructure (e.g. wireless masts, cell towers, phone lines, etc.)
Increased intensity of heavy precipitation, flooding	• Disruption to operation • Damage to facilities • Disruption to fuel transportation • Disruption to fuel extraction/ production	• Inundation of facilities • Increased demand on drainage system • Increased demand on wastewater treatment facilities • Lower treatment efficiency due to rapidly changing water quality • Damage to storage facilities	• Inundation of roads and rail tracks • Damage to roads • Increased risk of landslips and erosion • Increased risk of bridge scour • Damage and closure of runways due to flooding	• Increased flooding of low-lying and underground facilities • Exposed cables and trunks due to erosion • Reduced quality and strength of wireless signal during heavy raining
Increased temperature, heatwaves	• Reduced generation efficiency due to warming of cooling water	• Lower water quality due to increased content of bacteria, algal bloom, etc.	• Faster deterioration of road surfaces due to increased heat	• Location/density of wireless masts may become sub-optimal

Continued

Table 7.2 Examples of Climate Change Impacts on Infrastructure Systems—cont'd

Climate Change Impacts	Infrastructure Systems: Energy	Water/ Wastewater	Transport	ICT
	• Increased power demand • Increased expansion and reduced efficiency of transmission lines	• Increased load on existing infrastructure due to higher water demand • Higher water losses during storage	• Buckling of rail tracks • Overheating of rail electrical systems	• Increased need in air-conditioning
Longer and more frequent droughts	• Limitations on power generation due to lack of water for cooling • Reduced hydropower	• Shortage of fresh water • Increased demand for storage facilities • Increased demand for water quality treatment and recycling	• Reduces access to ports due to low water level • Reduced river navigability due to low water level	• Increased role in drought preparedness and mitigation
Rising sea level and coastal erosion	• Increased risk of flooding (incl. permanent) to facilities near the coast • Increased risk of damage or permanents loss of assets due to erosion	• Increased risk of inundation (incl. permanent) to facilities near the coast permanents loss due to erosion • Saltwater contamination of groundwater aquifers	• Increased risk of permanent inundation and loss of facilities/assets near the coast • Flooding of underground tunnels	• Permanent loss of facilities near the coast • Increased inundation and salt water corrosion of near-coast infrastructure

not straightforward and may result from human decisions. Since all infrastructure systems operate under certain regulations, they are also influenced by policies and laws, security and safety concerns, public health and welfare factors, as well as social and political trends. It should be noted that some interdependencies are not visible in a normal operational scenario but may emerge in some critical situations so that an outage in one of IIS may unexpectedly cause a significant harm to others and, eventually, disruption of people's lives.

Zimmerman (2004) has combined the interdependencies based on functional and geographical dependencies of IIS and proposed, using historical data, to adapt a different metric, which among other parameters has indicators of IIS that frequently cause failures of others and IIS that are frequently affected by failures of others. Their ratio may be thought of as how great the impact of a failure in one infrastructure on the others. As an example, Zimmerman (2004) has showed that failure of water mains has the highest ratio of 3.4 indicating that the effect of this failure on other IIS will be 3.4 times greater than the water mains will have to sustain from failures of the other IIS.

A slightly different approach has been proposed by Rinalidi et al. (2001) where the classification of failures was based on the type of disruption. Cascade failures have been identified as events occurring in one infrastructure that trigger failures of some elements in a dependent second infrastructure that will eventually cause failure of this entire infrastructure or a large part of it (of course, cascade failures are also possible within the same infrastructure). Escalating failures are those that can aggravate existing disruptions in other infrastructure systems; for example, failure of a traffic light control may escalate exiting traffic congestion that will eminently delay the arrival of a service team. Common cause failures are identified as two or more infrastructure systems disrupted simultaneously due to a common cause.

A single infrastructure system can be represented by a single or several layers that provide an abstract representation of subsystems divided into layers according to one of the above four types, for instance, based on Rinalidi et al. (2001). Within each sublayer, there may be several entities that are modelled by nodes. Nodes often, but not always, represent physically existing entities such as factories, power plants, refineries, storages, electrical and pumping stations, transport hubs, network servers, control and communication centres, satellites, etc. Depending on the objective of the modelling and level of details, a large power plant and a light pole can both be modelled as a node. These nodes are connected by physical or virtual links such as pipelines and roads, wires and optical fibres, etc. A number of IIS, thus, can be represented by an interconnected multilayered system with nodes and links (Fig. 7.1). Each node may have unidirectional or bidirectional links from/to one or many other nodes. For instance, a traffic light needs a unidirectional input from both electrical and ICT (control) infrastructure systems.

Often these links can be modelled using open or closed loops, feedback or feedforward paths. Due to the complexity of infrastructure networks, majority of the existing publications has been focused on linear interactions and, therefore, remained within the linear theory framework. However, it has been suggested by a number of researchers that some interdependencies may be regarded as linear only until a certain critical value is reached, beyond which the interdependencies become nonlinear, whereas some others cannot be considered as linear ones from the very beginning (Dudenhoeffer et al., 2006).

7.3.2 Modelling Techniques

Over the last 15 years, a number of different methodologies has been proposed or adapted to analyse the complex network of IIS. Methods used for modelling IIS can be broadly divided into the following categories:

(a) agent-based (AB) approach,
(b) input-output (IO) approach,
(c) network-based (NB) approach,

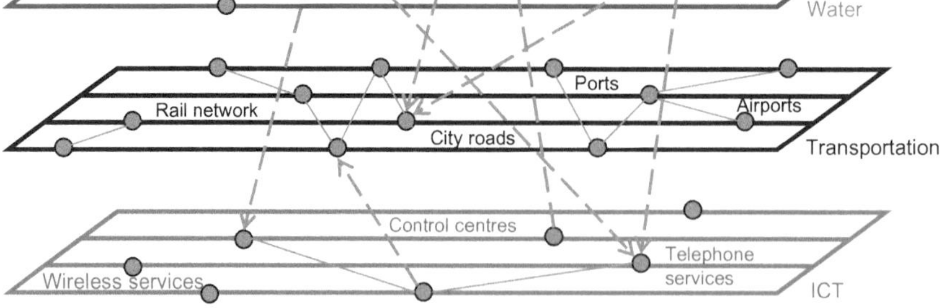

Fig. 7.1 Schematic presentation of interdependent infrastructure systems.

(d) system dynamics (SD) approach, and
(e) other approaches.

There is no commonly recognised and adapted unified methodology for modelling IIS. There are several reasons to it, for instance, different infrastructure systems may have different internal structures and regulations to operate in normal and emergency situations. The modelling scale (e.g. district, city, or country level) is important as well since at different levels there may be different types of interdependencies. Other important factors are objectives of modelling and computational capabilities since high level of details increases the computational costs enormously.

7.3.2.1 AB Approach

The AB approach originated as a branch of artificial intelligence. The main feature of this methodology is that a model comprises individual smart agents that mimic the behaviour of particular individuals or units. Each agent behaves according to a set of rules and interacts with other agents and its own environment based on perceived information and its experience. Each agent has its own specific set of goals and acts to fulfil them. Thus, this approach is often called *bottom-up* modelling because the entire problem and complex interactions between IIS come from individual interactions at different levels. Originally, this approach was mostly used in social and economic sciences where other types of models failed due to a number of factors. However, lately it has been adapted in engineering science, in particular to model complex IIS. The AB approach is relatively versatile and can be used at different levels of modelling so that at a low-level agents can represent physical entities like people (users and pedestrians), units of transport (cars, ships, trains, and planes), storages, pumping stations, etc. At a higher level, the agents may represent groups like costumers, passengers, small companies, and their commodities, whereas as the highest level the agents can represent global companies competing with each other for resources and clients.

As can be deduced from the basic idea behind the AB methodology, this approach is not 'governed' or 'controlled' centrally and every agent, even within a given set of agents, can make its own decisions. This is an important feature because, in some sense, this is how different levels within one infrastructure or different infrastructure systems may behave in a case of failure or emergency. Moreover, it provides researchers with an opportunity to model various levels of infrastructure systems by assigning them different characteristics and yet remaining within the AB paradigm. As a result, this approach allows observing some hidden behavioural patterns that cannot be observed in other models.

A number of software packages have been developed over the last 20 years utilising the AB approach. The Sandia National Lab developed the first AB software named ASPEN, which later was improved by including power infrastructure agents and called ASPEN-EE (Barton et al., 2004). Since then, the software has been further improved by incorporating economical, communication, and cyber interdependencies to study threats associated with cyberattacks and cybercrimes. The Argonne National Lab also developed their own AB modelling software SMART focused on the power systems, power market and electricity generation. This software has later been used as a basis for more modern version of Flexible Agent Simulation Toolkit (FAST) (North, 2001). There are a number of companies developing the AB software such as: AgentSheets, AnyLogic, MASON, NetLogo, StarLogo, and others.

In simple supply-demand modelling, the agents may mimic the behaviour of small supplying companies, intermediate shipping and planning units, as well as intermediate agents contributing or making an output product to the top-level consumer manufacturers. It has been observed that large chains create a number of interesting effects like delays and period doubling, even when the low-level supply and high-level demand remain constant due to intermediate players, with some certain trends appearing when the number of intermediate players grows. To simulate a more realistic environment or effects of various failures, probabilistic models can be used or shortage of certain supplies can be imposed (Sichman et al., 1998).

Nevertheless, this approach has a number of shortcomings, starting from a very simple question: how many agents should be used to observe a certain trend or to model behavioural features of, say, a million people? Should that number be the same or of the same order? Initial assumptions built into agents significantly influence the outcome of the modelling, whereas some assumptions are hard to validate or justify. Moreover, the model calibration becomes very challenging in the absence of sufficiently large amount of data and its relevance.

7.3.2.2 NB Approach

In the NB approach, the IIS are modelled as a network of nodes connected by links. Nodes usually represent and mimic the behaviour of real physical objects, whereas links represent connections between them. Works in this area can roughly be separated into two groups: *topology-based methods* and *flow-based methods*. In *the topology-based methods*, the IIS are modelled based on their topology with nodes and links usually having one of two states—normal or failed. Nodes and links, called vertices and edges correspondingly, have different properties that can be quantified. The vertices are characterised by the vertex degree and degree distribution, where the former describes the number of edges connected to a given vertex. According to Albert et al. (2000), complex networks can be divided into two major groups based on the connectivity distribution $d(k)$, where k is the number of edges connected to a given vertex. The first group consists of networks with the distribution tail decaying exponentially for large values of k. The other group comprises networks having the distributions with the tail behaviour proportional to a power law, that is, equivalent to k^p for large k. Two other parameters used to describe a network are the path length and the clustering coefficient. The path length is the shortest length between two vertices so that when the two vertices become disconnected the shortest length becomes infinite. The clustering coefficient refers to the connection of neighbouring vertices between themselves. Another important parameter is the redundancy ratio that can be thought of as the number of existing independent ways to the vertex, which usually is normalised to the maximum number of paths (Dueñas-Osorio et al., 2007a). Therefore, for a completely disconnected vertex within a network, the redundancy ratio is zero, whereas for a network with all vertices connected to each other, the redundancy ratio is equal to unity. One may think of this parameter as an indicator of whether it is possible to connect to the vertex when one edge or adjacent vertex fails.

Failures in this approach are modelled by simply removing either some vertices or some edges and by that reducing the number of possible paths to deliver or receive goods/services to/from the vertices. For IIS it is reasonable to introduce the conditional probability of failure of an element in a network A_i based on a failure of an element in another network A_j–Pr $(failure\ A_i \lfloor failure\ A_j)$ so that when the networks are independent $\Pr(failure\ A_i | failure\ A_j) = \Pr(failure\ A_i)$. Another parameter that can be used is the interdependency ratio that is the absolute difference between the interdependent and independent responses normalised by the maximum independent response attained at any removal fraction (Dueñas-Osorio et al., 2007b).

Having introduced a number of parameters describing IIS, different parametric studies have been performed to understand the complex behaviour of the networks due to failures. For instance, the number of vertices, the number of IIS nodes, the number of vertices removed due to failures, the distribution functions can be different in each network, and all these parameters may have different effects on the networks. However, in an analytical analysis, the number of vertices and the distributions for every network are usually taken the same (Buldyrev et al., 2010), whereas the number of removed vertices in one network may directly related to that in the others (coupled). It should be noted that the analytical analysis is possible only for very large number of vertices (one may think of k going to infinity), whereas all other cases should be investigated numerically including the cases of isolated and cascade failures (Hernandez-Fajardo and Duenas-Osorio, 2011). Due to the obvious topological nature of the approach, it has been shown that it may lack of information on the flow performance in the IIS and, therefore, should be integrated with other approaches.

The *flow-based* methods consider both nodes and edges that can supply/transduce/demand, send/tranship/deliver commodities or services to/from the nodes via the edges according to their capacities. This approach seems to be very general and provides a uniform network flow model for different IIS. However, it should be applied with caution since different infrastructure systems have diverse internal structures and management mechanisms. For instance, the study made by Ouyang et al. (2009) showed that the results obtained by the topology- and flow-based modelling were significantly different. The traditional flow-based methods have other shortcomings, especially when used for modelling IIS. Two important limitations are related to their inability of considering a flow that may change in time and the fact that only balanced networks can be considered. Some modifications to overcome these and other limitations have been proposed by a number of authors.

Mathematically speaking, this method in its classical linear formulation is focused on the equilibrium in nodes so that the sum of inflow and outflow for a node at a normal state is equal to zero. Nodes and edges in this approach are subjected to some constrains due to their physically limited capacities. In an extended formulation (Holden et al., 2013), nodes representing storage are naturally limited by the number/volume of commodities they can store, similarly, nodes representing manufacturing or production entities can make a fixed number of goods or deliver a fixed quantity of services. Edges, in turn, when representing pipes, electrical lines, roads, etc., can only transmit a limited amount of flow. Thus, a problem of studying the behaviour of IIS using such a method reduces to a problem of linear programming with given constrains and an objective function to minimise (or maximise).

Recently, a number of new ideas have been tested by incorporating probabilistic and stochastic approaches to the flow-based modelling. The model proposed by Val et al. (2014) allows one to make a time-variant analysis as well as analyse unbalanced networks due to failed or damaged infrastructure assets. This method also allows modelling the recovery of IIS after failure, although very few works have been focused on this phenomenon.

7.3.2.3 IO Approach

The IO approach is an economic model based on a simple relationship between the commodities required to manufacture and produce certain goods. In the model proposed by Leontief (1951), this relationship is assumed to be linear so that the output is linearly proportional to the weighted sum of input components. In mathematical language, it means that to produce an output x_i it is required $a_{ij}x_j$ quantity of input components so that $\mathbf{x} = \mathbf{Axe} + \mathbf{c}$, where $\mathbf{A}$ is a matrix consisting of a_{ij} elements, $\mathbf{x}$ and $\mathbf{c}$ are vectors, where $\mathbf{c}$ is the consumption by users or customers of the final product. In the case of failure, the system becomes inoperable and cannot produce a required level of output commodities.

In the case of IIS modelling, the *inoperability* is referred to the inability of the system to accomplish its functions, which may result in partial or complete disruption. Then, the vector $\mathbf{x}$ represents the risk of inoperability and the matrix $\mathbf{A}$ consists of the probability elements a_{ij}, which describe the inoperability probability of the system i in the case of failure in the system j. For instance, $a_{13} = 0.6$ means that failure of the third system will make the first system inoperable with 60% probability. Thus, the IO approach can model and predict how disturbances in one infrastructure can propagate and influence other IIS, which can help in the formulation of procedures to mitigate these disturbances and avoid global failures.

As with any other approach, there are shortcomings in the IO model, although some of them have been addressed. For instance, the proposed model is static and does not consider the time dependence of the inoperability, when its solution exists. Consequently, it is hard to understand how the system has evolved to reach the final state. To answer this question, the original linear system was transformed into the system of ordinary linear differential equation with the left-hand side being substituted by the derivative of the vector $\mathbf{x}$. Another deficiency of this approach is the liner dependencies between infrastructure systems, whereas in reality some of the dependencies are naturally nonlinear. This will lead to a set of nonlinear algebraic or differential equations that are much harder to deal with. In these cases, such systems are often analysed numerically using the computable general equilibrium approach (Rose, 1995). One also should note that the coefficients of **A** are often available for normal operational conditions but not for emergency conditions, which means that in the case of failure the matrix **A** may be different leading to somewhat unpredictable result.

7.3.2.4 SD Approach

The SD approach has been proposed by an electrical engineer Jay Wright Forrester (Forrester, 1961). Since then the SD method has been adopted in a number of areas including engineering, social, and ecological sciences. The main feature of the SD approach is very similar to the NB approach, where, real processes are represented by a stock of various entities (people, material, capital equipment, orders, money, etc.) and flows between the stocks. The method relies on two types of diagrams: *casual loop diagram* and *stock and flow diagram*. The first demonstrates the influence of one variable on another so that it often leads to a positive or negative feedback loop. A good example of a negative loop is the ratio between potential buyers and sellers since the number of potential buyers should decrease with more items being sold to them. In the ideal case, the numbers of buyers and items left should drop to zero simultaneously, which unfortunately is not the case in the real life. The information about the decline rate of the potential buyers is unknown in this case and, thus, it is almost impossible to predict how many more items will be required. In this case, the stock and flow diagram may help, where the flow rate of sales can be adjusted. The SD approach takes an aggregated view on the processes and flows and, thus, can be called a *top-down* approach, in opposite to the AB methodology. The SD approach allows taking into account nonlinear interaction of variables, model time evolution, and delays.

In this approach, the stock variables are governed by the initial value and the difference between inflows and outflows. To account for the overall inflows and outflows, an integral of the difference is usually considered, where inflows are taking with a positive sign and outflows with a negative one. A flow equation has also to be written for each variable and there are several ways to define the flow. The flow can be proportional to the stock value or independent of it, as well as can be proportional to a deficit of the corresponding stock. Monitoring and managing flow variables basically allow regulating the stock variables and the network in general. Stock variables are usually slowly changing despite rapid changes of flow, therefore, stocks may be considered as entities with delays and absorbers. In the case of failure, usually some flows are stopped and some stocks are disconnected.

The SD approach relies on numerical simulations and a number of software packages have been proposed for this purpose over the years. There are over 15 software packages that can be used to address a problem using the SD approach, for example: ASCEND, NetLogo, iThink, Semantics, and VenSim PLE. For modelling IIS, another tool called CIS/DSS has been developed, which includes a nonlinear optimisation package. This software has been used for studying the behaviour of seven critical IIS, defined by the Homeland Security Directive, at an aggregate level (Bush et al., 2005).

The SD approach has some shortcomings although, in general, it describes the interaction by a set of nonlinear differential equations. The SD approach cannot deal with changes in the infrastructure topology, cannot describe low-level behaviour and provides information only on the aggregate system level.

7.3.2.5 Other Approaches

Apart from the AB approach, practically all other methods are somewhat similar in their idea of describing the IIS mathematically as a set of linear, in the first approximation, algebraic or differential equations with/without additional conditions/constrains. There are other approaches used for analysis of IIS, some of which are very different from the above, like *empirical risk analysis*, others are similar in their core idea or combine some of the above methodologies with new features.

A *system of systems* (SoS) comprises heterogeneous, distributed, and independent existing systems embedded in networks at multiple levels (DeLaurentis, 2007). The key words here are independent and existing systems, where the former implies that the system may have very different engineering structures, may belong and be managed by different owners/executives. It also means that these systems have been built for a purpose and their integration into a network will bring new (inter)dependences and new functional requirements under the original capabilities leading to new behaviour. Analysis of such systems is the focus of the SoS approach as well as understanding how to achieve a common goal or synergy between the systems operating together. Originally, this method has been proposed in defence-related applications to improve the interaction between different branches of armed forces, which provide a perfect example of existing and independently managed units. Later, the SoS approach became popular within the engineering community in general and for studying the IIS in particular. The SoS approach distinguishes four pyramid-shape categories–policy, economics, resources, and operations, that are divided into a number of levels, providing a certain hierarchy within each category. Three dimensions–autonomy, connectivity, and system, are also introduced to help guiding the analysis of the SoS. This approach along with network-of-networks methodology has been used to study large-scale IIS (Thacker et al., 2017).

The *Petri-Net* (*PN*) approach has been proposed by German mathematician Carl Petri who suggested to describe a network by a set of four entities (P, T, I, O), where P represents a set of stands, T represents a set of transitions, and I and O represent a set of input and output functions, correspondingly (Petri, 1962). Graphically, the stands are represented by circles, transitions by vertical lines and tokens, which are other elements of the PN, are marked by dots inside the circles. The transition from one state of the network to another is associated with the tokens transfer throughout the network. An important feature of the PN approach is that it permits two independent and allowed processes to occur simultaneously. In the traditional PN approach, the problem is reduced to a system of linear algebraic equations; however, whether the PN can be optimised is another nontrivial question. A number of researchers used the PN approach to study the behaviour of IIS (Laprie et al., 2007), whereas some complex geographical features of nodes have been considered in GeoPN approach developed by Ge et al. (2010).

Hierarchical methods have been developed in the early 1980s as an effort to capture the hierarchical structure of very large and incommensurable systems with uncertainties and even conflicting objectives. The hierarchical holographic modelling (HHM) method appeared as a generalisation of the hierarchical overlapping coordination methodology, which was successfully applied in the water resources field. The main concept of the HHM approach is "overlapping among various holographic models with respect to the objective functions, constraints, decision variables, as well IO relationships of the basic system" (Haimes, 1981). In other words, the main driver for this idea was the desire to overcome the limitation of a single model that could not describe a complex large-scale system. The name *holographic* is taken from optics where it implies three-dimensional (3D) image rather than a conventional two-dimensional (2D) photography, thus, looking at a system from a different point of view. The HHM approach suits well for studying the risk and uncertainties in large-scale systems, properly capturing their heterogeneous properties. Risk here is used as a measure of wellness and/or performance. The HHM considers a large-scale system as a set of smaller subsystems and analyses them separately from different perspectives as well as the impact of subsystems onto each other. The linear programming methodology is used in the approach with multiple objectives at different levels.

The *empirical approach* is mainly based on historic data available from various types of accidents and disasters. As it has been mentioned, some critical infrastructure interdependencies cannot be foreseen in normal operational conditions but may appear in emergency situations caused by extreme events. This has been shown in some cases by analysing the data records and proliferation of failures throughout the IIS for various events, especially the recent ones that have been comprehensively recorded. These data can help to identify not only a number of 'hidden' interdependencies but also create a pattern of failures associated with certain infrastructure or connected infrastructure systems, identify weak links, the rate of proliferation and the direction of cascade failure within a given infrastructure and between IIS, the severity of impact onto people and industry as well as help in the future to prevent similar failures by promptly recognising a certain pattern or chain of events. To find trends or patterns, a huge database is required and specific methodology like *big data* analysis to work with available data. Databases have been developed for existing national-level infrastructure in a number of countries to store information on failures, disruptions, and incidents. For instance, in the United States, the Office of Infrastructure Protection has been developing and maintaining a National Asset Database that contains information on over 77,000 individual assets, including dams, hazardous materials sites, power plants including nuclear ones, and others (Moteff, 2007). In the United Kingdom, the Infrastructure Transitions Research Consortium has proposed and created an integrated national-scale infrastructure database that is enhanced with modelling and analytic tools for analysis. Luiijf et al. (2009) have analysed existing EU data on cascading failures and found that they happen fairly often, highly asymmetric and focused. Another interesting thing reported was the fact that the majority of cascade failures were originated in energy and ICT sectors but interdependencies were not strong enough to allow these failures to propagate deep into other infrastructure systems. Later, Van Eeten et al. (2011) conducted similar study based on empirical data available from media reports. Other databases have been proposed and studied by McDaniels et al. (2007), Wallace et al. (2003), and others. Apparently, this approach based on historical data is meant to benefit from thoroughly studying the real events but has a number of shortcomings. For instance, some confidential data on certain failures may not be released to general public but may contain information crucial for their understanding. Similarly, important data may have been unrecorded or unreported intentionally due to someone's thought of its insignificance or possible liability implications. There are also no rules or standards on the data collection for such databases. Thus, researchers may use any combination of data, based on their preferences, from different databases that may lead to erroneous or misleading results.

7.3.3 Risk Analysis Framework for Single and Multiple Hazards

As noted above, due to complexity and high interconnection of the present infrastructure systems, special attention should be paid to dependencies and cascading effects. The dependencies among systems imply that a perturbation in an initial single system is likely to cause stress in other connected systems, triggering sometimes an uncontrolled feedback process. Thus, analysing critical infrastructure systems without considering this relation would lead to underestimating both the risks and the consequences (Nogal and O'Connor, 2019). In fact, cascading effect hazards, which are characterised by a high impact low probability nature, have started to be recognised in technical standards and legislation concerned with the control of major accident hazards, such as the Council Directive 96/82/EC (EU Seveso-II Directive) and the European Parliament and Council Directive 2012/18/EU.

The EU-funded project Risk Analysis of Infrastructure Networks (RAIN) in response to extreme weather proposes an approach to estimate the probability associated with single events and cascading effects of single or multiple hazard events and identify the most effective adaptation measures to climate change. It is based on four levels, the first of which deals with the identification of the potential threats on the critical infrastructure. On the second parallel level, their vulnerabilities are analysed, which allows for the simplification of the problem by reducing the number of elements to be included in the analysis. In this level, the mitigation measures are introduced. The third level assesses the impact that any potential failure would bring from the security, economic, and societal perspectives (Nogal et al., 2016). The last level introduces different adaptation strategies to extreme weather events. The four components, namely, threats, vulnerabilities, impacts, and strategies are all combined in the framework presented in Van Gelder et al. (2016).

The Bayesian probabilistic model proposed by Van Gelder et al. (2016) can be split into two phases: an inference phase that quantifies the probabilities of the possible outcomes as a function of the actions that a decision maker might take; and the decision phase, which determines the optimal actions to minimise the already identified risks. More interesting aspect of this approach is that it considers not only the most likely scenario (i.e., expectation value) but also the worst- and best-case scenarios that are lower and upper bounds. According to the Allais paradoxes of behavioural economics, usually, decisions are not only based on the most likely scenario but also the worst- and best-case scenarios are expected to be also considered in the decision process (Nogal et al., 2018).

7.3.4 Risk Analysis Framework for Collateral Impacts of Cascading Effects

The cascading effects are introduced in the first level mentioned above. The estimation of the threats considers the probabilities of occurrence related to both extreme weather events and other hazards triggered by the weather events. The use of time-dependent Markov processes for the estimation of the latter seems a sound approach. Nevertheless, the challenge arising here is the estimation of the transition matrices, which for the presented problem would be intractable (in) homogeneous matrices. To tackle the problem, the probability sort algorithm (Van Erp et al., 2017) has been proposed, whose output is time-dependent probability maps of failure of the overall infrastructural systems.

In modelling this inhomogeneous Markov process, the transition probabilities change as a function of time. The time dependency is caused by physical processes and/or by human intervention measures in the system. For instance, when a landslide due to exposure to intensive rainfall occurs, then there will be an initial increase in the probability of a knock-on landslide occurring in the surrounding areas. It is assumed that this danger will quickly dissipate overtime if the knock-on effect fails to materialise. If we assume that the landslide implies several connected areas, the cascading events will propagate with that physical connection and then the corresponding inhomogeneous probability map can be estimated, $\mathbf{f}(\mathbf{s}^{(t-1)}, \mathbf{k}^{(t-1)})$, where t is the time step, $\mathbf{s}$ is the damage state vector, and $\mathbf{k}$ is the counter vector where the elements k_i are the number of time steps that an object is in a failed state.

Fig. 7.2 shows the graphical representation of the analysis of two landslides starting simultaneously and evolving overtime, while Table 7.3 presents the estimated probability map for the primary event at $t=3$, when assuming known probabilities of propagation for these physical connections,

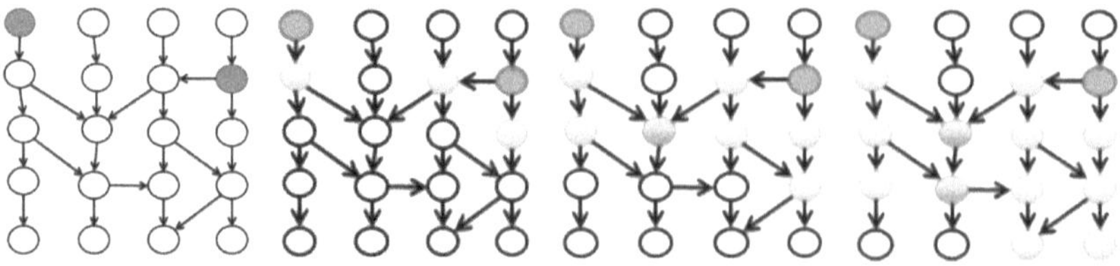

Fig. 7.2 Graphical representation of two cascading effect analysis using the probability sort algorithm in Van Erp et al. (2017). The darker the colour, the larger the probability of the event. (A–D) The initial state and the states at the following time steps, respectively.

Table 7.3 Estimated Probability Map for the Event at $t=3$, Where the Cell Position Stands for the Associated Node Position

1	0.0279	0.028	0.028
0.9124	0.0448	0.5437	1
0.8326	0.7985	0.499	0.9125
0.7407	0.8488	0.4409	0.8866
0.0521	0.0577	0.593	0.7418

for example, $P=0.9$ for the vertical connection, $P=0.7$ for the lateral and diagonal connections, and a general landslide probability of $P=0.01$.

An interesting advantage of this approach is that it allows a sort of exact evaluation of even exponentially large event trees of the system states as long as the information entropy in that event tree is low enough. The reader is referred to Van Erp et al. (2017) for the formulation and mathematical foundation of the approach.

7.4 Climate Change Adaptation of Infrastructure Systems Including Case Studies

7.4.1 Major Coastal Flooding Exposes Infrastructure Vulnerabilities

In January 2005, an extreme storm surge event affected Great Britain, Denmark, Sweden, Estonia, and the most populated area of Finland, including Helsinki and Stoos. The land areas in southwest Finland and in the west Uusimaa region were threatened by record values of the sea level (in

Helsinki, +151 cm over its normal value) combined with high wind speeds (with gusts of 30 m/s), intense rainfall, and snow and sleet in central Finland.

In Finland, the early-warning systems and the close coordination of the national and local authorities, governmental agencies, and private sector, which owns most of the critical infrastructure (e.g. railway, energy, and maritime transport) enabled a number of preparatory measures to be put in place. Among the measures, they built 1-m-high dams of paper bales, sealed off sewer overflows to avoid seawater to enter buildings through drainage systems and offered sandbags to members of the public. During the event and afterwards, water was pumped to avoid larger damages and permit a prompt recovery. Water was pumped during days from the underground critical infrastructure (heating, electricity, phone, water, and sewage lines), metro transformer booths, tunnels, and cellars of private houses.

Although large-scale flooding and storm damage were avoided, the storm surge event resulted in partial road closures due to flooding, causing important traffic disruptions in Helsinki and road closures in various coastal regions of Uusimaa. Passenger ferries between Helsinki and Tallinn were cancelled resulting in thousands of passengers having to remain in Tallinn overnight. Suomenlinna Island was isolated because the ferry could not carry passengers from the submerged wharf and the maintenance tunnel to Suomenlinna had been barricaded with sandbags. Flooding water entered cellars and garages and surrounded many of the private houses. Those residents subjected to flooding were displaced from their homes for months. As a result of the overload of the sewer network, 63,000 m^3 of untreated wastewater was dumped into the sea. A total of 886 flood damage cases were declared, worth in total 7.2 million euros, of which only about 1 million has been paid to date. The reader is referred to Carey et al. (2017) for a detailed description of the event.

On Hästholm Island, 90 km east from capital Helsinki, the Loviisa power plant is situated. According to the plant regulations, when the sea level is +200 cm rise, the power plant needs to be secured and shutdown, although the preparation for shutdown should be started when sea level rises over +175 cm. During the event, the sea level in the area reached +173 cm above normal levels. Closing down the nuclear power plant would have affected the primary power supply of a very large area, affecting many residential areas and critical services and infrastructure such as railway, water/wastewater treatment plants, and health services.

This case is an interesting one from the perspective of the cascading effects. The blackouts might impact the pumping capacity, which was definitely an important tool of flooding control. Few more centimetres of sea-level rise might have created a complex and multidimensional emergency situation, which is very likely in the future given that the sea level of the Baltic Sea is predicted to rise up to 40 cm during the next 100 years due to climate change. The case clearly shows interdependencies between various infrastructure systems (e.g. electricity network and flood defences) and highlights an urgent need in climate adaptation measures.

7.4.2 Designing for Resilience in Energy and Telecommunication Infrastructure Networks

Energy and telecommunications networks are especially vulnerable to extreme weather events due to their extensive coverage across the urban and rural countryside. The leading causes of failures due to extreme weather events are wind storms, ice/wet snow storms, extreme heat, lightning, flash floods, wildfires, and sandstorms.

Some remediation measures to mitigate the effects of these events on energy and telecommunications networks are related to maintenance operations. Network operators claim that the introduction of unified infrastructure protection guidelines would be beneficial and provide a standard to which maintenance operation bodies can adhere. These guidelines should include aspects such as regular tower inspections and maintenance and the rights of way maintenance, for example, of vegetation management.

The increasing threat of extreme weather events due to climate change also calls for protection plans allowing for adaptation, with consideration of, for instance, deicing and anti-icing measures and prevention of line sagging. The variability of weather events needs the ability to adapt and cope with mitigation and remediation techniques in design, construction, and operational phases of infrastructure management. A detailed description of the protection elements and methods in energy and telecommunication infrastructure networks can be found in Halat et al. (2015).

Energy and telecommunication infrastructure networks are considered as resilient if exposed to any disruption they guarantee an adequate level of performance according to the current situation, maintaining the critical functions (Nogal and O'Connor, 2017). This is enabled by the capacity of the communities to face a stress, which is rooted in its organisational capabilities and flexibility to change. Thus, the human actors, including users, operators and managers, and the information-related aspects, such as how the information is translated to and how it is understood and used, are the key factors when dealing with any type of system resilience (Nogal and O'Connor, 2018).

For instance, in the case of energy systems, the effects of outages are usually quantified in economic and engineering terms. Governmental regulations specify a number of 'reliability indices' such as the number of affected people, the value of lost load (VoLL) and the duration of the outage. These measurements provide information about the disruption and are clearly defined. Nevertheless, they do not consider some impacts on the population, for example, the loss of power supply will have a different impact when occurring to a hospital, a factory, a residential area, or to a public transportation system. Thus, social response, including the consideration of the special context during emergency situations caused by extreme weather disasters should be integrated into the resilience assessment. As pointed out in Nogal et al. (2018), resilience can be boosted by stakeholders in a shorter-term timeframe. The participation of the operators and managers in the identification of the strengths and vulnerabilities of the energy and telecommunication systems in the face of extreme weather can improve their resilience.

7.4.3 Risk-Based Adaptation of Local Electricity and Water Networks

This case study investigates the performance of two infrastructure networks—electricity and water, which provide services to a town during a weather-related hazard, namely flooding. The entire local network, which also includes customers, is shown in Fig. 7.3. The electricity network

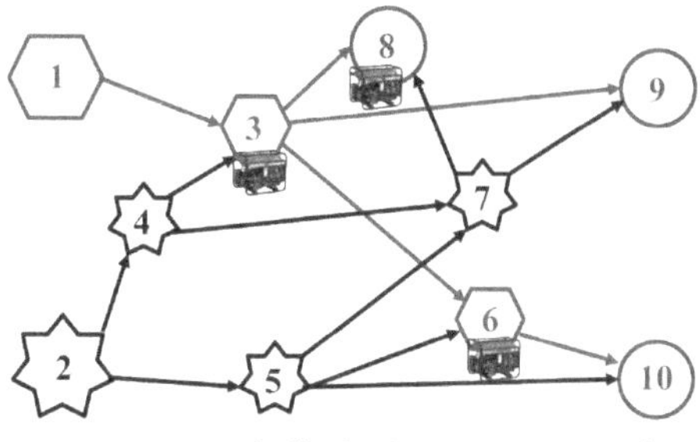

Fig. 7.3 Local infrastructure network (1, 3, 6—water network, 2, 4, 5, 7—electricity network, and 8, 9, 10—customers).

consists of four substations: a primary substation '2' and three secondary substations '4', '5', and '7'. The water network includes a water treatment plant '1', a water tower '3', and a water pumping station '6'; the network depends on the electricity network since pumps at the tower and the pumping station need power for their operation. The customers are represented by a hospital '8' and two residential areas '9' and '10', with a care home located in '9'. This arrangement of the network allows to take into account that loss or shortage of services (i.e., electricity and water) will have different impacts on different customers (i.e., hospital, care home, and residential area).

The performance of the entire network is analysed during a 1-in-100-year flood, which is usually used to represent a design event. A probabilistic flood map is available for the town so that the probabilities of flooding of the electricity and water facilities can be calculated. Of course, the probabilities depend on flood defences, which should be planned for such facilities in flood prone areas. In this particular case, the probabilities of flooding of the primary electricity substation and the water facilities are negligibly small, that is, only secondary substations are at risk of flooding. If a secondary substation is flooded, it will be shutdown. The infrastructure network will then function in a partially damaged state until the secondary substation will return to operation, which may take several days. The analysis is carried out using an extended network flow model presented in Val et al. (2014). The probabilities of service losses to customers are calculated depending on the time required to restore the secondary substations.

As noted above, one of the essential adaptation measures is to equip water pumping stations with backup generators in order to ensure that they will continue to operate in the case of power outage. Thus, it is considered that the water tower and the water pumping stations have backup generators; the hospital should have a backup generator as well. The analysis has shown that the availability of backup generators significantly improves the performance of the water network in the case of flooding. It has been checked that with a relatively small amount of fuel stored for the generators, uninterrupted water supply to the customers can be provided for up to 11 days, when the secondary substations are out of operation. Applying the model, the required amount of fuel for backup generators can be estimated depending on the expected duration of power outage and the corresponding probabilities of service losses. More details about the model and analysis can be found in Val et al. (2014).

7.4.4 Examples of Adaptation Measures

The development of adaptation strategies is significantly complicated by major uncertainties associated with climate change and its impacts. However, taking into account that many infrastructure assets have design working life of 50 years and longer proactive adaptation measures may need to be implemented now or in the nearest future. Although there is no single adaptation solution that would suit all infrastructure systems/networks and geographic areas there are some common measures. One of them is updating of codes/standards for the design and construction of infrastructure assets with long working lives by taking into account climate change impacts. Such changes may increase the initial costs of infrastructure assets but should help to avoid much larger costs of their maintenance, repair, and replacement in the future; even if the climate change predictions used for the code updating are not accurate, this will still lead to improvement of the asset resilience and eventual reduction of the above costs. Of course, these documents may need to be updated again when new information related to climate change will be available.

There are many other adaptation measures which may be relevant depending on a particular infrastructure network and geographical location of its assets; examples of such measures will briefly be described in the following. *Energy networks*: power plants, nuclear ones in particular, are often located on the coast or in estuaries for access to cooling water. Protective systems may need to be constructed/upgraded against storm surges and coastal erosion due to sea-level rise. Electricity substations located in flood prone areas may require additional protection, for example, installing flood defence barriers, permanent high-volume pumps, raising the level at which new substations are installed. Similar measures can be used for other energy infrastructure facilities at risk of flooding such as refineries and gas terminals. To protect overhead power transmission lines against high winds, proactive cutting of tree branches should be implemented. In areas prone to tropical cyclones, power transmission lines can be placed underground. To account for additional sagging of overhead power lines during heat waves, the height of poles/towers can be increased. *Water/wastewater networks*: for pumping stations and water or wastewater treatment plants at risk of flooding, measures similar to those for electricity substations in flood prone areas can be adopted. To prevent overflow of drainage systems due to heavy rainfall sustainable drainage systems (SuDS), which slow down runoff and keep water on the surface similar to natural systems, can be constructed. To mitigate adverse effects of drought, measures like the construction of additional reservoirs, effluent reuse, use of new or alternative water sources, for example, building a desalination plant, can be implemented. In addition, measures aimed at decreasing the water demand, that is, reduction of household consumption and leakage losses, can be undertaken. *Transport*: to use new pavement materials and design to expand the temperature range in which pavements remain functional. To install new road drainage systems and water absorbing pavements to reduce risk of flash flooding. To upgrade rail tracks for new ones, which will be operable at higher temperatures without being susceptible to buckling. To strengthen embankments to increase their stability in cases of heavy raining, improve coastal defences (when roads/rail tracks are located close to the coast), increase bridge protection against scour. *ICT*: to improve protection of assets such as cables, masts, towers, and data centres against flooding, including relocation of key facilities to areas not at risk of flooding. Increase redundancy of the networks. The list of adaptation measures, of course, can be continued.

7.5 Summary

The chapter addressed the issue of adaptation of critical infrastructure systems such as energy, water/wastewater, transport, and ICT, to climate change, with emphasis on their interdependencies. First, risks posed by a changing climate to these infrastructure systems were considered, with the main attention to extreme weather events and associated with them hazards. In particular, it has been shown that the frequency and intensity of various extreme weather events have increased and may continue to increase in the future due to climate change. Examples illustrating devastating effects of extreme weather events and weather-related hazards on the infrastructure systems have been presented and cases, when the infrastructure interdependencies had a major influence on consequences of these events, have been discussed. Potential adverse impacts of climate change on the performance of the infrastructure systems have been summarised. Since modelling the performance of infrastructure systems subject to weather-related hazards can provide data and support decisions regarding adaptation strategies, methods that were proposed and applied for this purpose have been overviewed. In particular, a risk-based framework for analysis of IIS subject to single or multiple hazards have been presented. Finally, possible measures for climate adaptation of the critical infrastructure systems have briefly been described, including three more detailed case studies.

7.6 Potential Design and Practice Evolutions

Many infrastructure assets have a long service life, for example, 50 years and longer. Therefore, it is essential to account for potential impacts of climate change in planning new infrastructure. For example, planning a new drainage system (or a major upgrade of an existing one) in a low-lying coastal city without taking into account sea-level rise will most likely result in this expensive investment becoming unfit to serve its purpose long before the end of its intended design life. Thus, there is a clear need in updating standards and guidelines for design of infrastructure facilities, especially the ones with long lifetime, to ensure that these facilities will remain fit for their purpose, without costly upgrading/retrofitting, during their design life in conditions of climate change. Of course, this concerns not only new infrastructure but also existing one as well. Many key existing infrastructure assets with long operational lifetime (e.g. bridges, reservoirs, power plants, water/wastewater treatment plants, etc.) are already experiencing harmful impacts of a changing climate but very costly to replace and in need of upgrading/retrofitting. Hence, standards/guidelines for their retrofitting, repair, and maintenance also require revision and updating to account for impacts of climate change. Large uncertainties associated with the prediction of future climate conditions significantly complicate this standards/guidelines updating. In this context, a decisive move from prescriptive standards to performance-based ones is clearly needed.

As shown in this chapter, infrastructure systems are increasingly interdependent. For example, water/wastewater treatment plants and pumping stations require power to operate, power plants need water for cooling, all these facilities rely on the transportation network for access of workers and emergency crews (in case of failure), the three infrastructure systems (i.e., water, power, and transport) depend on ICT for control and monitoring, while the latter needs power, etc. The importance of the interdependencies has been widely recognised. However, the current regulatory structure in practically all countries is based on a 'silo' approach, which concentrates on individual infrastructure systems without addressing their interdependencies. This represents a major barrier to developing efficient strategies for infrastructure adaptation to climate change, in particular for improving its resilience to extreme weather events and other weather-related hazards. Further efforts are needed to ensure that regulators, planners, and operators across various infrastructure sectors work together to address challenges posed by climate change.

7.7 Open Research Questions

The adaptation of IIS to a changing climate is an extremely complex and challenging problem. To develop efficient strategies for its solution significant gaps in our knowledge need to be addressed, for example:

- Impacts of climate change on extreme weather events (i.e., storms, tornados, lightning, etc.) in different regions of the world are still not well understood. Furthermore, there are major uncertainties associated with modelling these events and their effects on IIS.
- Modelling large complex infrastructure systems remains a major challenge. In particular, this concerns modelling cascading failures in interdependent systems, which are governed by different physical laws and have different internal structures. Various methods have been proposed for that so far but there are practically no benchmark studies, which would allow to compare the efficiency of these methods for particular applications.
- Not all infrastructure interdependencies have been identified and/or sufficiently well understood. There is a lack of data for that; in particular, there are very few databases that contain information on infrastructure failures, including cascading failures. No standards or rules on data collection for such databases currently exist.
- There are major uncertainties associated with weather-related hazards and responses of IIS to such hazards. However, only a few methods that have been proposed for analysis of IIS are capable of taking these uncertainties into account.
- Of course, the list can be continued.

References

Albert, R., Jeong, H., Barabasi, A.L., 2000. Error and attack tolerance of complex networks. Nature 406, 378–381.

Añel, J.A., Fernández-González, M., Labandeira, X., López-Otero, X., de la Torre, L., 2017. Impact of cold waves and heat waves on the energy production sector. Atmosphere 8, 209. https://doi.org/10.3390/atmos8110209.

Barton, D.G., Eidson, E.D., Schoenwald, D.A., Cox, R.G., Reinert, R.K., 2004. Simulating Economic Effects of Disruptions in the Telecommunications Infrastructure. Sandia report SAND-2004.

Buldyrev, S.V., Parshani, R., Paul, G., Stanley, H.E., Havlin, S., 2010. Catastrophic cascade of failures in interdependent networks. Nature 464, 1025–1026.

Bush, B., Dauelsberg, L., LeClaire, R., Powell, D., Deland, S., Samsa, M., 2005. Critical Infrastructure Protection Decision Support System (CIP/DSS) Overview. LANL report LA-UR-05-1870.

Carey, M., Clarke, J., Corbally, R. Connolly, L., et al., 2017. Report on benefits of critical infrastructure protection. Deliverable 6.3 of RAIN (risk analysis of infrastructure networks in response to extreme weather) project. http://rain-project.eu/wp-content/uploads/2017/08/6.3-Final.pdf.

Climate Institute, 2012. Coming Ready or Not: Managing Climate Risks to Australia's Infrastructure. The Climate Institute, Sydney.

DeLaurentis, D., 2007. Role of human in complexity of a system-of-systems. In: Digital Human Modelling, LNCS 4561. Springer-Verlag, pp. 363–371.

DoC, 2014. Service Assessment. The Record Front Range and Eastern Colorado Floods of September 11–17, 2013. The U.S. Department of Commerce, National Oceanic and Atmospheric Administration, National Weather Service. http://www.weather.gov/media/publications/assessments/14colorado_floods.pdf (accessed 20.08.18).

DoE, 2009. Comparing the impacts of the 2005 and 2008 hurricanes on U.S. energy infrastructure. The U.S. Department of Energy http://www.oe.netl.doe.gov/docs/HurricaneComp0508r2.pdf (accessed 20.08.18).

DoE, 2013. Comparing the impacts of northeast hurricanes on energy infrastructure. The US Department of Energy. Available from: https://www.energy.gov/sites/prod/files/2013/04/f0/Northeast%20Storm%20Comparison_FINAL_041513b.pdf (accessed 20.08.18).

Dosio, A., Mentaschi, L., Fisher, E.M., Wyser, K., 2018. Extreme heat waves under 1.5°C and 2°C global warming. Environ. Res. Lett. 13054006.

Dudenhoeffer, D.D., Permann, M.R., Manic, M., 2006. CIMS: A framework for infrastructure interdependency modeling and analysis. In: Perrone, L.E. et al., (Ed.), Proceedings of the 2006 winter simulation conference, pp. 478–485.

Dueñas-Osorio, L., James, I.C., Barry, J.G., Ann, B., 2007a. Interdependent response of networked systems. J. Infrastruct. Syst. 13 (3), 185–194.

Dueñas-Osorio, L., James, I.C., Barry, J.G., 2007b. Seismic response of critical interdependent networks. Earthq. Eng. Struct. Dyn. 36, 285–306.

Dupigny-Giroux, L.A., 2000. Impacts and consequences of the ice storm of 1998 for the North America north-east. Weather 55, 7–15.

Environment Agency, 2016. The Costs and Impacts of the Winter 2013 to 2014 Floods. ReportSC140025/R1, Environment Agency, Bristol.

Forrester, J.W., 1961. Industrial Dynamics. MIT Press, Cambridge.

Ge, Y., Xing, X., Cheng, Q., 2010. Simulation and analysis of infrastructure interdependencies suing a petri net simulator in a geographical information system. Int. J. Appl. Earth Obs. Geoinf. 12, 419–430.

Gibson, J.R., 2017. Built to Last: Challenges and Opportunities for Climate-Smart Infrastructure in California. Union of Concerned Scientists. https://www.ucsusa.org/sites/default/files/attach/2017/11/gw-whitepaper-smart-infrastructure.pdf (accessed on 20.08.18).

Gleick, P., 2015. Impacts of California's Ongoing Drought: Hydroelectricity Generation. Pacific Institute, Oakland, CA. http://pacinst.org/publication/impacts-of-californias-ongoingdrought-hydroelectricity-generation. Accessed 20.08.18.

Haimes, Y.Y., 1981. Hierarchical holographic modelling. IEEE Trans. Syst. Man Cybern. 11 (9), 606–617.

Halat, M., Clotet, X., Gaitán, V., Bechrakis, D., 2015. Protection elements and methods in energy and telecom infrastructure. In: Deliverable 4.2 of RAIN (Risk Analysis of Infrastructure Networks in Response to Extreme Weather) Project. http://rain-project.eu/wp-content/uploads/2015/11/D4.2-Protection-on-Electrical-and-Telecomunication.pdf.

Hanson, S., Nicholls, R., Ranger, N., Hallegate, S., Corfee-Morlot, J., Herweijer, S., Chateau, J., 2011. A global ranking of port cities with high exposure to climate extremes. Clim. Chang. 104 (1), 89–111.

Hernandez-Fajardo, I., Duenas-Osorio, L., 2011. Sequential propagation of seismic fragility across interdependent lifeline systems. Earthquake Spectra 27 (1), 23–43.
Holden, R., Val, D.V., Burkhard, R., Nodwell, S., 2013. A network flow model for interdependent infrastructures at the local scale. Saf. Sci. 53, 51–60.
Kunkel, K.E., Karl, T.R., Brooks, H., Kossin, J., Lawrimore, J., Arndt, D., Bosart, L., Changnon, D., Cutter, S.L., Doesken, N., Emanuel, K., Groisman, P.Y., Katz, R.W., Knutson, T., O'Brien, J., Paciorek, C.J., Peterson, T.C., Redmond, K., Robinson, D., Trapp, J., Vose, R., Weaver, S., Wehner, M., Wolter, K., Wuebbles, D., D., 2013. Monitoring and understanding trends in extreme storms: State of knowledge. Bull. Am. Meteorol. Soc. 94 (4), 499–514.
Laprie, J.C., Kanoun, K., Kaaniche, M., 2007. Modelling interdependencies between the electricity and information infrastructures. In: Proceedings of the 26th Int. conference on computer safety, reliability and security, SAFECOMP07.
Leavitt, W.M., Kiefer, J.J., 2006. Infrastructure interdependencies and the creation of a normal disaster: the case of hurricane Katrina and the City of New Orleans. Public Works Manag. Policy 10 (4), 306–314.
Leontief, W.W., 1951. Input-output economics. Sci. Am. 185, 15–21.
Luiijf, E., Nieuwenhuijs, A., Klaver, M., van Eeten, A., Cruz, E., 2009. Empirical findings on critical infrastructure dependencies in Europe. In: Critical Information Infrastructure Security. Springer, Berlin.
McDaniels, T., Chang, S., Peterson, K., Mikawoz, J., Reed, D., 2007. Empirical framework for characterizing infrastructure failure interdependencies. J. Infrastruct. Syst. 13 (3), 175–184.
Mendelsohn, R., Emanuel, K., Chonabayashi, S., Bakkensen, L., 2012. The impact of climate change on global tropical cyclone damage. Nat. Clim. Chang. 2 (3), 205–209.
Miles, S.B., Jagielo, N., Gallagher, H., 2015. Hurricane Isaac power outage impacts and restoration. J. Infrastruct. Syst. 22 (1) ASCE. 05015005.
Miller, R., 2006. Hurricane Katrina: communication & infrastructure impact. In: Tussing, B.B. (Ed.), Threats at our Threshold. U.S. Army War College, PA, pp. 191–205.
Moteff, J., 2007. Critical Infrastructure: The National Asset Database. (CRS report for congress).
New York State 2100 Commission, 2013. NYS2100 Commission Report. Recommendations to Improve the Strength and Resilience of the Empire State's Infrastructure. http://www.governor.ny.gov/assets/documents/NYS2100.pdf Accessed 20.08.18.
Nogal, M., O'Connor, A., 2017. Cyber-transportation resilience. Context and methodological framework. In: Resilience and Risk. Springer International Publishing, pp. 415–426.
Nogal, M., O'Connor, A., 2018. Resilience assessment of transportation networks. In: Sustainable and Resilient Infrastructure. Routledge Chapter 11.
Nogal, M., O'Connor, A., 2019. Transportation. Considerations on resilience management. In: Florin, M.-V., Linkov, I. (Eds.), Resource Guide on Resilience. EPFL International Risk Governance Center (IRGC), Lausanne.
Nogal, M., O'Connor, A., Caulfield, B., Brazil, W., 2016. A multidisciplinary approach for risk analysis of infrastructure networks in response to extreme weather. Transp. Res. Proc. 14, 78–85.
Nogal, M., O'Connor, A., Groenemeijer, P., Luskova, M., Halat, M., Clarke, J., Gelder, P., Gavin, K., 2018. Assessment of the impacts of extreme weather events upon the Pan-European infrastructure to the optimal mitigation of the consequences. Proceedings of Transport Research Arena, Vienna, Austria.
North, M.J., 2001. Toward strength and stability: agent-based modelling of infrastructure markets. Soc. Sci. Comput. Rev. 19 (3), 307–329.
Ouyang, M., Hong, L., Mao, Z.J., Yu, M.H., Fei, Q., 2009. A methodological approach to analyse vulnerability of independent infrastructures. Simul. Model. Pract. Theory 17, 817–828.

Pall, P., Aina, T., Stone, D.A., Stott, P.A., Nozawa, T., Hilberts, A.G.J., Lohmann, D., Allen, M.R., 2011. Anthropogenic greenhouse gas contribution to flood risk in England and Wales in autumn 2000. Nature 470, 382–385.

Petri, C.A., 1962. Communication With Automata. PhD thesis, University of Darmstadt.

Rinalidi, S.M., Peerenboom, J.P., Kelly, T., 2001. Identifying, understanding and analyzing critical infrastructure interdependencies. IEEE Control. Syst. Mag. 21 (6), 11–25.

Rose, A., 1995. Input-output economics and computable general equilibrium models. Struct. Chang. Econ. Dyn. 6, 295–304.

Sichman, J.S., Conte, R., Gilbert, N., 1998. Multi-agent systems and agent-based simulations. In: Proceedings of 1st Int. Workshop MABS'98.

Thacker, S., Pant, R., Hall, J.W., 2017. System-of-systems formulation and disruption analysis for multi-scale critical national infrastructure. Reliab. Eng. Syst. Saf. 167, 30–41.

Trenberth, K.E., 2011. Changes in precipitation with climate change. Clim. Res. 47, 123–138.

Val, D.V., Holden, R., Nodwell, D., 2014. Probabilistic analysis of interdependent infrastructures subjected to weather-related hazards. Civ. Eng. Environ. Syst. 31 (2), 140–152.

Van Eeten, M., Nieuwenhuijs, A., Luiijf, E., Klaver, M., Cruz, E., 2011. The state and the threat of cascade failure across critical infrastructures: the implications of empirical evidence from media incident reports. Public Adm. 89, 381–400.

Van Erp, N., Linger, R.O., Khakzad, N., van Gelder, P., 2017. Deliverable 5.2 report on risk analysis framework for collateral impacts of cascading effects. http://rain-project.eu/wp-content/uploads/2017/08/D5_2_Final_merged.pdf.

Van Gelder, P., van Erp, N., O'Connor, A., Nogal, M., et al., 2016. Deliverable 5.5 RAIN Workflow Integration. http://rain-project.eu/wp-content/uploads/2017/08/D5.5_final-draft-version.pdf.

Vose, R.S., Applequist, S., Bourassa, M.A., Pryor, S.C., Barthelmie, R.J., Blanton, B., Bromirski, P.D., Brooks, H.E., DeGaetano, A.T., Dole, R.M., Easterling, D.R., Jensen, R.E., Karl, T.R., Katz, R.W., Klink, K., Kruk, M.C., Kunkel, K.E., MacCracken, M.C., Peterson, T.C., Shein, K., Thomas, B.R., Walsh, J.E., Wang, X.L., Wehner, M.F., Wuebbles, D.J., Young, R.S., 2014. Monitoring and understanding changes in extremes: extratropical storms, winds, and waves. Bull. Am. Meteorol. Soc. 95 (3), 377–386.

Wallace, W.A., Mendonca, D.M., Lee, E.E., Mitchell, J.E., Chow Wallace, J.H., 2003. Managing disruptions to critical interdependent infrastructures in the context of the 2001 WTC attack. In: Beyond September 11th: An account of Post-Disaster Research. Special Publication 39, University of Colorado, pp. 165–198.

Walsh, J., Wuebbles, D., Hayhoe, K., Kossin, J., Kunkel, K., Stephens, G., Thorne, P., Vose, R., Wehner, M., Willis, J., Anderson, D., Doney, S., Feely, R., Hennon, P., Kharin, V., Knutson, T., Landerer, F., Lenton, T., Kennedy, J., Somerville, R., 2014. Ch. 2: Our changing climate. In: Melillo, J.M., Richmond, T.C., Yohe, G.W. (Eds.), Climate Change Impacts in the United States: The Third National Climate Assessment. U.S. Global Change Research Program, Washington, pp. 19–67. https://doi.org/10.7930/J0KW5CXT.

Weerakoon, R., Kumar, A., Desha, C., 2012. Sustainability in post disaster road infrastructure recovery projects in Queensland, Australia. In: Proceedings of the Central Region Engineering Conference 2012. Engineering Australia, Queensland Division, Rockhampton, QLD, pp. 63–68.

Zimmerman, R., 2004. Decision-making and the vulnerability of interdependent critical infrastructure. In: Proceeding of 2004 IEEE International Conference on Systems, Man and Cybernetic. vol. 5, pp. 4059–4063.

Zimmerman, R., Zhu, Q., de Leon, F., Guo, Z., 2017. Conceptual modeling framework to integrate resilient and interdependent infrastructure in extreme weather. J. Infrastruct. Syst.. 23 (4) ASCE. 04017034.

Further Reading

DoE, 2013. Comparing the Impacts of Northeast Hurricanes on Energy Infrastructure. The U.S. Department of Energy, http://energy.gov/sites/prod/files/2013/04/f0/Northeast%20Storm%20Comparison_FINAL_041513c.pdf (accessed 20.08.18).

PART III

Case Studies for Buildings

CHAPTER EIGHT

Adaptation of Residential Buildings to Coastal Floods: Strategies, Costs and Efficiency

Axel Creach*, Emilio Bastidas-Arteaga†, Sophie Pardo‡, Denis Mercier*

*Department of Geography and Planning, Faculty of Arts and Humanities, Sorbonne Université, Paris, France
†Research Institute in Civil and Mechanical Engineering, UMR CNRS 6183, Université de Nantes, Nantes Cedex, France
‡Laboratoire d'Economie et de Management de Nantes-Atlantique, Université de Nantes, Nantes, France

8.1 Introduction

The storm Xynthia (28 February 2010) had major impacts along the Atlantic coast of France, especially between the Loire and Gironde estuaries. The analysis of fatalities provided by Vinet et al. (2011, 2012) highlighted how buildings can trap people during floods. They showed that: (i) 100% of the drowning occurred when the flood level exceeded 1 m, (ii) 90% of the drowning occurred inside buildings located within 400 m of the flood defences that failed during the storm, and (iii) 78% of the drowning occurred inside the single-storey houses.

In response, the French government implemented the 'black zone' policy, which identified 1628 buildings in the areas of extreme flood risk for demolition at a cost of €315.7 million (Cour des Comptes, 2012). This policy was severely criticised because of its lack of methodological transparency, high costs, and lack of engagement of local communities (Mercier and Chadenas, 2012). In addition, this policy did not result in any appreciable reduction in human vulnerability to flood hazards along the French coast as it was only implemented in the areas that were flooded in 2010 (Pitié and Puech, 2010; Pitié et al., 2011). In summary, this policy was perceived as a 'knee-jerk' reaction, lacked a cost–benefit analysis and devoid of any scientific underpinning.

The French Atlantic coast is particularly susceptible to floods. For instance, it is predicted that a centennial flood could submerge 354,079 ha (CETMEF et al., 2009). In this area there are 535,500 permanent inhabitants (MEDDE, 2012b) and 136,711 residential buildings (CETMEF et al., 2009),

https://doi.org/10.1016/B978-0-12-816782-3.00008-5

22% of which are single-storey constructions (MEDDE, 2012b). Flood risks will probably increase under the effects of global warming that will result in sea-level rise (Chapter 1, IPCC, 2013; Nicholls, 2011). The official projection for the French Atlantic coast announces a rise of 0.60 m of sea level for 2100 (MEDDTL, 2011). Under a projection of sea-level rise of 1 m the French Atlantic coast will comprise 58% of potentially flooded areas (CETMEF et al., 2009). There is a trend for an increase of storms frequency along the French Atlantic coast due to climate change (Goeldner-Gianella, 2013; Vinet, 2010).

Two consequences could be related with future sea-level rise:

- A flood similar to Xynthia could be produced by a less intense storm.
- A storm similar to Xynthia could generate a more dangerous flood.

In this context of potential increase of coastal floods under a changing climate in zones where existing residential buildings are vulnerable for people, its adaptation becomes essential to prevent death by drowning inside the buildings. According to Nicholls (2011), adaptation strategies are well known: protection, accommodation and retreat. However, these strategies have different costs and efficiency to reduce vulnerability of buildings for people. To address these problems, in this chapter we propose a global framework to evaluate potential fatalities in relation to coastal floods and, on the basis of these results, determine the cost-efficiency of adaptation strategies. The framework implements the VIE index methodology to identify buildings where people could be trapped and drowned during floods (Creach et al., 2015). Thus, the proposed methodology will help the decision-makers to establish the most cost-effective adaptation strategies to protect the human life from flood risks in the future.

Section 8.2 will present the methodology to estimate the VIE index. This methodology will be applied to evaluate the vulnerability of existing buildings and the risk reduction related to the implementation of the adaptation strategies presented in Section 8.3. The proposed framework will be illustrated by two study cases in Section 8.4.

8.2 VIE Index: A Tool to Identify Buildings Which Could Expose Their Occupants to Death

8.2.1 Methodology Description

The VIE index is based on four major criteria which contributed to the people being trapped inside the residential buildings during the storm Xynthia (Vinet et al., 2011, 2012) (Fig. 8.1):

- CR1: Potential water depth inside buildings.
- CR2: Distance to flood defences.
- CR3: Architectural typology of buildings, since single-storey constructions are more likely to trap people than multi-storey buildings where people could escape upstairs.
- CR4: Proximity to a rescue point to facilitate evacuation.

Each criterion is rated from 0 (no vulnerability) to 4 (high vulnerability). Creach et al. (2015) proposed Eq. (8.1), validated through statistical analysis (Section 8.2.2) to identify highly vulnerable buildings (Fig. 8.1).

$$\mathrm{VIE} = \begin{cases} 0 & \text{if } \mathrm{CR1} = 0 \\ \frac{2}{3}\mathrm{CR1} + \frac{1}{3}\mathrm{CR4} + \mathrm{CR2} + \mathrm{CR3} & \text{elsewhere} \end{cases} \tag{8.1}$$

The results range from 0 (no vulnerability) to 12 (maximum vulnerability). The results of the index evaluation are classified into four levels of vulnerability:

- Class A (green, VIE = 0): buildings are not exposed to floods and so do not endanger people.
- Class B (orange, VIE ∈ [1–5[): buildings are of a suitable design to reduce risk to people during floods. The level of vulnerability for people is low.
- Class C (red, VIE ∈ [5–8[): the vulnerability of buildings to floods is high but nonlethal excepting for children, older, or disabled people.
- Class D (black, VIE ∈ [8–12]): the vulnerability of buildings to floods is very high and could result in fatalities in the case of floods.

8.2.2 Statistical Validation

The method was validated by comparing the locations of deaths during storm Xynthia for La Faute-sur-Mer town in France (Creach et al., 2015). La Faute-sur-Mer was the most impacted municipality during this event with 29 deaths, all inside the buildings. The calculation of the VIE index shows that, for a total of 2695 buildings, 48%, 33%, 8%, and 11% were, respectively, in Classes D (black), C (red), B (orange), and A (green) (Creach et al., 2015). Thus, about 50% of residential buildings were identified as posing a lethal risk to occupants in case of flood.

The results presented in Fig. 8.2 are in agreement with the location of deaths. Fig. 8.2 shows that 83% of fatalities were inside the buildings in Class D (black), while 17% of the fatalities were in Class C (red). This confirms that buildings in the black or red classes could expose their occupants to a risk of death in case of floods. The number of deaths during storm Xynthia (29 persons) seems to be small considering that 1305 buildings were

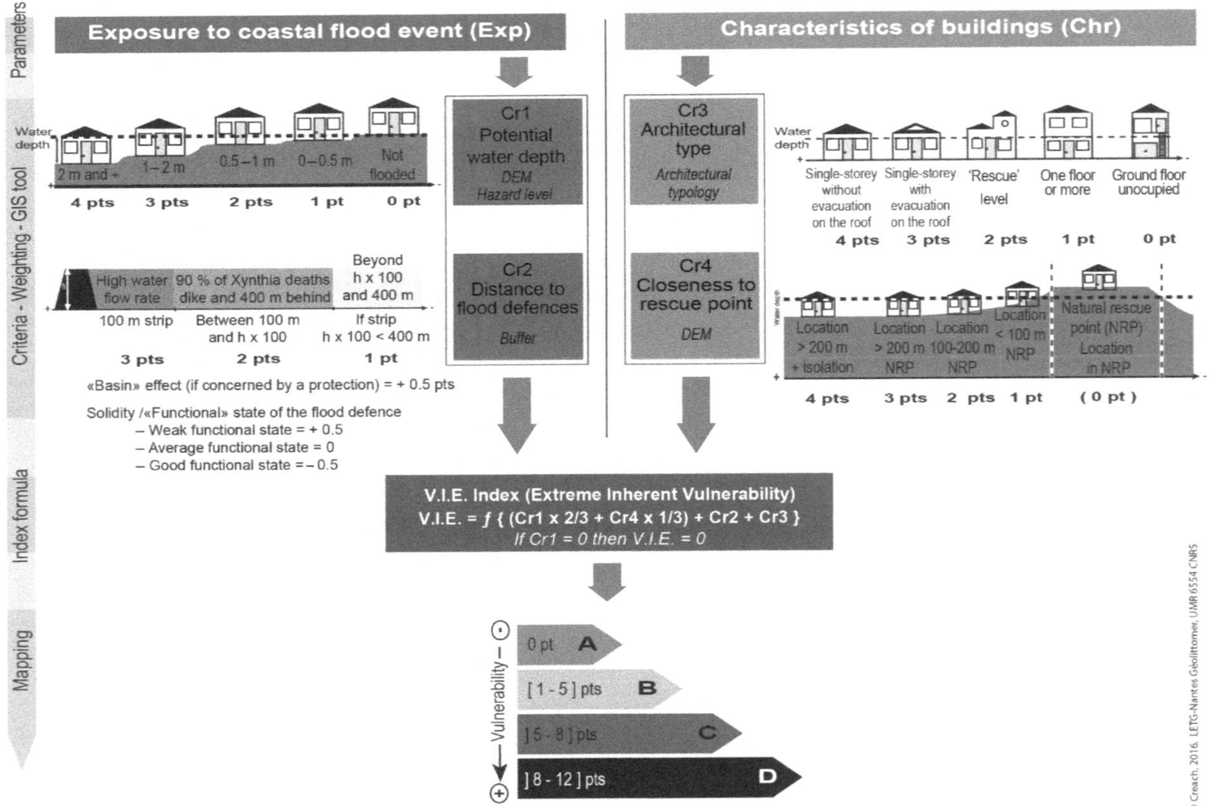

Fig. 8.1 The V.I.E. index methodology. *(Reprinted by permission from Natural Hazards/Springer, The use of a micro-scale index to identify potential death risk areas due to coastal flood surges: lessons from Storm Xynthia on the French Atlantic coast, Creach et al., © 2015).*

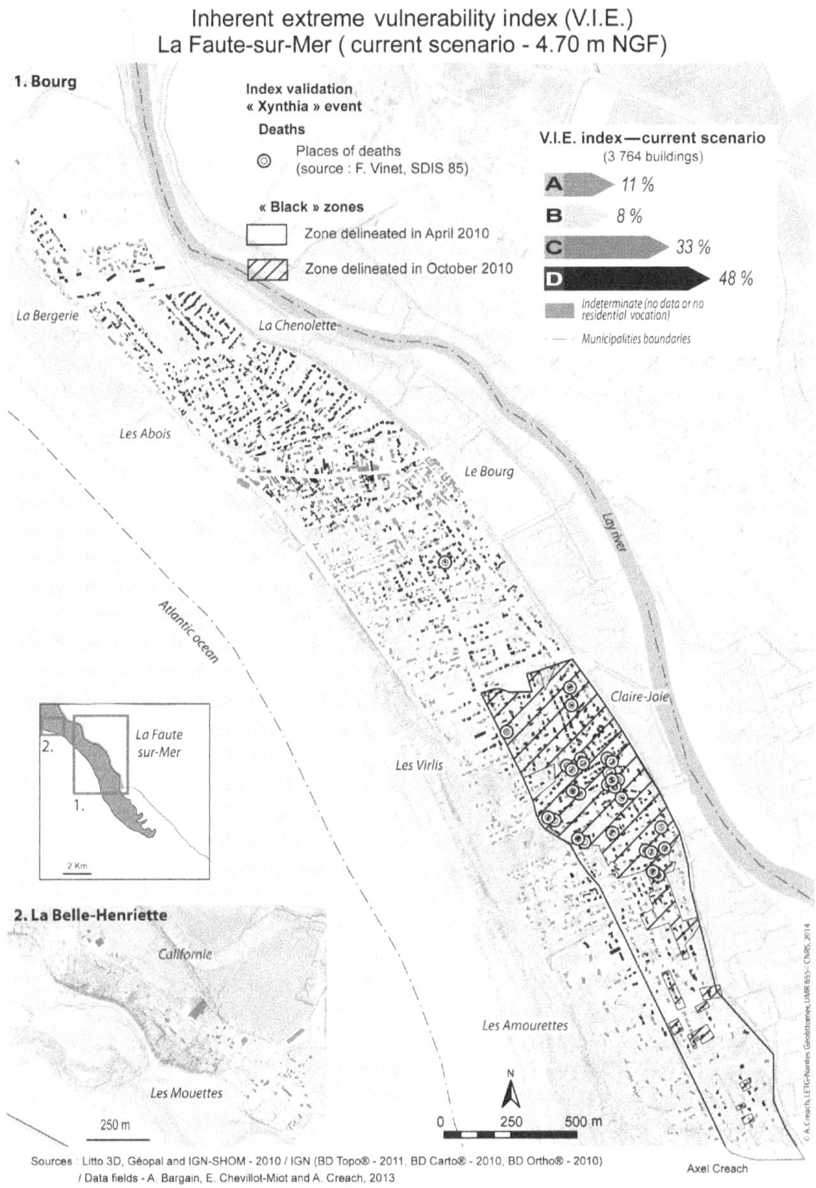

Fig. 8.2 Results of the VIE index for La Faute-sur-Mer municipality and the location of deaths linked to storm Xynthia. *(Reprinted by permission from Natural Hazards/Springer, The use of a micro-scale index to identify potential death risk areas due to coastal flood surges: lessons from Storm Xynthia on the French Atlantic coast, Creach et al., © 2015).*

identified in the Class D by the VIE index. This is explained by the fact that 85% of residential buildings are secondary houses (INSEE, 2014) and were probably unoccupied during the storm.

Fig. 8.2 also shows two areas of high vulnerability located in the south (where the deaths were located) and in the north. During Xynthia, the northern area was not subjected to high water depths thus reducing the number of fatalities compared to the southern area. However, the northern area was still identified as potentially dangerous by experts, because of its proximity to the Lay River (Pitié and Puech, 2010). Surprisingly, this north area was not included in the black zone policy (Pitié and Puech, 2010).

8.3 Adaptation Strategies

Various adaptation strategies could be implemented to reduce the vulnerability of houses during coastal flood events. Table 8.1 summarises the classification of structural adaptation strategies found in the literature which could be applied to the buildings (Klein et al., 2001; Nicholls, 2011; Paskoff, 2004). Moreover, preventive warning and evacuation could be considered as an accommodation strategy crucial to prevent fatalities (Priest et al., 2007; Lumbroso and Vinet, 2011; Kolen et al., 2013; Lumbroso and Davison, 2016). In this way, this chapter considers *protection*, *retreat* and two accommodation strategies (*housing architectural adaptation* and *preventive warning*

Table 8.1 Adaptation Strategies for Housing Subjected to Coastal Flood Risks

Adaptation Strategy	**Scheme**
Protection: Construction of a protective structure that reduces the probability of floods in a particular zone.	
Retreat: Moving an exposed building to safe zones (MEDDE, 2012a). This strategy was adopted after storm Xynthia via the "black areas policy"	
Accommodation: Minimisation of flood impact by renovating housing at the original location	

Adapted from Nicholls, R., 2011. Planning for the impacts of sea level rise. Oceanography 24(2), 144–157.

and evacuation). The following sections will describe these adaptation strategies including their costs and effects on VIE index assessment.

8.3.1 Protection: Dike Reinforcement and Raising

8.3.1.1 Objectives and Implementation

This strategy aims at artificially maintaining the position of the shoreline by the construction of dikes or protective walls that reduce the likelihood of floods (Klein et al., 2001; Nicholls, 2011; Paskoff, 2004). Nevertheless, there are three failure mechanisms of these protection structures: (i) structural failure, (ii) overtopping, and (iii) degradation. Despite criticisms regarding its high costs (Jonkman et al., 2013) and the false sense of security that they provide to people living behind these structures (Vinet, 2010), this strategy is popular in developed countries (Klein et al., 2001).

This chapter considers only the adaptation of existing protection facilities to overcome sea-level rise due to climate change. In France, since 2015 and the introduction of the GEMAPI law (GEstion des Milieux Aquatiques et Prévention des Inondations, i.e., Water Environment and Floods Prevention Management), local authorities are now responsible for management of flood defences. This poses a significant challenge to the local authorities because prior to 2010, 3000 km of existing flood defences were not managed or maintained properly (Lumbroso and Vinet, 2011).

8.3.1.2 Costs

Dollet and Sylla (2015) quantified reinforcement and raising costs for a segment of a dike located in Noirmoutier Island in France. A 120 cm raise of the dike crest was estimated as sufficient to overcome the maximum expected sea level for actual conditions of 4.20 m NGF ('nivellement général de la France', i.e., ordnance datum). This raise also includes an extra protection (80 cm) to mitigate future sea-level rise. The new altitude of the dike will be then 5 m NGF.

The total cost of this work was estimated at €500 per m of dike, plus an annual maintenance allowance equivalent to 2% of the investment (€10 per m of dike). Although this value is similar to others found in the literature (Goeldner-Gianella et al., 2015), it is important to highlight that these costs vary depending on the initial state of the structure, required dike raise, and/or availability and proximity of construction materials.

8.3.1.3 Impact on VIE Index Calculation

This strategy affects the second criterion CR2 of the VIE index which concerns the distance between flood defences and buildings. This study considers that flood defences are reinforced and therefore its probability of collapse is less important. Consequently, a value of 0.5 is subtracted to the mark of CR2 by considering that the defence is in a good functional state. In contrast, raising flood defences could increase the surprise effect in case of failure and therefore increase flood vulnerability for houses and their occupants.

8.3.2 Retreat

8.3.2.1 Objectives and Implementation

In some cases, implementation of protective strategies for high-risk exposure levels becomes very expensive and accommodation strategies cannot guarantee a significant decrease in vulnerability. Consequently, it is necessary to relocate vulnerable buildings outside of high-level hazard areas.

This measure has been previously implemented for housing and facilities placed in zones where coastal erosion is unavoidable (Klein et al., 2001). Regarding the marine flood risk, this approach is still new in France, that is, the "black areas policy" was the first example of the application of this strategy after the Xynthia event. In the context of sea-level rise due to climate change, this strategy should become widely employed in the future, as proposed in the French national strategy for shoreline management (MEDDE, 2012a).

The implementation of this type of strategy requires the purchase and destruction of the more exposed buildings and facilities. Although retreat is expensive in the short-term, it should eliminate future flood consequences. Like protection measures, this strategy also concerns mainly public authorities and is carried out by friendly takeovers or expropriations.

8.3.2.2 Costs

Housing retreat costs after the Xynthia event are not considered in this study. Therefore, the costs for this adaptation strategy are estimated based on available information as the sum of the purchase and demolition costs. The purchase costs in coastal zones are mainly influenced by land pressure that differs for each construction. To account for these variations, this study estimates a mean cost per m^2 for La Guérinière town in France from information available in three specialised websites (http://www.meilleursagents.com/prix-immobilier, http://prix-immobilier.drimki.fr, and http://Efficity.com). Data for seven towns located around La Guérinière zone indicate a mean purchase cost of

€2366/m^2 with a standard deviation of €555/m^2. The surface area of the buildings to be demolished is multiplied by this mean cost to estimate the purchase cost.

Demolition costs are estimated at €20,000 per building considering the information available in the website: http://www.renovationettravaux.fr/cout-demolition-maison-prix-devis. Considering the additional costs such as land restoration or installation in new retreat areas could complete retreat costs but they are not considered herein.

8.3.2.3 Impact on VIE Index Calculation

This strategy has no direct impact on any of the VIE index criteria. In this study, buildings identified in Class D (black) are considered too dangerous and consequently, it is supposed that they will be demolished.

8.3.3 Accommodation 1: Housing Architectural Adaptation

8.3.3.1 Objectives and Implementation

This strategy focuses on adapting buildings to reduce occupants' vulnerability to coastal flood events. The different types of solutions can be classified into two categories:

- *measures that reduce or avoid potential flood zones*: for example, the increase of the first floor level by using of stilts, construction of mounds around the building, use a floating foundation, etc.; and
- *measures that provide a safe zone in case of a flood*: for example, the addition of a refuge floor, conversion of loft, etc.

The implementation of these solutions depends on both the use and temporality of the construction. For example, the first category would be more suitable for new buildings while the second one could be more appropriate for the existing or secondary constructions. The present study only considers architectural adaptation of existing buildings. It will then focus specifically on the construction of a refuge floor, as proposed by the French repository for flood prevention in existing houses (MEDDE and METL, 2012) (Fig. 8.3).

Considering that 17% of deaths occurred in single-storey buildings during the Xynthia event (Vinet et al., 2011, 2012), the implementation of this adaptation strategy could be cost-effective. However, it should be noted that it does not completely eliminate the risks for occupants in case of a coastal flood. In some cases they could be trapped on the ground floor despite the presence of a refuge. The application of the strategy concerns mainly private individual initiatives but it may also be supported by public financial incentives.

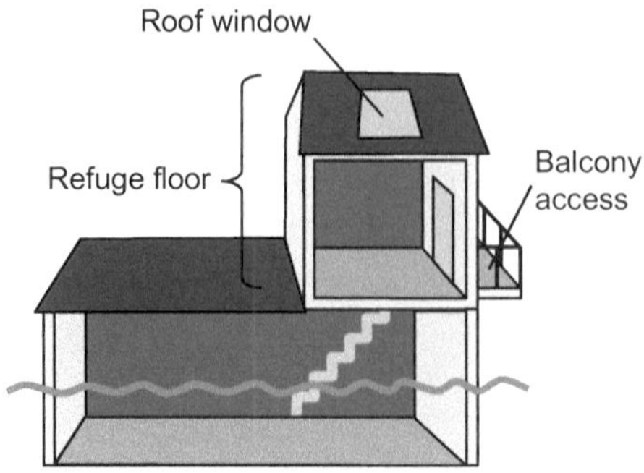

Fig. 8.3 Example of creation of a refuge floor. *(According to MEDDE and METL, 2012).*

8.3.3.2 Costs

The solution takes into account the recommendations of the French ministry for flood prevention in relation to the existing housing (MEDDE and METL, 2012). It consists of constructing a refuge floor with the following characteristics: (i) surface area varying from 9 to 20 m^2, (ii) a floor level higher than 4.2 m above the sea level, and (iii) external access (roof, window, or balcony) allowing evacuation by helicopter or boat (Fig. 8.3). On the basis of these requirements, Festuot and Guillaume (2015) estimated that the minimum construction cost of a refuge floor of 20 m^2 is €19,000. However, this cost may vary depending on the type of existing foundation and structure as well as the selection of construction materials. Potential benefits include the use of the new space as an active part of the house and the increase of the economic value of the residential building.

8.3.3.3 Impact on VIE Index Calculation

The measure of housing architectural adaptation has impacts on the third criterion CR3 of the VIE index (architectural configuration of the house). The single-storey buildings identified in red and black classes in the initial results will be adapted by building a refuge floor.

8.3.4 Accommodation 2: Preventive Warning and Evacuation

8.3.4.1 Objectives and Implementation

This strategy aims to facilitate preventive evacuation in case of an alert of a coastal flood. Kolen et al. (2013) affirmed that such an evacuation would significantly reduce human fatalities during storm Xynthia. Moreover, Chadenas et al. (2013) showed that prevention could be greatly improved for coastal floods in France; however, no real-time monitoring of coastal surges and warning was available during storm Xynthia (Lumbroso and Vinet, 2011).

Nevertheless, the effectiveness of an evacuation depends on: (i) the number of people who effectively evacuate, which can be extremely variable (Priest et al., 2007); and (ii) the time available between the beginning of the alert and the occurrence of the storm event as well as the evacuation logistics (Parker et al., 2009).

A successful warning and evacuation, in case of a coastal flood, requires people to have awareness of flood risk and preliminary preparation has been undertaken. This necessitates that appropriate guidelines are issued regarding evacuation procedures and drills conducted frequently. The difficulty with such a measure is maintaining risk awareness as an ongoing process.

This strategy could be implemented at two levels:

- *Implementation of preventive measures from population information and education*: this task could be carried out by disseminating regulatory documents for information and prevention (e.g., risk prevention plans) and conducting evacuation drills regularly. These actions would facilitate the successful evacuation of inhabitants in case of an alert.
- *Deployment of evacuation equipment and/or construction of refuge structures*: since the total evacuation of an area is unrealistic, provision of compulsory household equipment such as the 'standby emergency kit' (http://www.developpement-durable.gouv.fr/Kit-d-attente-des-secours-pendant.html) are useful to localise and provide timely help to people during a flood event. On the other hand, specific structures like shelter platforms could also be installed for use as a refuge. Various types of refuge platforms have been already implemented elsewhere (Creach, 2015).

The application of this strategy depends primarily on public authorities. Nevertheless, several private facilities such as holiday centres have already built refuge platforms (Creach, 2015).

8.3.4.2 Costs

This study considers the implementation of information and educational measures as well as the construction of a refuge shelter platform. Preventive actions are defined at the municipal level. The costs of the production and dissemination of documents and evacuation training for the Flood Prevention Action Plan for La Rochelle (France) are €357,000 (CdA La Rochelle, 2013). An update every five years is envisaged to maintain risk awareness.

The installation of the shelter platforms is compulsory in certain holiday centres and campgrounds. According to one company that constructed and installed them in France (TDEM company), the cost of a platform with a capacity of 150–200 people is about €80,000. This cost varies depending on the size and additional equipment of the platform (staircases, lifts, etc.). Finally, the cost of the 'standby emergency kit' is estimated at €250 taking into account the list of equipment recommended by the Ministry of Ecology.

8.3.4.3 Impact on VIE Index Calculation

This adaptation measure impacts the fourth criterion (CR4) of the VIE index (proximity to a rescue point to facilitate evacuation). The number of platforms to install is estimated by supposing that each house should be located at least at 200 m of a natural or artificial shelter or safe haven.

8.4 Case Studies

The main objective of this section is to illustrate how the VIE index could be used to identify houses with higher flood risk and then to use this information to estimate the cost-efficiency of the adaptation strategies. The first case study (Section 8.4.2) will focus on the assessment of cost-efficiency for the four adaptation strategies described previously. For this case study, it is supposed that the adaptation strategies are applied separately and compared for La Guérinière town. The second case study (Section 8.4.3) will illustrate how these adaptation strategies could be combined to improve its cost-efficiency for La-Faute-Sur-Mer town.

8.4.1 Basic Considerations

8.4.1.1 Considered Coastal Flood Scenarios

This study analyses the vulnerability of houses for people and costs of the adaptation strategies for four scenarios of coastal floods: frequent, medium, medium including sea-level rise in 2100, and extreme. These scenarios have been defined on the basis of studies made by the French administration to formulate and implement the Flood Directive (European Commission, 2007). Table 8.2 provides the sea water levels for each scenario estimated in (DREAL Pays de la Loire, 2014) for Fromentine town. These values are considered herein because Fromentine town is also placed in the Atlantic coast close to La Guérinière and La-Faute-Sur-Mer towns.

Table 8.2 Coastal Flood Scenarios (DREAL Pays de la Loire, 2014)

Scenario	Return Period	Sea Water Level at Fromentine Town
Frequent	10 years	3.6 m NGF
Medium	100–300 years	4.2 m NGF
Medium + SLR[a]	Same as above +0.6 m	4.8 m NGF
Extreme	1000 years or more	5.2 m NGF

[a]SLR: Sea-level rise.

8.4.1.2 Flood Modelling, Assumptions, and Limitations

Based on the sea water levels presented in Table 8.2, the potentially flooded areas were determined employing a static flood model (Breilh et al., 2013). This model uses high-resolution topographic datasets (Lidar Litto-3D) (IGN and SHOM, 2012) and considers that inland areas located below sea level are inundated within a range of 100 m of the seashore. The difference between sea level and the ground level of the house is used to estimate the potential water level in case of floods. This method is relatively simple but has two main limitations: (i) the house ground level is derived from LiDAR datasets; however Cooper et al. (2013) identified several uncertainties for LiDAR datasets particularly in marshes and urbanised areas, and (ii) it is assumed that the LiDAR ground level corresponds to the ground level of the house. If both the aspects are accounted for, there will be in some cases a difference of 20–60 cm with respect to the real ground level of the house inducing some errors for the assessment of the real water level. Consequently, infield data would be necessary to improve the accuracy of the vulnerability assessment.

8.4.1.3 Discount Rates

There is some uncertainty about the discount rate, particularly for climate change economic assessments [see Chapter 1 or Bastidas-Arteaga and Stewart, 2015, 2016 for a further discussion]. France used a discount rate of 8% to evaluate public investments from 1985 to 2005. However, following the 2005 Lebègue report (Lebègue et al., 2005), the 'Commissariat Général au Plan' has recommended a 4% discount rate for short-term investments and a lower discount rate of 2% for cash flows occurring after more than 30 years (Gollier, 2012). These discount rates were revised in 2013 by the 'Commissariat Général à la Strategie et à la Prospective' recommending 2.5% and 1.5% discount rates for short-term (lifetime less than 70 years) and long-term investments, respectively (Quinet, 2013). Quinet (2013) also recommends carrying out a sensitivity analysis with a 4.5% discount rate to compare new and old approaches. The present study considers a 2% discount rate as recommended by the 'Commissariat Général à la Strategie et à la Prospective' in 2005 in the Lebègue report (Lebègue et al., 2005).

8.4.2 Case Study 1: VIE Index for Adaptation Measures Applied Separately

8.4.2.1 Site Description

Noirmoutier is a French Atlantic island (area of 49 km^2) located in the Vendée department (Fig. 8.4). The island is in a low-lying area currently protected by 24 km of flood defences on the east coast and by a sand barrier on the west coast (Fattal et al., 2010). Without this protection, around 68% of the island would have been under water during storm Xynthia (Creach et al., 2016).

Since 1968, the population of the island has increased by 19%, meanwhile, the urbanisation has increased by +162% during the same period (Creach et al., 2017). This important growth is mainly due to the construction of leisure residential houses which represented 50% of the urbanised area in 2011 (INSEE, 2014). Leisure housing is mainly located in low-lying areas (Creach et al., 2016) and today one-half of these constructions are located in potential flood areas.

Noirmoutier island was not severely affected by storm Xynthia in 2010 with only 3% of the territory flooded (Devaux et al., 2012). Since the 1950s,

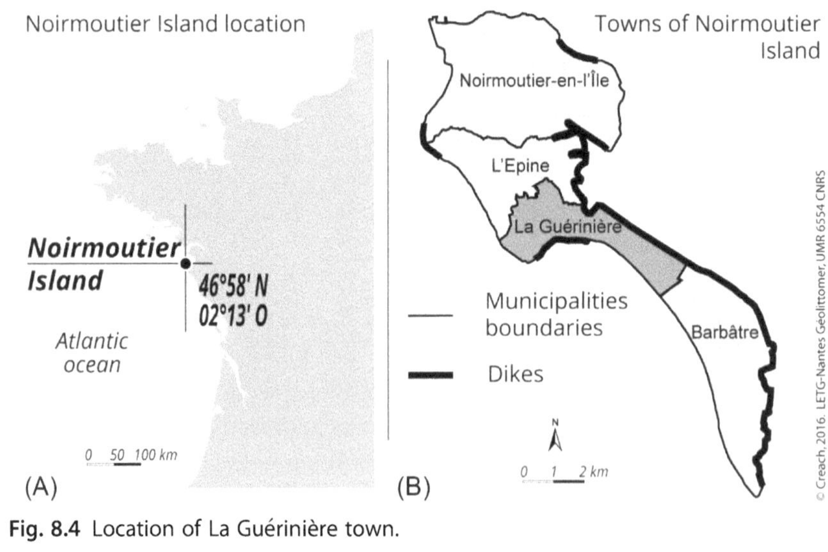

Fig. 8.4 Location of La Guérinière town.

no major floods have occurred. A large part of the island was flooded in 1937; however, the relatively unpopulated nature of the island at that time resulted in minimal impacts on human life and infrastructure (Garnier et al., 2012). Taking into account the natural risk of floods for this island, the additional risk related to sea-level rise and its present urbanisation, it is important to evaluate the potential consequences of extreme events occurring today and in the future.

This section focuses on one of the most vulnerable towns of the island called La Guérinière. La Guérinière covers 7.8 km^2 (16% of the island) with 80% of its territory located below the sea water level attained during storm Xynthia (4.20 m NGF) (Devaux et al., 2012). The town is in the narrowest part of the island (less than 800 m between the west coast and the east coast) which increases the risk of coastal floods (see Fig. 8.4). On the east coast, La Guérinière is protected by flood defences which are 5 m high (following renovation in 2014) (Fig. 8.4). On the west coast, the town is isolated from the Atlantic Ocean by a sand barrier which is high and wide on the north and the south, but narrow and less elevated in the central part of the town (approximately 5 m high and 50 m wide) and supported by a seawall.

The town had 2667 houses with 1460 inhabitants in 2011 (INSEE, 2014). The number of houses has increased by 144% since 1968 and presently 73% are located under the seawater level reached during storm Xynthia (Creach et al., 2017).

8.4.2.2 VIE Index Without Adaptation Measures

The VIE index was calculated for each flood scenario for 2817 houses identified in 2012. Results are showed in Table 8.3. As expected, vulnerability of houses for people depends on the intensity of the hazard. The proportion of non-floodable houses (Class A) decreases from 47% to 26% from the frequent to the extreme scenario. On the other hand, the proportion of houses that could expose their occupants to a fatality risk (Class D) increases from 16% to 29% from a frequent to an extreme event.

8.4.2.3 VIE Index for the Adaptation Strategies

Tables 8.4–8.7 summarise the VIE indexes for all adaptation strategies. It is noted that each adaptation strategy has a different performance depending on the flood scenario. In a global manner, retreat is the most efficient measure because all the houses considered as the most dangerous (Class D) are demolished. Therefore Table 8.4 shows that Class D is empty. However, from political, economical, and social points of view, it is harder to

Table 8.3 VIE Index without Adaptation Measures for La Guérinière

	Frequent Event (3.60 m NGF)		Medium Event (4.20 m NGF)		Medium Event + SLR (4.80 m NGF)		Extreme Event (5.20 m NGF)	
Class A	1,312	47%	1,034	37%	890	32%	690	26%
Class B	304	11%	329	12%	313	11%	267	10%
Class C	732	26%	885	31%	908	32%	911	35%
Class D	469	16%	569	20%	706	25%	758	29%
Total of identified houses	2,817	100%	2,817	100%	2,817	100%	2,626	100%
Non-identified[a]	1,251	31%	1,251	31%	1,251	31%	1,442	35%
Total	4,068	100%	4,068	100%	4,068	100%	4,068	100%

[a]Indeterminate: no data or no residential vocation.

Table 8.4 VIE Index for Retreat Adaptation Strategy for La Guérinière

	Frequent Event (3.60 m NGF)		Medium Event (4.20 m NGF)		Medium Event + SLR (4.80 m NGF)		Extreme Event (5.20 m NGF)	
Class A	1,312	56%	1,034	46%	890	42%	690	37%
Class B	304	13%	329	15%	313	15%	267	14%
Class C	732	31%	885	39%	908	43%	911	49%
Class D	0	0%	0	0%	0	0%	0	0%
Total of identified houses	2,348	100%	2,248	100%	2,111	100%	1,868	100%
Non-identified[a]	1,251	35%	1,251	36%	1,251	37%	1,442	44%
Total	3,599	100%	3,499	100%	3,362	100%	3,310	100%

[a]Indeterminate: no data or no residential vocation

Table 8.5 VIE Index for Housing Architectural Adaptation Strategy for La Guérinière

	Frequent Event (3.60 m NGF)		Medium Event (4.20 m NGF)		Medium Event + SLR (4.80 m NGF)		Extreme Event (5.20 m NGF)	
Class A	1,312	47%	1,034	36%	890	32%	690	26%
Class B	631	22%	701	25%	618	22%	588	22%
Class C	848	30%	975	35%	966	34%	906	35%
Class D	26	1%	107	4%	343	12%	442	17%
Total of identified houses	2,817	100%	2,817	100%	2,817	100%	2,626	100%
Non-identified[a]	1,251	31%	1,251	31%	1,251	31%	1,442	35%
Total	4,068	100%	4,068	100%	4,068	100%	4,068	100%

[a]Indeterminate: no data or no residential vocation.

Table 8.6 VIE Index for Protection Adaptation Strategy for La Guérinière

	Frequent Event (3.60 m NGF)		Medium Event (4.20 m NGF)		Medium Event + SLR (4.80 m NGF)		Extreme Event (5.20 m NGF)	
Class A	1,312	47%	1,034	37%	890	31%	690	26%
Class B	364	13%	382	14%	320	11%	268	10%
Class C	889	31%	1,015	35%	1,085	39%	1,106	42%
Class D	252	9%	386	14%	522	19%	562	22%
Total of identified houses	2,817	100%	2,817	100%	2,817	100%	2,626	100%
Non-identified[a]	1,251	31%	1,251	31%	1,251	31%	1,442	35%
Total	4,068	100%	4,068	100%	4,068	100%	4,068	100%

[a]Indeterminate: no data or no residential vocation.

implement when considering the destruction of 16% (frequent event) to 29% (extreme event) of residential buildings (King et al., 2014).

The second most efficient measure is housing architectural adaptation (Table 8.5). It reduces the number of dangerous houses (Class D) from 95% (for a frequent event) to 42% (for an extreme event) with respect to the reference values provided in Table 8.3. The protection measure (Table 8.6) decreases the number of dangerous houses from 46% (frequent event) to 26% (extreme event). Finally, the warning and evacuation measures appear to be less efficient (Table 8.7), but their efficiency increases with the intensity of the coastal flood hazard. For example, for a frequent event, it reduces the quantity of Class D houses by 2.3%, whereas the reduction reaches 14% for an extreme event.

Table 8.7 VIE Index for Preventive Warning and Evacuation Adaptation Strategy for La Guérinière

	Frequent Event (3.60 m NGF)		Medium Event (4.20 m NGF)		Medium Event + SLR (4.80 m NGF)		Extreme Event (5.20 m NGF)	
Class A	1,312	47%	1,034	37%	890	32%	690	26%
Class B	304	11%	362	13%	378	13%	314	12%
Class C	743	26%	914	32%	931	33%	970	37%
Class D	458	16%	507	18%	618	22%	652	25%
Total of identified houses	2,817	100%	2,817	100%	2,817	100%	2,626	100%
Non-identified[a]	1,251	31%	1,251	31%	1,251	31%	1,442	35%
Total	4,068	100%	4,068	100%	4,068	100%	4,068	100%

[a]Indeterminate: no data or no residential vocation.

8.4.2.4 Cost-Efficiency of Adaptation Strategies

Costs of adaptation (C_{adapt}) are also calculated including investment and maintenance costs for a lifetime of 100 years. The reduction in risk caused by climate adaptation strategies (ΔR) is estimated in terms of the ability of the adaptation measure to reduce the number of dangerous houses in Class D as:

$$\Delta R = \frac{nb_{\text{with adap}} - nb_{\text{without adap}}}{nb_{\text{without adap}}} \times 100\% \tag{8.2}$$

where $nb_{\text{with adap}}$ and $nb_{\text{without adap}}$ are respectively the number of houses in Class D with and without considering the adaptation strategy.

Fig. 8.5 shows costs and corresponding risk reduction for the considered adaptation strategies and flood scenarios for La Guérinière. It is observed that the costs of adaptation strategies are very different: from about €2 million (for preventive warning and evacuation) to more than €100 million (for retreat). Protection is around €4 million while housing architectural adaptation is between €16 and €22 million.

Considering the risk reduction, it can be noted that retreat is the most efficient adaptation strategy (ΔR = 100%) for all flood scenarios. The efficiency of housing architectural adaptation and protection strategies decreases for larger hazard intensities. The ΔR for housing architectural adaptation diminishes from 95% to 42%. This means that this strategy is appropriate for dealing with

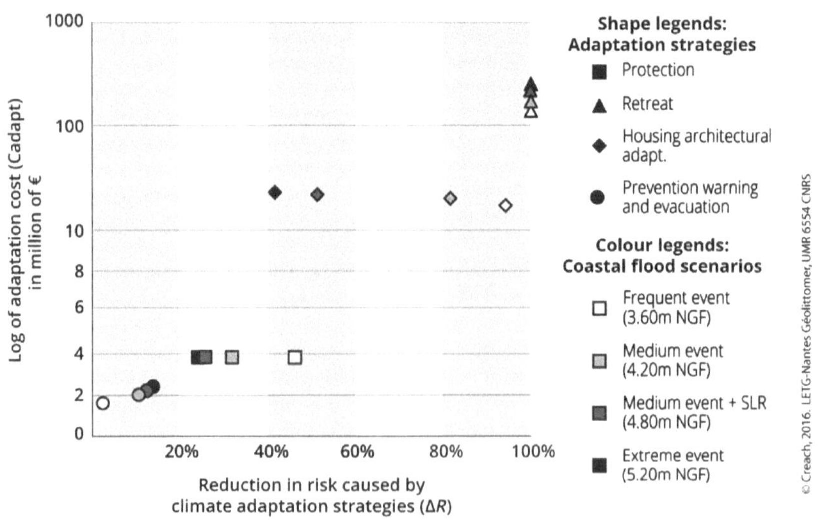

Fig. 8.5 Adaptation costs and corresponding risk reduction for adaptation strategies applied to La Guérinière.

current flood scenarios but it is not effective for reducing future risks related with climate change. The efficiency of the protection strategy is smaller than 50% for frequent events and could decrease up to 23% for extreme events. For the preventive warning and evacuation strategy ΔR is very low but increases from 2% (frequent scenario) to 14% (extreme scenario). This could be explained by the fact that additional refuge platforms are implemented for extreme events. This will reduce the distance between exposed population and houses and then increase the efficiency of the measure. Overall results show that, independently from the flood scenario, the strategies could answer to different political choices: saving money (economical choice, it concerns protection) to safety first (safety choice, it concerns retreat).

8.4.3 Case Study 2: VIE Index for Combined Adaptation Measures

8.4.3.1 Site Description

Hinkel et al. (2013) stated that climate change adaptation cannot be based on a unique solution but on a mix of strategies. In Section 8.4.2 we found that each strategy has a different cost and risk reduction performance. Nevertheless, what would be the efficiency of preventive warning and evacuation and housing architectural adaptation together to protect humane life? Or would this mix of strategies reduce the number of houses to be relocated? To provide some partial responses to these challenges, this section includes combined adaptation strategies for La Faute-sur-Mer town.

La Faute-sur-Mer is located on a small strip of coastal land and 80% of its 6.8 km^2 territory could potentially be flooded (Creach, 2015); and some low-lying areas are protected from the Lay River by a dike (Fig. 8.2). In addition, urbanised areas have grown by 262% between 1968 and 2006 (Creach, 2015). In 2006, there were 3737 residential buildings (INSEE, 2014) among which 85% are leisure residential houses (INSEE, 2014) and 70% are single-storey houses (Creach, 2015).

La Faute-sur-Mer was the most impacted municipality by storm Xynthia (29 people died). After storm Xynthia, the French government bought and demolished 552 houses for a total cost of €160 million (Cour des Comptes, 2012). This 'black zones' policy could be referred to as a retreat of the most dangerous buildings (Class D).

8.4.3.2 Results of VIE Index

Fig. 8.2 showed that about 50% (1305 over 2695 houses) of the houses in La Faute-sur-Mer could expose their occupants to extreme vulnerability

(Class D) in case of coastal flood similar to Xynthia. To reduce this vulnerability, the four adaptation strategies presented in Section 8.3 were considered. This analysis also includes the implemented 'black zones' policy and two alternatives mixing the adaptation strategies proposed above:

- Alternative 1: it combines preventive warning and evacuation and housing architectural adaptation. The objective is to avoid the retreat of houses in the flood prone area but facilitating evacuation of people during a flood.
- Alternative 2: it combines preventive warning and evacuation, housing architectural adaptation and retreat. It is the same scenario described above but the most dangerous houses are relocated. The goal is to minimise the risk of death for Class D houses in cases where housing architectural adaptation is not efficient.

The VIE index for each adaptation strategy and a medium coastal flood hazard is given in Table 8.8. It is observed that adaptation strategies decrease the percentage of houses in Class D from 48% up to 0%. The higher reductions correspond to the cases of retreat and both alternatives combining the different adaptation strategies.

8.4.3.3 Cost-Efficiency of Adaptation Strategies

Costs of adaptation (C_{adapt}) and the reduction in risk caused by climate adaptation strategies (ΔR) are estimated by following the same considerations given in Section 8.4.2.4. The results of the analysis are presented in Table 8.9. It is noted that protection and preventive warning and evacuation are the less interesting measures because its risk reduction is less than 7% for an investment of €3.9 million. Retreat is the most expensive measure (€320.5 million), two times higher than 'black zones' policy (€160 million), but with a $\Delta R = 100\%$. This could be explained by the fact that 'black zones' policy concerned only the south part of La Faute-sur-Mer while there are also dangerous houses on the north part of the town (see Fig. 8.2). Housing architectural adaptation provides a ΔR higher than 70% with a cost 10 times cheaper than retreat.

Concerning combined adaptation strategies, Alternative 1 is 11% more expensive than architectural adaptation alone but provides a larger ΔR compared to this single strategy. Alternative 2 provides the same ΔR than retreat but is 76% less expensive. These results indicate that combining adaptation strategies is paramount to reduce the adaptation costs and increase the risk reduction.

Table 8.8 VIE Index for Different Adaptation Strategies, Including Two Alternative Scenarios, on La Faute-sur-Mer Municipality for a Medium Coastal Flood

	VIE Index (Medium Event–4.70 m NGF)		'Black Zones' Policy		Protection		Housing Architectural Adaptation		Preventive Warning and Evacuation		Retreat		Alternative 1		Alternative 2	
Class A	289	11%	289	13%	289	11%	289	11%	289	11%	289	21%	289	11%	289	11%
Class B	216	8%	216	10%	172	6%	399	15%	230	9%	216	16%	422	16%	422	17%
Class C	885	33%	749	35%	1,016	38%	1,626	60%	920	34%	885	64%	1,811	67%	1,811	72%
Class D	1,305	48%	891	42%	1,218	45%	381	14%	1,256	47%	0	0%	173	6%	0	0%
Total of identified houses	2,695	100%	2,145	100%	2,695	100%	2,695	100%	2,695	100%	1,390	100%	2,695	100%	2,522	100%
Non-identified[a]	1,069	28%	964	31%	1,069	28%	1,069	28%	1,069	28%	1,069	43%	1,069	28%	1,069	30%
Total	3,764	100%	3,109	100%	3,764	100%	3,764	100%	3,764	100%	2,459	100%	3,764	100%	3,591	100%

[a]Indeterminate: no data or no residential vocation.

Adapted from Creach, A., 2015. "Cartographie et analyse économique de la vulnérabilité du littoral atlantique français face au risque de submersion marine (2 volumes)." Université de Nantes.

Table 8.9 Adaptation Costs and Corresponding Risk Reduction for Adaptation Strategies Applied to La Faute-sur-Mer Under a Medium Coastal Flood

Adaptation Strategy	Adaptation Cost (in € million)	Δ*R* (%)
'Black zones' policy	160	31.7
Protection	3.9	6.7
Housing architectural adaptation	31.9	70.8
Preventive warning and evacuation	3.9	3.7
Retreat	320.5	100
Alternative 1	35.7	86.7
Alternative 2	76.4	100

8.5 Summary

Sea-level rise and vulnerability of the existing residential buildings would increase the risk of death in case of future floods. To prevent these fatalities, adaptation is an essential solution. However, the cost and efficiency of available adaptation strategies are difficult to evaluate. Therefore, this chapter presented a global framework that aims to assess vulnerability of buildings for people under different flood scenarios including climate change. It is based on the VIE index and is useful to identify buildings which need to be adapted in priority. Four different adaptation strategies were illustrated in the chapter; protection, retreat, housing architectural adaptation, and preventive warning and evacuation. The proposed methodology was used to compare their costs and risk reduction.

Two case studies showed that retreat is one the most efficient but expensive alternatives to reduce vulnerability of buildings for people. In addition, it was observed that a combination of adaptation strategies could reduce risks and adaptation costs. The results highlighted that the proposed framework could be used as a decision-making tool to compare or optimise the performance and costs of adaptation strategies under various flood scenarios.

8.6 Potential Design and Practice Evolutions

Adaptation of the coastal areas subjected to flood risks is not an easy task for decision-makers. For example, after storm Xynthia, the lack of standardised procedures including a comprehensive methodology to evaluate and compare various adaptation strategies conducted to the implementation of a costly, ineffective and criticised 'black zones' policy.

The standardised procedures should include building adaptation to reduce vulnerability for people in a context of sea-level rise due to climate change. Building adaptation could reduce the vulnerability of buildings but there will be a remaining risk. For instance, 5 of the 41 deaths related to storm Xynthia occurred inside multistorey houses where people were trapped at the ground floor (Vinet et al., 2011).

Human factors including flood risk awareness and acceptability should also be considered when formulating adaptation strategies. The acceptability of the 'black zones' policy for the local population is an example in France where people rejected to leave their houses after the storm Xynthia (Mercier and Chadenas, 2012). Acceptability could be created by explaining to local communities about the potential consequences of flood for various scenarios and proposing to them several adaptation strategies to deal with.

A final aspect that should be considered when formulating 'acceptable' adaptation strategies is to propose alternative economical models. Nowadays governments or insurers provide the main source of funding to implement adaptation measures after a flood event. However, one could imagine that all the involved actors (governments, insurers and owners) participate preventively for cofunding the implementation of adaptation strategies. For instance, a cofunding scheme with financial incentives from the government and reduction of insurance fees could motivate the owners to invest in the construction of a refuge floor. This solution in turn will increase the living space of the house generating cobenefits for the owner.

8.7 Open Research Questions

The proposed methodology is a useful decision-making tool to reduce the vulnerability of buildings and to protect people from future floods. To be fully operational the framework needs to be improved in the following directions:

- Consideration of dynamic effects on flood modelling and ageing and failure mechanisms for defences.
- Quantification and propagation of uncertainties related to the hazard, the adaptation strategies and the cost-efficiency analysis.
- Calibration of the VIE index with databases different from the storm Xynthia. For instance, databases in the Netherlands (Di Mauro and Lumbroso, 2008), UK (Gerritsen, 2005) and the USA (Jonkman et al., 2009)

- Consideration of acceptability on the formulation of adaptation strategies.
- Formulation of innovative funding schemes where all the involved actors could participate actively and preventively in the whole process.

Acknowledgements

We would like to thank the following persons and institutions: Prof. Freddy Vinet from Université Montpellier III for providing data regarding the circumstances under which fatalities occurred during Storm Xynthia; the Departmental Service of Fire and Rescue of Vendée and Charente Maritime for helpful discussions; and all those who helped us to collect data in the field, especially the researchers of LETG UMR CNRS 6554, and the "Communauté de Communes de l'île de Noirmoutier". This research work was supported by the COSELMAR research program funded by the Regional Council of Pays de la Loire (2012–2017): www.coselmar.fr.

References

Bastidas-Arteaga, E., Stewart, M.G., 2015. Damage risks and economic assessment of climate adaptation strategies for design of new concrete structures subject to chloride-induced corrosion. Struct. Saf. 52, 40–53.

Bastidas-Arteaga, E., Stewart, M.G., 2016. Economic assessment of climate adaptation strategies for existing reinforced concrete structures subjected to chloride-induced corrosion. Struct. Infrastruct. Eng. 12 (4), 432–449.

Breilh, J.-F., Chaumillon, E., Bertin, X., Gravelle, M., 2013. Assessment of static flood modeling techniques: application to contrasting marshes flooded during Xynthia (western France). Nat. Hazards Earth Syst. Sci. 13 (6), 1595–1612.

CdA La Rochelle, 2013. Programme d'Action de Prévention des Inondations «Agglomération Rochelaise». In: Communauté d'agglomération de La Rochelle.

CETMEF, CETE Méditerranée, CETE Ouest, 2009. Vulnérabilité du territoire National aux risques littoraux. CETMEF/DLCE.

Chadenas, C., Creach, A., Mercier, D., 2013. The impact of storm Xynthia in 2010 on coastal flood prevention policy in France. J. Coast. Conserv. 18 (5), 529–538.

Cooper, H.M., Fletcher, C.H., Chen, Q., Barbee, M.M., 2013. Sea-level rise vulnerability mapping for adaptation decisions using LiDAR DEMs. Prog. Phys. Geogr. 1–22.

Cour des Comptes. (2012). Les enseignements des inondations de 2010 sur le littoral atlantique (Xynthia) et dans le Var. Rapport public thématique, Cours des Comptes, 299.

Creach, A., 2015. Cartographie et analyse économique de la vulnérabilité du littoral atlantique français face au risque de submersion marine (Ph.D. thesis) (2 volumes). Université de Nantes.

Creach, A., Pardo, S., Guillotreau, P., Mercier, D., 2015. The use of a micro-scale index to identify potential death risk areas due to coastal flood surges: lessons from storm Xynthia on the French Atlantic coast. Nat. Hazards 77 (3), 1679–1710.

Creach, A., Chevillot-Miot, E., Mercier, D., Pourinet, L., 2016. Vulnerability to coastal flood hazard of residential buildings on Noirmoutier Island (France). J. Maps 12 (2), 371–381.

Creach, A., Pardo, S., Mercier, D., 2017. Diagnostic préventif de la vulnérabilité des constructions résidentielles pour leurs occupants face au risque de submersion marine appliqué à l'île de Noirmoutier (Vendée, France). VertigO - la revue électronique en sciences de l'environnement. 17(1). https://doi.org/10.4000/vertigo.18457.

Devaux, E., Désiré, G., Boura, C., Lowenbruck, J., Bérenger, N., Rouxel, N., Romain, N., 2012. La tempête Xynthia du 28 février 2010—Retour d'expérience en Loire-Atlantique et Vendée-Volet hydaulique et ouvrages de protection. Retour d'expérience, CETE-Ouest/ DREAL Pays-de-la-Loire/DDTM Loire-Atlantique/ DDTM Vendée. 78.
Di Mauro, M., Lumbroso, D.M., 2008. Hydrodynamic and Loss of Life Modelling for the 1953 Canvey Island Flood. Keble College, Oxford, UK, pp. 1117–11126.
Dollet, V., and Sylla, O. (2015). Techniques de renforcement et coûts d'adaptation des digues face à des nouveaux risques en lien avec le changement climatique—Application au cas de l'Ile de Noirmoutier. Master thesis, Université de Nantes - M2 GC CRB, Saint-Nazaire, 53.
DREAL Pays de la Loire, 2014. Rapport de présentation de la cartographie du risque de submersion marine sur le secteur Noirmoutier - Saint Jean de Monts (Directive inondations Bassins Loire-Bretagne). Rapport final, DREAL Pays de la Loire, p. 123.
European Commission/Commission Européenne, 2007. Directive 2007/60/CE du Parlement Européen et du Conseil du 23 octobre 2007 relative à l'évaluation et à la gestion des risques d'inondation.
Fattal, P., Robin, M., Paillart, M., Maanan, M., Mercier, D., Lamberts, C., Costa, S., 2010. Effets des tempêtes sur une plage aménagée et à forte protection côtière: la plage des Éloux (côte de Noirmoutier, Vendée, France). Norois 215 (2), 101–114.
Festuot, R., and Guillaume, L. (2015). R&D: Etude technico économique sur la réduction de la vulnérabilité des constructions de plain pied soumise à des risques de submersions. Master thesis, Université de Nantes—M2 GC CRB, Saint-Nazaire, 56.
Garnier, E., Henry, N., Desarthe, J., 2012. Visions croisées de l'historien et du courtier en réassurance sur les submersions: recrudescence de l'aléa ou vulnérabilisation croissante? Gestion des risques naturels—Leçons de la tempête Xynthia. Quae, Versailles, 23.
Gerritsen, H., 2005. What happened in 1953? The big flood in the Netherlands in retrospect. Philos. Transact. A Math. Phys. Eng. Sci. 363 (1831), 1271–1291.
Goeldner-Gianella, L., 2013. *Dépoldériser en Europe occidentale. Pour une géographie et une gestion intégrées du littoral.* «Territoires en mouvement». Editions du CNRS, Paris.
Goeldner-Gianella, L., Bertrand, F., Oiry, A., Grancher, D., 2015. Depolderisation policy against coastal flooding and social acceptability on the French Atlantic coast: The case of the Arcachon Bay. Ocean Coast. Manag. 116, 98–107.
Gollier, C., 2012. Pricing the Planet's Future: The Economics of Discounting in an Uncertain World. Princeton University Press, New Jersey.
Hinkel, J., van Vuuren, D.P., Nicholls, R.J., Klein, R.J.T., 2013. The effects of adaptation and mitigation on coastal flood impacts during the 21st century. An application of the DIVA and IMAGE models. Clim. Change 117 (4), 783–794.
IGN, SHOM, 2012. Litto3D®—v 1.0—Spécifications techniques. IGN/SHOM, p. 16.
INSEE, 2014. Recensement général de la population 2011. (Publié le 26 juin 2014). Insee).
IPCC, 2013. Climate change 2013: the physical science basis. In: Stocker, T.F., Qin, D., Plattner, G.-K., Tignor, M., Allen, S.K., Boschung, J., Nauels, A., Xia, Y., Bex, V., Midgley, P.M. (Eds.), Contribution of Working Group I to the Fifth Assessment Report of the Intergovernmental Panel on Climate Change. Cambridge University Press, Cambridge, United Kingdom and New York, NY, USA.
Jonkman, S.N., Hillen, M.M., Nicholls, R.J., Kanning, W., van Ledden, M., 2013. Costs of adapting coastal Defences to sea-level rise- new estimates and their implications. J. Coast. Res. 29 (5), 1212–1226.
Jonkman, S.N., Maaskant, B., Boyd, E., Levitan, M.L., 2009. Loss of life caused by the flooding of New Orleans after hurricane Katrina: Analysis of the relationship between flood characteristics and mortality. Risk Anal. 29 (5), 676–698.

King, D., Bird, D., Haynes, K., Boon, H., Cottrell, A., Millar, J., Okada, T., Box, P., Keogh, D., Thomas, M., 2014. Voluntary relocation as an adaptation strategy to extreme weather events. International Journal of Disaster Risk Reduction 8, 83–90.

Klein, R.J.T., Nicholls, R.J., Ragoonaden, S., Capobianco, M., Aston, J., Buckley, E.N., 2001. Technological options for adaptation to climate change in coastal zones. J. Coast. Res. 17 (3), 531–543.

Kolen, B., Slomp, R., Jonkman, S.n., 2013. The impacts of storm Xynthia February 27–28, 2010 in France: Lessons for flood risk management. J. Flood Risk Manage. 6 (3), 261–278.

Lebègue, D., Baumstark, L., Hirtzman, P., 2005. Révision du taux d'actualisation des investissements publics. Commissariat Général du Planp. 112.

Lumbroso, D., Davison, M., 2016. Use of an agent-based model and Monte Carlo analysis to estimate the effectiveness of emergency management interventions to reduce loss of life during extreme floods. J. Flood Risk Manage. 16.

Lumbroso, D., Vinet, F., 2011. A comparison of the causes, effects and aftermaths of the coastal flooding of England in 1953 and France in 2010. Nat. Hazard. Earth Syst. Sci. 11 (8), 2321–2333.

MEDDE, 2012a. Stratégie nationale de gestion intégrée du trait de côte—Vers la relocalisation des activités et des biens. Ministère de l'Écologie, du Développement durable et de l'Énergie, p. 11.

MEDDE. Première évaluation nationale des risques d'inondation: Principaux résultats—EPRI 2011. MEDDE, 2012b, 9.

MEDDE, METL, 2012. Référentiel de travaux de prévention du risque d'inondation dans l'habitat existant, 81.

MEDDTL, 2011. Circulaire du 27 juillet 2011 relative à la prise en compte du risque de submersion marine dans les plans de prévention des risques naturels littoraux.

Mercier, D., Chadenas, C., 2012. La tempête Xynthia et la cartographie des «zones noires» sur le littoral français: analyse critique à partir de l'exemple de La Faute-sur-Mer (Vendée). Norois 222 (1), 45–60.

Nicholls, R., 2011. Planning for the impacts of sea level rise. Oceanography 24 (2), 144–157.

Parker, D.J., Priest, S.J., Tapsell, S.M., 2009. Understanding and enhancing the public's behavioural response to flood warning information. Meteorol. Appl. 16 (1), 103–114.

Paskoff, R.P., 2004. Potential implications of sea-level rise for France. J. Coast. Res. 20 (2), 424–434.

Pitié, C., Bellec, P., Maillot, H., Nadeau, J., Puech, P., 2011. Expertise des zones de solidarité Xynthia en Charente-Maritime. Rapport principal, CGEDD/MEEDDM, p. 186.

Pitié, C., Puech, P., 2010. Expertise complémentaire des zones de solidarité délimitées en Vendée suite à la tempête Xynthia survenue dans la nuit du 27 au 28 février 2010. CGeDD/MEEDDM, 80.

Priest, S., Wilson, T., Tapsell, S., Penning-Rowsell, E., Viavattene, C., Fernandez-Bilbao, A., 2007. Building a model to estimate Risk to Life for European flood events. Rapport Final, Middlesex University, p. 185.

Quinet, E., 2013. L'évaluation socioéconomique des investissements publics. Commissariat Général à la Strategie et à la Prospective, Paris, France.

Vinet, F., 2010. Le risque inondation. Diagnostic et gestion. Sciences du risque et du danger. Tec & Doc Lavoisier, Paris.

Vinet, F., Boissier, L., Defossez, S., 2011. La mortalité comme expression de la vulnérabilité humaine face aux catastrophes naturelles: deux inondations récentes en France (Xynthia, Var, 2010). VertigO-la revue électronique en sciences de l'environnement 11 (2), 28.

Vinet, F., Lumbroso, D., Defossez, S., Boissier, L., 2012. A comparative analysis of the loss of life during two recent floods in France: The sea surge caused by the storm Xynthia and the flash flood in Var. Nat. Hazards 61 (3), 1179–1201.

CHAPTER NINE

Climate Adaptation for Housing in Hurricane Regions

Sigridur Bjarnadottir*, Yue Li†, Mark G. Stewart‡
*VSÓ Consulting, Reykjavik, Iceland
†Case Western Reserve University, Cleveland, OH, United States
‡Centre for Infrastructure Performance and Reliability, School of Engineering, The University of Newcastle, Callaghan, NSW, Australia

9.1 Introduction

Hurricanes are one of the most expensive natural hazards to affect the United States. The 2005 Atlantic hurricane season was the costliest hurricane season in the US history, estimated to have caused $150 billion in damages (Pielke et al., 2008). The 2017 hurricane season surpassed previous seasons, causing over $200 billion in damages, mainly due to the devastating effects of Hurricanes Harvey, Irma, and Maria (Bloomberg Business, 2017). These are significant increases from damages in previous years, which were estimated to be $6 billion annually from 1989 to 1995 (Pielke and Pielke, 1997). The coastal population in the United States continues to grow despite this increase in hurricane damage losses. Approximately 50% of the US population lives in coastal regions (Culliton, 1998), and it is estimated that the populations of these coastal regions will increase by about 25% in the next 25 years (Boesch et al., 2000).

Sea surface temperatures (SST) are acknowledged to be a result of the global climate change (IPCC, 2007). There are varying classes of thought as to whether or not rising SST will alter hurricane intensity and/or frequency. First, there are studies that show a direct connection between altered hurricane activity in the Atlantic Ocean and rising SST (Elsner et al., 2008; Emanuel, 2005; White House, 2009). Estrada et al. (2015) estimate that 2%–12% of damage costs from the 2005 Atlantic hurricane season are directly attributed to rising SST. Other studies have found that the altered hurricane activity in the Atlantic Ocean is not the product of rising SST, but rather due to the natural climate variability (Klotzbach, 2006;

https://doi.org/10.1016/B978-0-12-816782-3.00009-7

Landsea et al., 2006; Landsea, 2007; Pielke et al., 2005). Independent of whether or not rising SST alters hurricane activity, changing hurricane patterns have led to increased damage costs.

Hurricane risk assessment models must be able to account for a nonstationary wind speed to accurately estimate damage costs. However, existing models assume that wind speeds are, in fact, stationary over periods of time (Stewart et al., 2003; Li and Ellingwood, 2006). Stewart and Li (2010) and Stewart et al. (2014) developed a framework for tropical cyclone risk assessment in Queensland, Australia. The framework assessed the impacts the changing global climate may have on damage costs and evaluated cost-effectiveness of various adaptation strategies. Bjarnadottir et al. (2011) developed a hurricane risk assessment framework for the United States considering climate change in hurricane damage estimates.

The chapter proposes a hurricane risk assessment framework for single-family residential housing in the United States. Chapter 1 highlights the importance of risk-based decision support in infrastructure assessments. The proposed framework is capable of considering various scenarios of climate change in its damage assessment. Furthermore, this chapter identifies various adaptation strategies that may aid to mitigate hurricane damage costs, and explores the cost-effectiveness of these adaptation strategies considering climate change. The purpose of the study is not to examine whether there is direct relationship between climate change and change patterns of wind hazard, nor to endorse any specific scenario of climate change (or lack thereof).

Wind speed will be used as the control variable for hurricane intensity as a starting point to look into this complex problem, consistent with appropriate hurricane wind and vulnerability models. The economic losses are determined by convolving the hurricane wind model and the hurricane vulnerability model. Emanuel (2005) and Elsner (2006) estimate that for every increase in SST of 1 °C, the peak wind speed of hurricanes could increase by approximately 5%. Knutson et al. (2010) estimates that hurricane wind speeds may increase by 2%–11% in the 21st century. In addition, it is important to explore the scenario of regional decrease in wind speed to account for the high variation in wind speed. Pryor et al. (2009) suggested that a decrease in annual mean wind speeds is possible. See Chapter 1 for more discussion on the potential impacts of climate change on infrastructure. Herein, a range of potential changes in wind speed (−5% to 10%) will be used to assess the impact of climate change on hurricane damage risks.

9.2 Hurricane Risk Assessment Framework

9.2.1 Hurricane Wind Models Considering Impact from Climate Change

The Weibull distribution is an appropriate model of the peak annual 3-s gust wind speed, at a height of 10 m on open terrain for hurricanes in the United States (Li and Ellingwood, 2006). Climate change may cause a change in hurricane wind speed; these dynamic changes should be reflected in hurricane wind speed models.

When nonstationary wind speed is assumed due to climate change, probability density function (PDF) of the Weibull distribution is:

$$f_v(v,t) = \frac{\alpha(t)}{u(t)} \cdot \left(\frac{v}{u(t)}\right)^{\alpha(t)-1} \exp\left[-\left(\frac{v}{u(t)}\right)^{\alpha(t)}\right] \tag{9.1}$$

where v is the 3-s gust wind speed, α and u are time-dependent parameters for the Weibull distribution, and t is time (years).

Using wind contour maps provided by Vickery et al. (2009), the Weibull parameters can be determined for any location in the United States. Subsequently, the wind speed can be modelled. For example, the current (year 1) Weibull parameters for Miami-Dade County, Florida, are $u(1)=27.36$ and $\alpha(1)=1.77$. The parameters correspond to a mean maximum annual wind speed of 24.3 m/s. The coefficient of variation (COV) of the maximum annual wind speed is 0.584.

As there is no projection on how the change of wind speed will occur overtime, it is assumed that the wind speed over a chosen time period (e.g., 50 years) changes linearly regardless of climate change scenario (Bjarnadottir et al., 2011). For example, if it is assumed that wind speed increases by 10% over a 50-year time period, the mean maximum annual wind speed increases from 24.3 m/s in year 1 to 26.7 m/s in year 50. The corresponding Weibull parameters at year 50 are $u(50)=30.0$ and $\alpha(50)=1.77$. The COV of 0.584 in year 1 was assumed to be constant for the time period.

9.2.2 Hurricane Vulnerability Models

A hurricane vulnerability model estimates the building damage caused by hurricanes. Various vulnerability models have been developed (Leicester et al., 1979; Sparks et al., 1994; Khanduri and Morrow, 2003). Huang et al. (2001)

developed a damage model for single-family housing units using Southeastern US insurance data from Hurricanes Hugo and Andrew. This particular model predicts the damage ratio (in percentage) which is the total amount paid by the insurer vs the total insured value. To develop the vulnerability model, herein, the damage ratio is related to the recorded average mean surface speed of the hurricane.

Huang's damage ratio model, in percentage, given as:

$$F_D(v_{\text{mean}}) = \exp(0.252v_{\text{mean}} - 5.823) \quad v_{\text{mean}} \leq 41.1\,\text{m/s} \tag{9.2}$$

$$F_D(v_{\text{mean}}) = 100 \quad v_{\text{mean}} \geq 41.1\,\text{m/s} \tag{9.3}$$

where v_{mean} is the mean surface wind speed in m/s.

9.2.3 Annual and Cumulative Damage

The expected annual damage can be determined by convolving the vulnerability function and the PDF of wind field model (Li and Ellingwood, 2006). The expected cumulative damage, $D(v, t)$, is found by summing the expected annual damage for a defined time frame:

$$D(v, t) = \int F_D(v_{\text{mean}}) \cdot c_1 c_2 \cdot f_v(v, t)\,dv \tag{9.4}$$

where c_1 and c_2 are conversion factors. The model in Huang et al. (2001) was developed based on the mean surface wind speed with a 10-min duration. The wind contour maps in ASCE-7 are for 3-s gust. Therefore, a conversion factor of $c_2 = 0.7$ was used to adjust the 3-s gust stipulated in wind speed maps to surface wind speed for all the exposure categories, according to ASCE-7 (2007).

The location of construction has an impact on how vulnerable construction is to wind damage. Therefore, Huang et al. (2001) identified three exposure categories: Within 1 km from coast, with a gradient-to-surface conversion factor of $c_1 = 0.90$ (Exposure Category 1); locations within 10 km inland, with a gradient-to-surface conversion factor of $c_1 = 0.80$ (Exposure Category 2); and all other locations further inland, with a gradient-to-surface conversion factor of $c_1 = 0.72$ (Exposure Category 3). These factors were accounted for in the damage model herein.

- Exposure Category 1: $c_1 = 0.90$, $c_2 = 0.70$
- Exposure Category 2: $c_1 = 0.80$, $c_2 = 0.70$
- Exposure Category 3: $c_1 = 0.72$, $c_2 = 0.70$

Huang's damage model and the Weibull distribution of maximum annual wind speeds are used to illustrate the probabilistic framework of estimating the damage hurricanes with the potential impact of climate change. The analyses conducted herein are primarily aimed at assessing the comparative effects of possible climate change scenarios, such as percentage change in wind damages. This suggests that although the wind field and vulnerability models may be relatively simple, their use for comparative analyses is instructive and suitable for initial climate impact and adaptation screening.

9.2.4 Regional Hurricane Risk Assessment

The annual hurricane damage risk (D_{annual}) can be estimated by combining the building inventory of a region and damage risk for building at each exposure site, as shown below:

$$D_{\text{annual}}(t) = \sum\nolimits_{j=1}^{3} \gamma_{Ej} M(t) D(v, t) \tag{9.5}$$

where j represents the exposure sites (1–3), γ_{Ej} is the proportion of houses located within each site, $M(t)$ is the number of houses, and $D(t)$ is the expected annual damage for a specific site at time t.

The insured value of a house (C_I) can be used to estimate the expected damage costs in monetary units. The land price is assumed to be approximately 50% of the median value (Zigomanis, 2007), so the replacement house value is 50% of the house value. Huang et al. (2001) found that because homeowners often hold contents insurance, the insured value of a household is 50% greater than the house replacement value. Therefore, it is assumed that the insured house value (C_I) is 150% that of the replacement value.

The cumulative regional loss (economic risk) $E(L)$ is found by summing the annual regional loss estimation (D_{annual}) over a time period of 50 years:

$$E(L) = \sum_{t=1}^{50} [D_{\text{annual}}(t)] \frac{C_I}{(1+r)^t} \tag{9.6}$$

where r is the discount rate at year t. The discount rate is assumed to be 3%, unless stated otherwise.

Fig. 9.1 shows a flowchart of the proposed risk assessment framework presented herein.

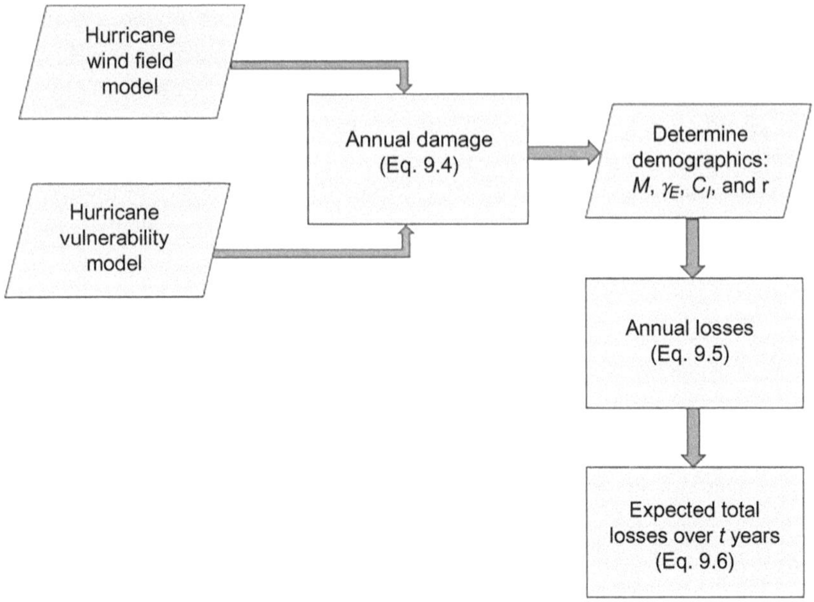

Fig. 9.1 Hurricane risk assessment framework.

9.3 Cost Assessment of Adaptation

Studies (e.g., Roth, 1997 and Stewart et al., 2003) emphasise the importance of retrofitting structures to comply with current building codes to reduce building vulnerability to hurricane damage. If the frequency and intensity of hurricanes may increase as a result of climate change, it becomes essential to identify and evaluate effective means to reduce building vulnerability. This study proposes five adaptation strategies in an effort to mitigate damage costs. For illustration purposes, the economic viability of the adaptation strategies is investigated under an increase in wind speed of 5% over 50 years.

Structural vulnerability can be affected by either strengthening or weakening the construction, and so the vulnerability models in Eqs. (9.2) and (9.3) are modified with an enhancement/reduction vulnerability (or risk reduction) factor (ΔR), expressed as a percentage of existing vulnerability, to reflect the variable structural vulnerability of the building stock within a region. As construction ages, the structural vulnerability of the construction increases.

Jain and Davidson (2007) divided the building inventory of North Carolina into four classes: pre-1970, 1970–85, 1985–95, and post-1995. Similarly, Pinelli et al. (2009) classified buildings in Florida into four groups: weak, constructed in 1970, medium, constructed in 1986, strong, constructed in 1998, and stronger, constructed after 2000. For consistency, all construction is classified into the following categories along with the corresponding enhancement/reduction vulnerability factor:

1. Weak: built before 1970 ($\Delta R = -15\%$)
2. Medium: built 1970–85 ($\Delta R = 0\%$)
3. Strong: built 1986–97 ($\Delta R = 20\%$)
4. Stronger: built after 1998 ($\Delta R = 30\%$)

The vulnerability of new construction can be further reduced by implementing strengthened construction that is resistant to higher wind speeds. This is enhanced construction is accounted for within the vulnerability function (Eq. 9.7) by further increasing the enhancement/reduction vulnerability factor (R): Strengthened Type I ($\Delta R = 50\%$), and Strengthened Type II ($\Delta R = 80\%$). Fig. 9.2 shows the impact varying the enhancement/reduction vulnerability (or risk reduction) factor (ΔR) has on the overall structural vulnerability of a structure as calculated with Eq. (9.7) for Miami-Dade County, Florida.

$$F_D(v_{\text{mean}}) = \left(\frac{100 - R}{100}\right) \exp\left(0.252 \cdot v_{\text{mean}} - 5.823\right) \quad F_D(v) \leq 100\% \quad (9.7)$$

To further illustrate the impact of the enhancement/reduction vulnerability factor (ΔR), Table 9.1 shows percentage change in the cumulative damage costs in Miami-Dade County for 0% and 5% wind speed change scenarios over 50 years. This was calculated with Eqs. (9.6) and (9.7). For example, if building vulnerability across the region is decreased by $\Delta R = -15\%$, the cumulative damage costs $E(L)$ for the region may increase by around 11% for both scenarios. In comparison, through the implementation of strengthened construction with $\Delta R = 50\%$, it is possible to potentially decrease the cumulative damage cost by around 37% for both climate change scenarios. The reduced cumulative damage is not linearly proportional to ΔR because maximum damage is limited to $F_D = 100\%$, see Eq. (9.7).

Five adaptation strategies for residential construction are proposed, see Table 9.2. To evaluate the potential cost-effectiveness of the adaptation strategies, a life-cycle cost (LCC) analysis is conducted. The LCC involves combining the reduction in structural vulnerability and the cost associated

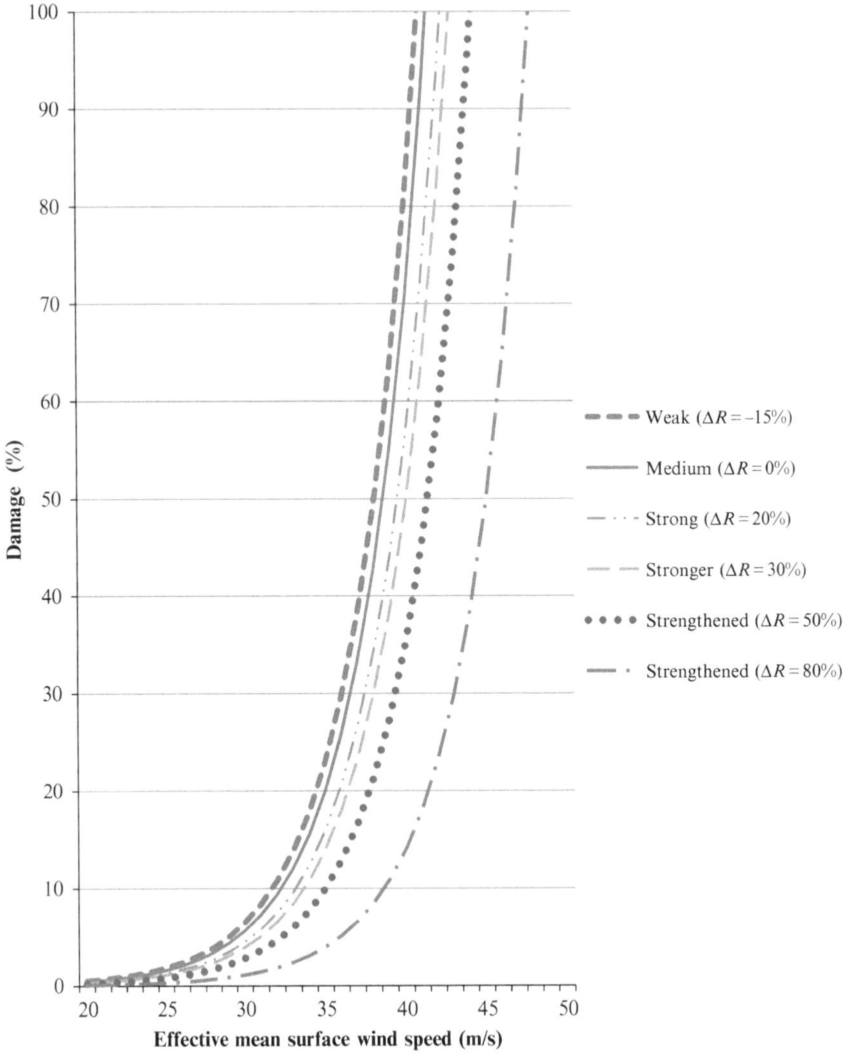

Fig. 9.2 Effect of enhancement/reduction vulnerability factor (ΔR) on hurricane vulnerability.

Table 9.1 Percentage Change in Cumulative Damage Cost considering Variable Enhancement/Reduction Vulnerability Factor (ΔR)

		Enhancement/Reduction Vulnerability Factor (ΔR)				
		−15%	**20%**	**30%**	**50%**	**80%**
Change in Wind Speed over 50 years	0%	11%	−15%	−21%	−37%	−59%
	5%	10%	−14%	20%	−36%	−57%

Table 9.2 Adaptation Strategies

Adaptation Strategy	Description
1	Strengthening a percentage of "weak" construction in Exposure Category 1
2	Strengthening a percentage of "weak" construction within the whole region (within all Exposure Categories)
3	Strengthening a percentage of "weak" and "medium" construction in Exposure Category 1
4	Strengthening a percentage of "weak" and "medium" construction within the whole region (within all Exposure Categories)
5	Reducing the vulnerability of new construction in Exposure Category 1

with strengthening construction (Stewart et al., 2003). To estimate the potential benefits of adaptation, the LCC for each adaptation strategy is compared to the "business as usual" case to estimate the potential benefits of adaptation. The "business as usual" case assumes no proactive retrofit of residential construction.

For Adaptation Strategies 1 and 2, "weak" buildings are retrofitted to "stronger" standards, while the "medium" and "strong" units are left as is. For Adaptation Strategies 3 and 4, however, both "weak" and "medium" structures are retrofitted to the standards of "stronger" buildings, while the "strong" units are left as is. All new construction is built to "stronger" standards for these four adaptation strategies.

For Adaptation Strategy 5, new construction ($t-1$) is built to enhanced standards, with $\Delta R = 50\%$ and $\Delta R = 80\%$, or Strengthened Construction Types I and II, respectively. This strategy is solely focused on new construction within Exposure Category 1. This strategy would cause no disruption to existing home owners, and has the potential to be highly viable because its ease of implementation.

For illustration purposes, the general LCC equations for Adaptation Strategy 1 are presented. These equations are developed on the LCC principles found in Chapter 1 and Stewart et al. (2003). For the sake of brevity, the equations for Adaptation Strategies 2–5 are not presented, see Bjarnadottir et al. (2011) for information relative to these adaptation strategies.

$$\mathrm{LCC}_{\mathrm{adapt}}(t,T)=\sum_{t=t_0}^{T}\sum_{j=1}^{3}\gamma_{Ej}\left(H_{\mathrm{weak}}\left[\left(\frac{100-nt}{100}\right)D_{\mathrm{weak}}(t)+\left(\frac{nt}{100}\right)D_{\mathrm{stronger}}(t)+\frac{n}{100}\left(\frac{C_{\mathrm{adapt}}}{1.5}\right)\right]+H_{\mathrm{medium}}D_{\mathrm{medium}}(t)+H_{\mathrm{strong}}D_{\mathrm{strong}}(t)+H_{\mathrm{stronger}}D_{\mathrm{stronger}}(t)+(H_{\mathrm{total}}m)tD_{\mathrm{stronger}}(t)\right)\cdot\frac{C_I}{(1+r)^t}\quad nt<100 \tag{9.8a}$$

$$\mathrm{LCC}_{\mathrm{adapt}}(t,T)=\sum_{t=t_0}^{T}\sum_{j=1}^{3}\gamma_{Ej}\left(H_{\mathrm{weak+stronger}}D_{\mathrm{stronger}}(t)+H_{\mathrm{medium}}D_{\mathrm{medium}}(t)+H_{\mathrm{strong}}D_{\mathrm{strong}}(t)+(H_{\mathrm{total}}m)tD_{\mathrm{stronger}}(t)\right)\cdot\frac{C_I}{(1+r)^t}\quad nt>100 \tag{9.8b}$$

where j represents the exposure sites (Exposure Categories 1–3), γ_{Ej} is the proportion of houses located within each site, H_{total} is the total number of single-family housing units within the area, H_{weak}, H_{medium}, H_{strong}, and H_{stronger} refer to the number of "weak", "medium", "strong", and "stronger" housing units within the area, $H_{\mathrm{weak+stronger}}$ refers summation of "weak" and "stronger" housing units in the region, respectively, $D_{\mathrm{weak}}(t)$, $D_{\mathrm{medium}}(t)$, $D_{\mathrm{strong}}(t)$, and $D_{\mathrm{stronger}}(t)$ are the expected annual damage for "weak", "medium", "strong", and "stronger" buildings, respectively, n is the percentage of houses being retrofitted each year, m is the annual growth rate of new housing, C_{adapt} is the additional cost of retrofit (adaptation) in terms of percentage of replacement value, and C_{I} is the median insured house value.

The cost of retrofit (C_{adapt}) is assumed to be 1%–50% (Stewart and Li, 2010) of the median insured house value, C_{I}, for this analysis, which is assumed to be an additional cost, and the discount rate is set as 3% as indicated before. The net present value (NPV) (or net benefit) is the difference between the adaptation LCC and the "business as usual" LCC:

$$\mathrm{NPV}=\mathrm{LCC}(t,T)-\mathrm{LCC}_{\mathrm{adapt}}(t,T) \tag{9.9}$$

where the "business as usual" LCC is:

$$\mathrm{LCC}(t,T)=\sum_{t=t_0}^{T}\sum_{j=1}^{3}\gamma_{Ej}\left(H_{\mathrm{weak}}D_{\mathrm{weak}}(t)+H_{\mathrm{medium}}D_{\mathrm{medium}}(t)+H_{\mathrm{strong}}D_{\mathrm{strong}}(t)+H_{\mathrm{stronger}}D_{\mathrm{stronger}}(t)+(H_{\mathrm{total}}m)tD_{\mathrm{stronger}}(t)\right)\cdot\frac{C_I}{(1+r)^t} \tag{9.10}$$

The cost of retrofitting (C_{adapt}) depends on the required reduction in vulnerability, structural configuration, current design and construction practices, and local labour cost and availability of retrofit material; therefore, it is difficult to estimate costs accurately (Li and Ellingwood, 2009). A number of studies have found that the additional cost to new housing for increased cyclone (hurricane) resistance is in the range of 1%–10% of the initial (replacement) value of the home (contents not included) (Reardon and Henderson, 1988). Stewart et al. (2003) estimated a cost of approximately 5% for installing Australian cyclone-resistant systems for new construction. There is very little data on the costs of retrofitting an existing house for increased hurricane resistance in the United States. If 5% is assumed for strengthening a new house in Miami-Dade County, then this equates to around $7500. In comparison, there is a relatively wide range of retrofit costs (C_{adapt}) for retrofitting existing housing construction. For example, according to Pinelli et al. (2009), the mitigation cost for existing gable roof timber residential construction built before the 1970s in Florida is around 15%–25% of replacement value, depending on the region and retrofit details (e.g., shingle, roof-to-wall connection, impact class, etc.).

9.4 Case Study

Miami-Dade County, Florida, New Hanover County, North Carolina, and Galveston County, Texas are used as case study locations to investigate the possible impacts climate change may have on hurricane damage risks and adaptation strategies, as well as to illustrate the effect of geographic difference on the assessment.

The demographic information from the US Census Bureau (2009) was compiled to estimate the hurricane damage risk in each of the three counties. The demographic information and how it influences wind field and economic characteristics are summarised in Table 9.3. The site-specific Weibull parameters at year 1, $u(1)$ and $\alpha(1)$, the mean maximum wind speed at year 1, the number of single-family housing units, the housing distribution (i.e., percentage of houses in Exposure Category 1% of houses in Exposure Category 2% of houses in Exposure Category 3), and the median insured house value are included in the table for each county.

The expected annual damage for the three case study locations is determined with Eq. (9.4). A comparison of the expected annual damage at Exposure Category 1, considering various scenarios of climate change over 50 years, is presented in Table 9.4. For example, assuming no climate change

Table 9.3 Demographic Information

Parameter	Description	Miami-Dade County	New Hanover County	Galveston County
$u(1)$	Weibull	27.4	19.9	23.4
$\alpha(1)$	Weibull	1.8	1.6	1.8
$\nu(1)$	Mean wind speed (year 1)	24.3 m/s	18.0 m/s	21.0 m/s
M	Single-family housing units	452,000	58,200	56,300
γ_E	Housing distribution	20-60-20	20-70-10	20-30-50
C_I	Insured house value	\$147,000	\$191,100	\$82,000

Table 9.4 Expected Annual Damage for Exposure Category 1 Considering Various Scenarios of Climate Change at 50 years

	Wind Speed Change			
	−5%	0%	5%	10%
Miami-Dade County	1.20%	2.68%	3.50%	4.45%
New Hanover County	0.49%	0.68%	0.92%	1.22%
Galveston County	0.62%	0.88%	1.21%	1.62%

(or 0% change in wind speed), the expected annual damage for Miami-Dade County is 2.7%. This increases to 3.5% and 4.5% for increases in wind speed of 5% and 10%, respectively, over 50 years. These are increases in annual damage risks ranging from 30% to 66% which could translate to significant increases in cumulative damage costs. Similar, increases can be expected for the other counties.

The expected cumulative damage for Exposure Category 1 over 50 years for each county is presented in Table 9.5, for various scenarios of climate change. For example, if no climate change is assumed, the expected cumulative damage for Miami-Dade County is 116%. If it is assumed that wind

Table 9.5 Expected Cumulative Damage for Exposure Category 1 Considering Various Scenarios of Climate Change over 50 years

	Wind Speed Change			
	−5%	0%	5%	10%
Miami-Dade County	116%	134%	155%	177%
New Hanover County	29%	34%	40%	47%
Galveston County	37%	44%	51%	61%

speeds will increase by 10% over 50 years, the cumulative damage for Miami-Dade County increases to 177%, in Exposure Category 1. The expected cumulative damage for Miami-Dade County is approximately 300% higher than the expected cumulative damage in New Hanover County and 200% higher than Galveston County, independent of climate change scenario.

The cumulative damage cost is determined for each county by inputting the demographic information found in Table 9.2 into Eq. (9.6). To normalise the damage risk and for comparative purposes, Fig. 9.3 shows the cumulative damage cost after 50 years per 100,000 residents for all three counties under various scenarios of climate change. From this figure, it can be seen that Miami-Dade County has a considerably higher expected cumulative damage cost per 100,000 residents than the other two counties. If no climate change is assumed, the cumulative damage costs is \$1600 million per 100,000 residents in Miami-Dade County, \$620 million per 100,000 residents in New Hanover County, and \$240 million per 100,000 residents in Galveston County. These figures increase to \$1840, \$720, and \$280 million for Miami-Dade County, New Hanover County, and Galveston County, respectively, under an assumed increase in wind speed of 5%.

Fig. 9.4 shows the percentage increase in cumulative damage costs for various scenarios of climate change from the no climate change scenario.

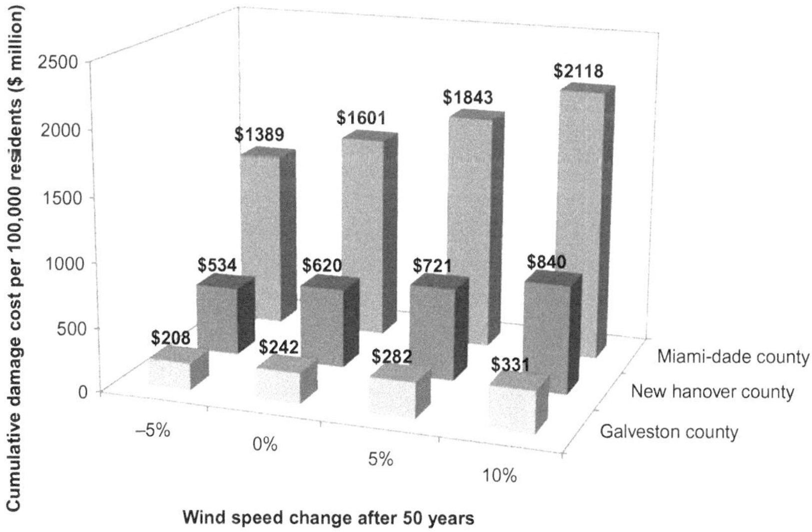

Fig. 9.3 Cumulative damage cost per 100,000 residents considering various scenarios of climate change over 50 years.

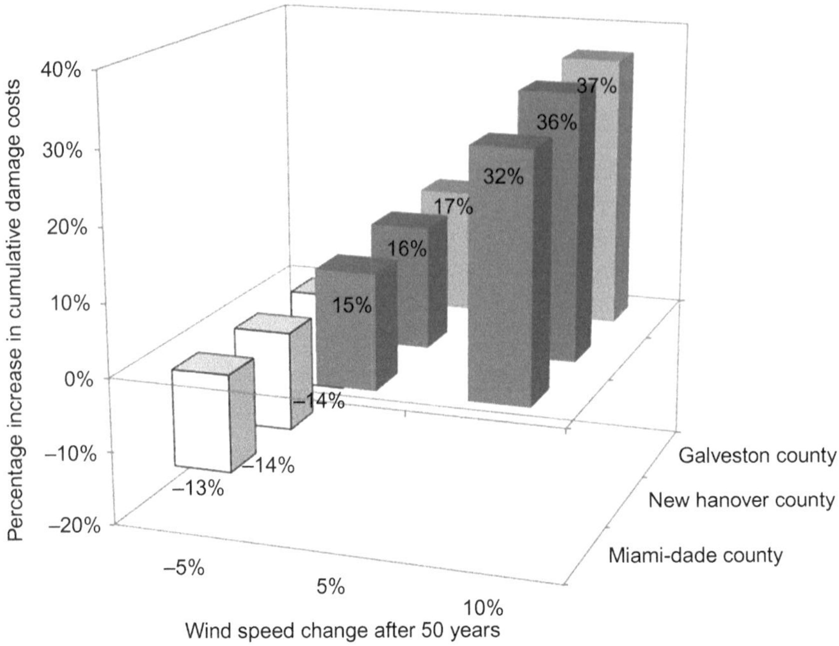

Fig. 9.4 Percentage change in cumulative damage cost considering various scenarios of climate change over 50 years.

Despite the considerable differences in potential cumulative damage costs seen in Fig. 9.3, the percentage change from no climate change to various changes in wind speed is similar for all the counties. Therefore, climate change may have a considerable effect on the cumulative damage cost in an individual county even though the damage cost seems insignificant in comparison with other more high-risk counties, such as Miami-Dade County.

The estimated cumulative damage costs per 100,000 residents in Miami-Dade County are considerably higher than that of New Hanover and Galveston Counties. This can be attributed to several factors. First, the mean maximum annual wind speed assuming no climate change for Miami-Dade County is approximately 24.3 m/s, in comparison to 18.0 m/s for New Hanover County, and 21 m/s for Galveston County. The number of single-family housing units exposed to hurricane events in Miami-Dade County is approximately 452,000 units, while in New Hanover County there are 58,500 units and in Galveston County there are 56,300 units.

The replacement house value is $98,000, $143,000, and $82,000 for Miami-Dade County, New Hanover County, and Galveston County, respectively. Even though the house replacement value in New Hanover

County is 45% higher than of Miami-Dade County, the expected damage cost in Miami-Dade County is still significantly higher. The number of exposed housing units has a considerable effect on the potential damage cost to a region, and Miami-Dade County has about seven times the number of exposed housing units than both other counties. Therefore, the expected damage cost for Miami-Dade County is significantly higher than those in the other two counties.

9.4.1 Uncertainty in Risk Assessments

Two groups of uncertainty, that is, aleatoric uncertainty and epistemic uncertainty can be identified within hurricane risk assessment. Aleatoric uncertainty arises because of the unpredictable, random nature of the physical system under study. On the other hand, epistemic uncertainty is due to the lack of knowledge of the system in respect to quantities and processes within the system. Aleatoric uncertainty cannot be reduced only identified and quantified, while epistemic uncertainty can be reduced through more comprehensive study. Epistemic uncertainties can be reduced through a thorough sensitivity analysis.

The Monte Carlo simulation (MCS) can be used to estimate the damage risk to incorporate the epistemic uncertainty in change in hurricane wind speed, that is, the wind increase is a variable. How climate change will affect regional climates and pertinent variables (epistemic uncertainty) is not well known, limiting the ability to predict consequential effects. Fig. 9.5 shows the results of the MCS for the annual damage in Miami-Dade County, Florida, assuming a mean increase in wind speed of 5% with COV of 0.50 in year 50. This figure is for illustration purposes only, as significant refinement is required to estimate probability distributions for future mean temperatures or other climatic variables (e.g., the COV for changes in wind speed) when new information is available (Solomon et al., 2007). A presentation like this figure can also be modified to represent the cumulative damage of a region to allow for make statements with certain confidence level, such as "I am 95% confident that the wind damage for this region will be under $1.0 billion for an increase in wind speed of 5% over 50 years". Such statements can be very useful in risk communication for decision-making.

9.4.2 Validation of Risk Assessment Framework

The HAZUS-MH Hurricane Model is used to validate the proposed risk assessment framework presented herein. The HAZUS-MH Hurricane Model was developed to estimate the potential damage of hurricanes to

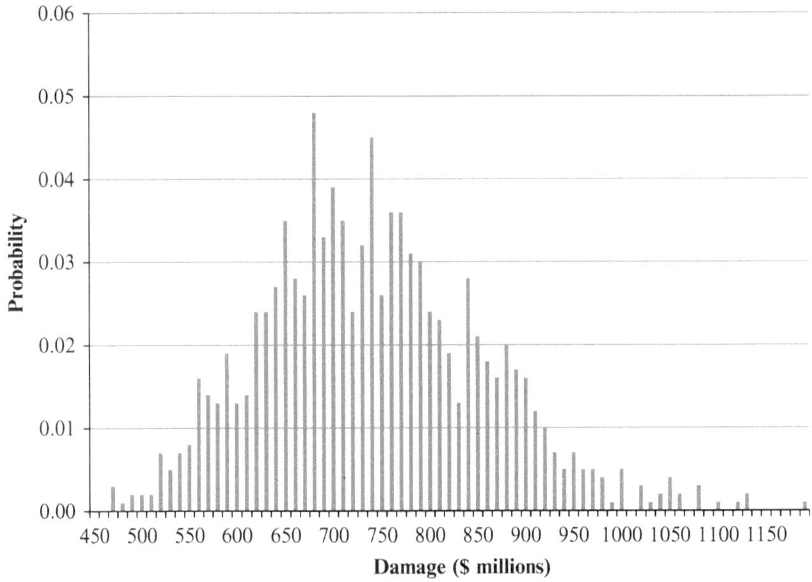

Fig. 9.5 Simulation histogram of annual damage losses in year 50 (Bjarnadottir et al., 2011).

buildings. The track and wind field of a hurricane is modelled using a hurricane hazard model, which has been validated with historic hurricane data (Vickery et al., 2000). The wind speed is calculated as a function of the following parameters: central pressure, surface roughness, and translation speed. The model classifies the building inventory of a region by the characteristics of the building envelope and frame, and has the capabilities to calculate damage loss to the building interior and contents (FEMA, 1999). However, the model assumes that wind speeds are stationary overtime, which cannot be used to estimate of hurricane losses when climate change is possible.

The HAZUS-MH Hurricane Model predicts the annual damage to single-family housing due to hurricanes for Miami-Dade County to be $666 million. In comparison, the framework proposed herein estimates the damage to single-family housing due to hurricanes as $966 million per year, assuming no climate change. The discrepancy of the values is explained with the assumption of the average single-family housing value. The proposed framework herein uses relatively current housing values, while the estimates from the HAZUS-MH Hurricane Model use 2000 US Census Data. More specifically, the proposed framework assumes an average value of $196,000 while HAZUS-MH Hurricane Model assumes a value of $124,000. The annual damage is recalculated with the proposed framework using the lower average housing value, and the annual damage estimate is approximately $613 million.

The difference between the estimated value and the prediction found in HAZUS is approximately 8%, which is not unreasonable given the large uncertainty associated with regional hurricane loss estimation (Li and Stewart, 2009). In addition, for the comparative purposes of this study and the assumption that the mean annual wind speed is applied to the whole region, the estimation using the proposed framework can be viewed as acceptable.

Chen et al. (2009) used the Florida Public Hurricane Loss Model (FPHLM) to estimate the damage hurricanes may potentially cause in Miami, Florida. The FPHLM is an interdisciplinary online tool designed to estimate hurricane losses in Florida. The model estimates that a Category 4 hurricane could potentially cause $12.8 billion in damages and a Category 3 could cause $3.1 billion in damages to the city of Miami, using 2003 data. In comparison, the framework of this study was modified with demographic data for the city of Miami for 2003, and found that a Category 4 hurricane, with wind speeds ranging from 58.3 to 69 m/s, could cause up to $12.9 billion in damage, and a Category 3 hurricane, with wind speeds ranging from 49.4 to 58 m/s, could cause up to $3.2 billion in damage. The results from the FPHLM are within the ranges established with the proposed framework.

9.4.3 Parametric Sensitivity Analysis

To identify the dominant factors that affect the cumulative damage costs, a sensitivity study was conducted. The effects of regional annual growth rate of houses, inventory of houses in different exposure categories, difference in median house prices in Exposure Category 1, and discount rates are investigated. Miami-Dade County was used to perform this sensitivity analysis. Fig. 9.6 shows a tornado plot of the results of the sensitivity analysis. A tornado plot is a useful tool in describing how sensitive a value of an output variable is to the input variables of the model.

- *Discount rate*:

  Not all damage will occur at the same point in time; therefore, it is necessary to discount future damage costs to present values. The results presented herein were calculated with a discount rate of 3% (Li and Ellingwood, 2009). Wen (2001) investigated the sensitivity of optimal

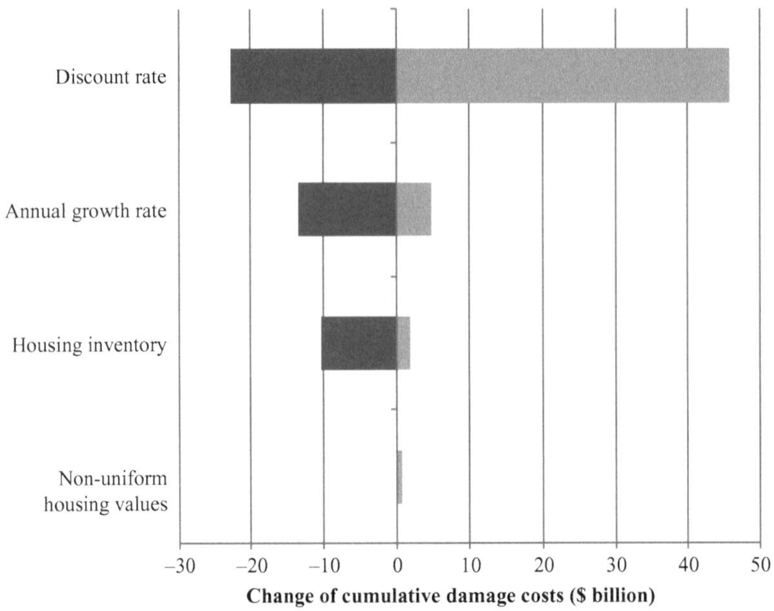

Fig. 9.6 Tornado plot of sensitivity analysis for Miami-Dade county ($ Billion).

design against multi-hazards to discount rates varying from 0% to 10%. Similarly, to explore the effects of the discounted rate, two cases where explored: $r=0\%$ and $r=8\%$, and then compared with $r=3\%$. The discount rate has a significant impact on damage cost estimates.

- *Regional annual growth rate*:
  Three regional growth rates other than $\mathrm{m}=2\%$ (US Census Bureau, 2009) were explored to reflect the uncertainty in population growth; $\mathrm{m}=0\%$, $\mathrm{m}=1\%$, and $\mathrm{m}=2.5\%$. As expected, the potential for higher damage to a region increases as the population of the coastal areas continues to grow, because more property is exposed to hurricane damage.
- *Inventory of houses in different exposure categories*:
  The results presented herein, for Miami-Dade County, were based on the following inventory distribution: 20%-Exposure Category 1; 60%-Exposure Category 2; 20%-Exposure Category 3, or 20%–60%–20% (Miami-Dade County Department of Planning and Zoning, 2003). To investigate the sensitivity of the inventory distribution on future damage estimates, three different cases for building inventory distributions were explored: 10%–30%–60%, 20%–50%–30%, and 30%–40%–30%. The results of these inventory distributions were then compared with the original results.

- *Nonuniform housing values*:
  Coastal properties are usually priced considerably higher than those located farther inland, due to higher property taxes and a higher demand. A report by RealtyTrac (2008), which compiled the median house value for single-family units in Miami-Dade County, found that the median house value for Miami Beach is \$246,000. This is an increase of 25% when compared to the county average house value of \$196,000. To explore the effect of nonuniform housing prices on damage estimates, increases in the median house value in Exposure Category 1 of 10%, 25%, and 40% are explored.

It was found that by increasing the annual growth rate, the cumulative regional damage increased as well. For example, if the annual growth rate was assumed to be 2.5% instead of 2%, the cumulative damage increased by 18%, under an assumed increase in wind speed of 10% over a 50-year time period. By modifying the building inventory from the original 20%–60%–20%, as seen in Table 9.3, it was found the concentration of houses located in Exposure Categories 1 and 2 controlled the cumulative damage to the region. This is because of the high damage risk in these two exposure categories.

On the other hand, it was found that increasing the median house value in Exposure Category 1 does not have a significant effect on the cumulative damage losses because of the relatively low percentage of houses in the exposure category. For example, if it was assumed that the median insured house value in Exposure Category 1 is increased by 25%, the cumulative damage cost only increased by 1%. Finally, a discount rate of 0% may result in a 119% increase of the present value of future losses over 50 years, while a discount rate of 8% will result in 60% decrease of the present value of future losses over a 50-year time period, when compared to the current case of $r=3\%$.

9.4.4 Cost-Effectiveness of Adaptation Strategies

The cost-effectiveness of the adaptation strategies varies significantly among the three counties, mainly due to the fact that the expected annual damage in Miami-Dade County is considerably higher than that of the two other case study locations. All adaptation strategies were found to be cost-effective for Miami-Dade County if the cost of retrofit is not >10%. In comparison, for New Hanover and Galveston Counties, the cost of retrofit would have to be as low as 3% for all strategies to be cost-effective.

Table 9.6 shows the NPV (Eq. 9.9) for Adaptation Strategies 1–4 considering an increase in wind speed of 5% and no change in wind speed.

Table 9.6 NPV ($ million) for Adaptation Strategies 1–4 Considering Increases in Wind Speed of 0% and 5% over 50 years

		Strategy 1		Strategy 2		Strategy 3		Strategy 4	
		0%	5%	0%	5%	0%	5%	0%	5%
Miami-Dade County									
Cost of Retrofit (C_{adapt})	1%	610	690	1,820	2,100	910	1,030	2,690	3,100
	2.5%	580	660	1,650	1,930	850	970	2,390	2,810
	5%	520	600	1,360	1,640	750	870	1,900	2,310
	10%	400	480	780	1,060	550	670	900	1,320
	25%	60	140	−950	−670	−40	70	−2,070	−1,650
New Hanover County									
Cost of Retrofit (C_{adapt})	1%	18.1	21.2	50	59.9	24.9	29.4	67.1	81.3
	2.5%	15.3	18.5	36.4	46.3	20.4	24.5	42.5	56.7
	5%	10.8	13.9	13.7	23.6	11.8	16.3	1.5	15.7
	10%	1.7	4.9	−31.6	−21.7	−4.6	−0.1	−80.5	−66.3
	25%	−25.5	−22.3	−167.7	−157.8	−53.8	−49.3	−326.5	−312.3
Galveston County									
Cost of Retrofit (C_{adapt})	1%	11.7	13.8	31	37.6	15.9	18.9	40.9	50.3
	2.5%	9.7	11.8	21.2	27.8	12.4	15.4	23.3	32.7
	5%	6.4	8.6	4.9	11.5	6.5	9.5	−6	3.3
	10%	−0.1	2	−27.8	−21.2	−5.2	−2.2	−64.7	−55.3
	25%	−19.7	−17.6	−125.9	−119.3	−40.4	−37.4	−240.7	−231.3

The NPV (net benefit) presented is calculated for a fixed rate of retrofit of $n=3\%$ (i.e., all houses are retrofitted within 34 years), as it is assumed that this is a reasonable annual rate of retrofit for the region. The cost of retrofit (C_{adapt}) varies from 1% to 25%.

From Table 9.6, if the cost of retrofit is 10%, Adaptation Strategy 2 could result in a NPV of $1.06 billion for Miami-Dade County over 50 years considering an increase in wind speed of 5%. Even if there is no change in wind speed, and rate and cost of retrofit are unchanged, Adaptation Strategy 2 could result in a NPV of $780 million over 50 years for Miami-Dade County. This implies that certain adaptation strategies could result in a "win-win" situation irrespective of climate change predictions. Similarly, adaptation strategies for New Hanover and Galveston Counties are found to be cost-effective even if there is no change in wind speed, see Table 9.6.

Fig. 9.7 shows the difference in cost-effectiveness between Adaptation Strategies 1–4 for Miami-Dade County. This figure is useful in risk-informed decision-making, by showing what combinations of cost of retrofit (C_{adapt}) and annual rate of retrofit (n) that result in a NPV of zero, that is, breakeven point, for a time period of 50 years, under an assumed increase in wind speed of 5%. The area above each line shows for what combinations the strategy is not cost-effective, while the area below represents the cost-effective combinations.

Adaptation Strategy 5 involved strengthening only new construction built in Exposure Category 1. Table 9.7 summarises the NPV for all three counties, under increases in wind speed of 5% and 0%, for Strengthened Construction Types I and II. From the table, it can be seen that reducing the vulnerability of new construction by $\Delta R=80\%$ is more cost-effective than reducing the vulnerability of new housing by $\Delta R=50\%$, if the cost of strengthening stays the same. From Table 9.7, it is evident that Adaptation Strategy 5 could be cost-effective if wind speeds in the three counties remain constant over a 50-year time frame. This implies that this adaptation strategy could result in a "win-win" situation for society.

Fig. 9.8 shows the various combinations of ΔR and cost of retrofit (C_{adapt}) for Adaptation Strategy 5 that result in a NPV of zero, that is, break-even point, for a time period of 50 years, under an assumed increase in wind speed of 5%. The area above each line shows for what combinations the strategy is not cost-effective, while the area below represents the cost-effective combinations. From this figure, it can be seen that Adaptation Strategy 5 is not cost-effective for reduction in vulnerability that is $<40\%$. Reducing the vulnerability of new construction at foreshore

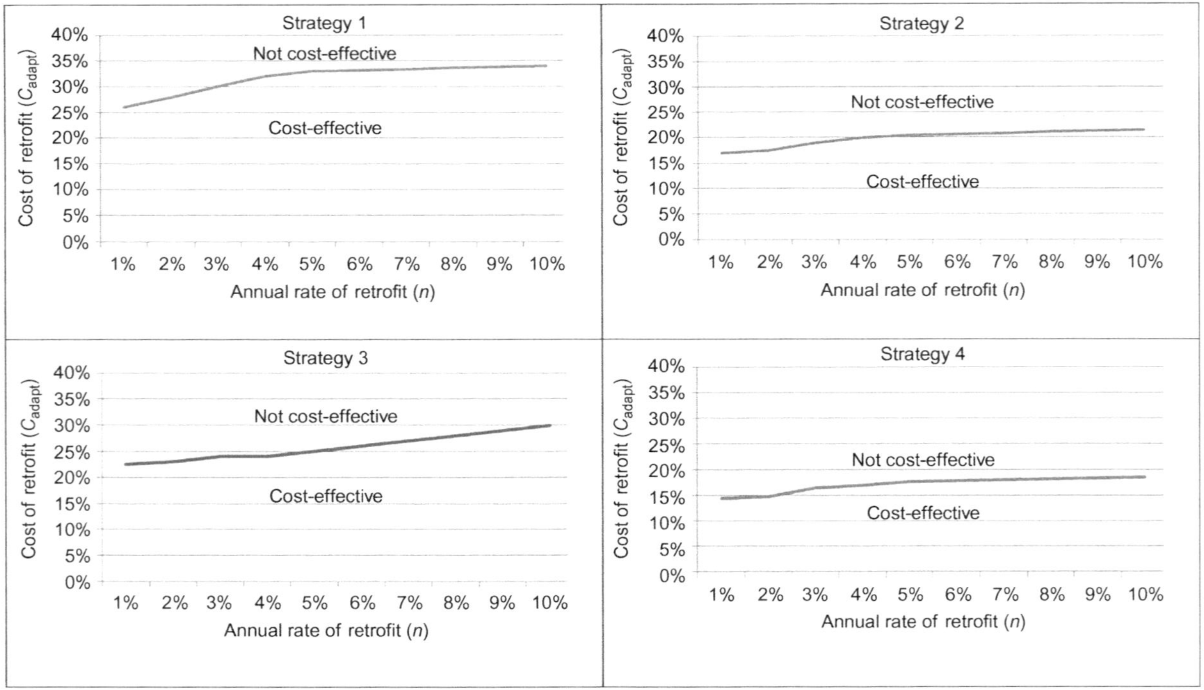

Fig. 9.7 Combinations of cost of retrofit (C_{adapt}) and annual rate of retrofit (n) in which the NPV is zero for adaptation strategies 1–4 after 50 years for Miami-Dade county (5% increase in wind speed).

Table 9.7 NPV ($ million) for Adaptation Strategy 5 Considering Increases in Wind Speed of 0% and 5% Over 50 Years

Strengthened Construction Type I (ΔR = 50%)

	Miami-Dade County		New Hanover County		Galveston County	
	0%	5%	0%	5%	0%	5%
Cost of retrofit **(C_{adapt})**						
1%	531	612	15	19	19	24
2.5%	463	544	6	10	15	19
5%	350	430	−8	−4	7	11
10%	124	204	−37	−33	−9	−5
25%	−555	−475	−123	−120	−57	−53
50%	−1,687	−1,607	−267	−264	−136	−132

Strengthened Construction Type II (ΔR = 80%)

	Miami-Dade County		New Hanover County		Galveston County	
	0%	5%	0%	5%	0%	5%
Cost of retrofit **(C_{adapt})**						
1%	1,600	2,110	50	68	45	59
2.5%	1,532	2,042	41	59	40	54
5%	1,419	1,929	27	45	32	46
10%	1,193	1,702	−2	16	16	31
25%	514	1,023	−88	−70	−32	−17
50%	−618	−109	−232	−215	−111	−96

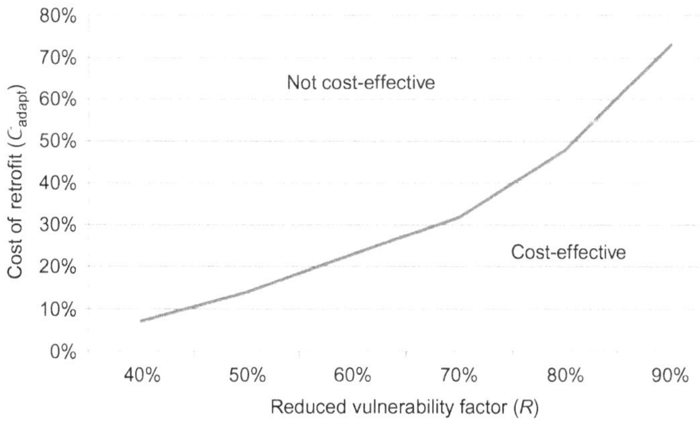

Fig. 9.8 Combinations of cost of retrofit (C_{adapt}) and annual rate of retrofit (n) in which the NPV is zero for Adaptation Strategy 5 after 50 years for Miami-Dade county (5% increase in wind speed).

Table 9.8 Ranking of the Adaptation Strategies

	Miami-Dade County	New Hanover County	Galveston County
1st Place	2 and 5	1 and 5	1 and 5
2nd Place	3	3	3
3rd Place	1	2	2
4th Place	4	4	4

locations by 40% could be cost-effective if the cost of retrofit is not >8%. In contrast, reducing the vulnerability of new construction by 70% could be cost-effective at a cost of retrofit as high as 32%.

The adaptation strategies were ranked individually for the three counties, see summary in Table 9.8. The rankings are based on how cost-effective the strategies are for each county and also how probable they are in implementation for each county. Adaptation Strategy 2 was found to be the most economically viable for Miami-Dade County, despite being an unlikely option for New Hanover and Galveston Counties. Adaptation Strategy 2 proposes retrofitting "weak" housing throughout the region. This is ineffective for New Hanover and Galveston Counties because the cost is outweighed by the benefits. But for Miami-Dade County, which has a considerably higher expected annual damage, the benefits outweigh the cost, and the adaptation strategy provides the highest NPV.

In contrast, Adaptation Strategy 1 was the most likely option for both New Hanover and Galveston Counties, because it can be undertaken at the highest cost of retrofit. Adaptation Strategy 4 was the least likely option for all three counties, because of the sheer number of houses the strategy requires to be retrofitted for implementation. Adaptation Strategy 5 is recommended in all counties, because it causes the least disruption while being cost-effective. Adaptation Strategy 5 was found to be cost-effective for the three counties under current climate conditions, that is, no climate change, which could result in a "win-win" situation for the regions.

9.5 Summary

The proposed framework includes probabilistic models of hurricane intensity and vulnerability model for typical US residential construction. It is possible that hurricane patterns may change as a result of climate change and so the proposed framework provides a tool to evaluate the potential impact of such change on hurricane damage risk assessment.

Notwithstanding the relative simple wind field and vulnerability models, mainly due to the proprietary nature of such models, the framework is appropriate for comparative studies to explore the potential impact of climate change on hurricane risk assessment.

The proposed framework was used to estimate hurricane damage risks for three hurricane-prone counties, Miami-Dade County, Florida, New Hanover County, North Carolina, and Galveston County, Texas. The annual damage risk for Miami-Dade County was found to be considerably higher than the annual damage risk for New Hanover County and Galveston County. This is attributed to the higher mean wind speed in Miami-Dade County. For example, under current climate conditions, the annual damage risk is 2.7% for Miami-Dade County considering Exposure Category 1 (foreshore), while the annual damage risk is 0.7%, and 0.9% for New Hanover County and Galveston County, respectively.

In terms of cumulative damage costs, climate change could have a significant impact. If wind speeds increase by 5% over 50 years, the expected annual damage costs may reach \$120 million in Miami-Dade County, \$4 million in New Hanover County, and \$2 million in Galveston County in 50 years' time. If this or other scenarios are believed likely then, it is necessary to assess the economic viability of mitigation (adaptation) strategies to strengthen or retrofit construction in hurricane-prone zones.

Five adaptation strategies were proposed herein and the cost-effectiveness of these strategies were evaluated for the three counties. For Miami-Dade County, assuming an increase in wind speed of 5% over 50 years, all adaptation strategies were found to be cost-effective if the cost of retrofit is <10%. In comparison, for New Hanover and Galveston Counties, the cost of retrofit would have to be as low as 3% for all strategies to be cost-effective, which is unrealistic because it is unlikely that retrofit can be undertaken at costs <10% of the insured house value. Furthermore, for Miami-Dade County, strengthening "weak" construction within the whole region proved to be more cost-effective than focusing solely on construction in Exposure Category 1 (foreshore), decreasing the expected cumulative damage to the region by 4.0% on average.

On the other hand, for New Hanover and Galveston Counties, it was most cost-effective to retrofit "weak" construction located solely in Exposure Category 1. The variation in the cost-effectiveness of the adaptation strategies for the three counties is due to the high annual damage risk in Miami-Dade County in comparison to New Hanover County and Galveston County. Finally, it was found that strengthening new construction in

Exposure Category 1 (foreshore) proved to be cost-effective for all three counties, while causing the least disruption to homeowners because it does not require the retrofit of older buildings.

9.6 Potential Design and Practice Evolutions

The current study provides a framework to explore the potential impact of climate change on hurricane risk assessment and adaptation strategies. The framework can be easily refined to address specific concerns of the user. For example, wind field models that consider local topographic factors, shielding effects present within the region, and terrain roughness can be incorporated. The framework can also be modified to investigate parameters of hurricanes other than wind speed, such as frequency and average translation speed (i.e., wind speed duration on structures).

Furthermore, the framework can be adapted with building vulnerability models that account for specific construction types, construction materials, age profiles, code specifications, etc. In addition, to evaluate the indirect losses and social disruption by hurricanes, it would be beneficial to develop a more thorough economic decision analysis, which would consider the effect of various factors (e.g., insurance premiums, discount rates, exposure periods, life safety, social issues, incentives, and hurricane mitigation). In short, the proposed framework provides a methodology that can be easily adapted to address specific concerns of decision-makers while incorporating the dynamic impacts of climate change in hurricane risk assessments.

9.7 Open Research Questions

There are some research questions that can be considered in the future, which include:

- How climate change will affect the rate of deterioration of strength/capacity of the houses? Which in turn will affect the risk assessment and adaptation of the houses?
- How to incorporate the indirect losses due to hurricane wind under the impact from climate change in the risk assessment and adaptation strategies?
- Is there a way to consider the confidence interval for the risk assessment results given the uncertainties associated with various climate change scenarios and wind field models?

- How can the rate and consequence of construction defects be quantified probabilistically, and then how does workmanship and/or construction defects affect expected losses and risks?

References

ASCE-7, 2007. Minimum Design Loads for Buildings and Other Structures. (ASCE Standard 7-05), Am. Soc. of Civil. Engineers, Reston.

Bjarnadottir, S., Li, Y., Stewart, M.G., 2011. A probabilistic-based framework for impact and adaptation assessment of climate change on hurricane damage risks and costs. Struct. Saf. 33, 173–185.

Bloomberg Business, 2017. The Most expensive US hurricane season ever: By the numbers. [Cited 2018 January 29]; Available from: https://www.bloomberg.com/news/articles/2017-11-26/the-most-expensive-u-s-hurricane-season-ever-by-the-numbers.

Boesch, D.F., Field, J.C., Scavia, D., 2000. The potential consequences of climate variability and change on coastal areas and marine resources. In: Report of the coastal areas and marine resources sector team, US National Assessment of the Potential Consequences of the Climate Variability and Change. US Global Change Research Program. National Oceanic and Atmospheric Administration Coastal Ocean Program Decision Analysis Series No. 21. Silver Spring, Maryland.

Chen, S.-C., Chen, M., Zhao, N., Hamid, S., Chatterjee, K., Armella, M., 2009. Florida public hurricane loss model: Research in multi-disciplinary system integration assisting government policy making. Gov. Inf. Q. 26, 285–294.

Culliton, T.J., 1998. Population: Distribution, Density and Growth. In: National Oceanic and Atmospheric Administration State of the Coast Report. Silver Spring, Maryland. Available: http://state-of-coast.noaa.gov/topics/html/pressure.html.

Elsner, J.B., 2006. Evidence in support of the climate change–Atlantic hurricane hypothesis. Geophys. Res. Lett. 33, 1–3.

Elsner, J.B., Kossin, J.P., Jagger, T.H., 2008. The increasing intensity of the strongest tropical cyclones. Nature 455, 92–95.

Emanuel, K., 2005. Increasing destructiveness of tropical cyclones over the past 30 years. Nature 436, 686–688.

Estrada, F., Wouter-Botzen, W.J., Tol, R.S.J., 2015. Economic losses from US hurricanes consistent with an influence from climate change. Nat. Geosci. 8, 880–884.

FEMA, 1999. HAZUS-MH hurricane model. [Cited 2009 April 7]; Available from: http://www.fema.gov/plan/prevent/hazus/hz_wind.shtm.

Huang, Z., Rosowsky, D.V., Sparks, P.R., 2001. Long-term hurricane risk assessment and expected damage to residential structures. Reliab. Eng. Syst. Saf. 74, 239–249.

IPCC, 2007. Climate change 2007: the physical science basis. In: Solomon, S., Qin, D., Manning, M., Chen, Z., Marquis, M., Averyt, K.B., Tignor, M., Miller, H.L. (Eds.), Contribution of Working Group I to the Fourth Assessment Report of the Intergovernmental Panel on Climate Change. Cambridge University Press, Cambridge.

Jain, V.K., Davidson, R., 2007. Forecasting changes in the hurricane wind vulnerability of regional inventory of wood-frame houses. J. Infrastruct. Syst. 13, 31–42.

Khanduri, A.C., Morrow, G.C., 2003. Vulnerability of buildings to windstorms and insurance loss estimation. J. Wind Eng. Ind. Aerodyn. 91, 455–467.

Klotzbach, P.J., 2006. Trends in global tropical cyclone activity over the past twenty years (1986–2005). Geophys. Res. Lett. 33.

Landsea, C.W., 2007. Counting Atlantic tropical cyclones back to 1990. EOS Trans. Am. Geophys. Union 88, 197–208.

Landsea, C.W., Harper, B.A., Hoarau, K., Knaff, J.A., 2006. Can we detect trends in extreme tropical cyclones? Science 313, 452–454.
Leicester, R.H., Bubb, C.T.J., Dorman, C., Beresford, F.D., 1979. An assessment of potential cyclone damage to dwellings in Australia. In: Cermak, J.E. (Ed.), Proc 5th Int. Conf. on Wind Engineering. Pergamon, New York, United States, pp. 23–36.
Li, Y., Ellingwood, B.R., 2006. Hurricane damage to residential construction in the US: Importance of uncertainty modeling in risk assessment. Eng. Struct. 28, 1009–1017.
Li, Y., Ellingwood, B.R., 2009. Risk-based decision making for multi-hazard mitigation for wood-frame residential construction. Aust. J. Struct. Eng. 9, 17–26.
Li, Y., Stewart, M.G., 2009. Effects of Climate Change on Cyclone Damage Estimation, the 10th International Conference on Structural Safety and Reliability (ICOSSAR). Osaka, Japan.
Miami-Dade County Department of Planning and Zoning, 2003. Housing in Miami-Dade County 2000. Planning Research Section, Miami.
Pielke, R.A.J., Pielke, R.A.S., 1997. Hurricanes: Their nature and impacts on society. In: Wiley. U.K, Chichester.
Pielke, R.A.J., Landsea, C., Mayeld, M., Laver, J., Pasch, R., 2005. Hurricanes and global warming. Bull. Am. Meteorol. Soc. 86, 1571–1575.
Pielke, R.A.J., Gratz, J., Landsea, C.W., Collins, D., Saunders, M.A., Musulin, R., 2008. Normalized hurricane damage in the United States: 1900–2005. Nat. Hazards Rev. 9, 29–42.
Pinelli, J.P., Torkian, B.B., Gurley, K., Subramanian, C., Hamid, S., 2009. Cost effectiveness of hurricane mitigation measures for residential buildings. In: The 11th Americas Conference on Wind Engineering (ACWE), San Juan, Puerto Rico.
Pryor, S.C., Barthelmie, R.J., Takle, G.S., 2009. Wind speed trends over the contiguous USA. In: IOP Conference Series: Earth and Environmental Science.
RealtyTrac Inc., 2008. Miami Beach. [Cited 2009 June 11]; Available from: http://www.realtytrac.com/states/Florida/Miami-Dade-County/Miami-Beach.html.
Reardon, G.F., Henderson, D., 1988. Cyclone risk assessment of houses in North Queensland. In: Butterworth, J. (Ed.), Australasian Structural Engineering Conference. 2. Structural Engineering Society of New Zealand, Auckland, pp. 1007–1014.
Roth, R.J., 1997. Insurable risks, regulation, and the changing insurance environment. In: Diaz, H.F., Pulwarthy, R.S. (Eds.), Hurricanes: Climates and Socioeconomic Impacts. Springer, Berlin, Germany, pp. 261–272.
Solomon, S., Qin, D., Manning, M., Chen, Z., Marquis, M., Averyt, K.B., Tignor, M., Miller, H.L. (Eds.), 2007. Climate change 2007: the physical science basis. Contribution of WG I to the 4th Assessment Report of the IPCC. Cambridge University Press, Cambridge.
Sparks, P.R., Schiff, S.D., Reinhold, T.A., 1994. Wind damage to envelopes of houses and consequent insurance losses. J. Wind Eng. Ind. Aerodyn. 53, 145–155.
Stewart, M.G., Li, Y., 2010. Methodologies for economic impact and adaptation assessment of cyclone damage risks due to climate change. Aust. J. Struct. Eng. 10, 121–136.
Stewart, M.G., Rosowsky, D.V., Huang, Z., 2003. Hurricane risks and economic viability of strengthened construction. Nat. Hazards Rev. 4, 12–19.
Stewart, M.G., Wang, X., Willgoose, G.R., 2014. Direct and indirect cost and benefit assessment of climate adaptation strategies for housing for extreme wind events in Queensland. Nat. Hazards Rev. 15 (4). 04014008(12).
US Census Bureau, 2009. State and county quick facts. [Cited 2009 April 14]; Available from: http://quickfacts.census.gov/qfd/states/12/12086.html.
Vickery, P.J., Skerlj, P.F., Twisdale, 2000. Simulation of hurricane risk in the US using empirical track model. J. Struct. Eng. 126, 1222–1237.

Vickery, P.J., Wadhera, D., Twisdale Jr., L.A., Lavelle, F.M., 2009. U.S. hurricane wind speed risk and uncertainty. J. Struct. Eng. 135, 301–320.

Wen, Y.K., 2001. Minimum lifecycle cost design under multiple hazards. Reliab. Syst. Saf. 73, 223–231.

White House, 2009. Global Climate Change Impacts in the United States. U.S. Global Change Research Program, Washington, D.C. http://downloads.globalchange.gov/usimpacts/pdfs/climate-impacts-report.pdf.

Zigomanis, A., 2007. Land Prices Still Challenge for New Housing Market. Property Council of Australia. [cited 2009 April 21]; Available from: http://www.propertyoz.com.au/Article/NewsDetail.aspx?id=15.

Further Reading

Knuston, T.R., McBride, J.L., Chan, J., Emanuel, K., Holland, G., Landsea, C., Held, I., Kossin, J.P., Srivastava, A.K., Sugi, M., 2010. Tropical cyclones and climate change. Nat. Geosci. 3, 157–163.

Pinelli, J.P., Simiu, E., Gurley, K., Subramanian, C., Zhang, L., Cope, A., Filliben, J.J., Hamid, S., 2004. Hurricane damage prediction model for residential structures. J. Struct. Eng. 130, 1685–1691.

Vickery, P.J., Skerlj, P.F., Lin, J., Twisdale Jr., L.A., Young, M.A., Lavelle, F.M., 2006. HAZUS-MH hurricane model methodology. II: Damage and loss estimation. Nat. Hazards Rev. 7, 94–103.

CHAPTER TEN

Housing Resilience to Wind-Induced Damage in Developing Countries

Kazuyoshi Nishijima
DPRI, Kyoto University, Kyoto, Japan

10.1 Introduction

There seems a general trend in the developing countries worldwide that traditional construction be replaced with "modern" construction that partially or totally utilise industrial products as construction materials. It is, however, doubtful whether such a trend leads to the improvement of local construction. Here, the improvement refers not only structural performance but also other performances related to, for example, indoor climate or daily use. As for the structural performance, poor construction work often prevents the structure from achieving the anticipated performance. Poor construction work can occur, since, for example, the new construction method is different from the method that has been utilised for traditional local construction; opportunities to learn the new construction methods are not offered; and tools required for the new construction methods are not available. Limited access to construction materials due to poor supply market and/or relatively high prices can result in insufficient use of the materials; for example, less nails are used, thinner iron sheets are used, less cement is mixed with more sand and water, little rebars are installed, etc. As for the indoor climate and daily uses, the new constructions do not always appreciate the architectural characteristics of traditional constructions, which are believed to have been empirically adapted considering their local climate and lifestyles as well as dominating natural hazards. For example, walls made with bamboo have been utilised in tropical regions, which enable gentle winds to pass through rooms in houses. This is not possible with walls with concrete blocks. A thick thatching roof helps in isolating heat from sun light during the day, whereas the galvanised iron (GI) sheet roof accelerates heating.

Climate Adaptation Engineering
https://doi.org/10.1016/B978-0-12-816782-3.00010-3

These do not mean that such a trend be stopped. Rather, it is to say that improvement of traditional construction especially for housing requires balanced improvement considering not only from a disaster risk reduction perspective but also from architectural and socioeconomical perspectives.

As stated, there is a shift from traditional construction to construction using industrial materials. The ultimate goal of the shift is to realise disaster resilient construction. The approach to achieve the goal by this way is indicated with the arrows at the bottom and right in Fig. 10.1. Note that as written in the figure, construction using industrial materials in developing countries is often built by nonexperts, hence, the performance of these constructions is often poor. An alternative approach to achieving resilient construction is to advance traditional construction methods. This approach is illustrated with the arrow at the left and top in the figure. Here, the advancement means that construction is improved maintaining the continuity of their architectural characteristics that reflect the local climate, culture, lifestyles, etc. One of the successful examples of this approach is the traditional houses in Japan. The traditional houses in Japan are made primarily of timber for main structure and clay tiles for roofing materials. These houses used to be vulnerable to earthquakes, fires, and strong wind. However, weak points have been overcome with scientific and technological development and currently these houses attain a certain level of resilience against natural hazards. An attempt is being made by the author's group under a Japan International Cooperation Agency (JICA) Grassroots project "Advancement of

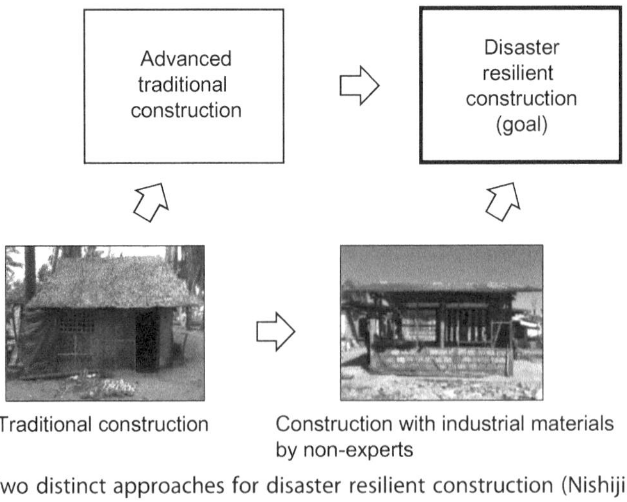

Fig. 10.1 Two distinct approaches for disaster resilient construction (Nishijima, 2016).

Vernacular Constructions in Tanna Island" in the Republic of Vanuatu focusing on the local traditional house called "Nimaleten", see Fujieda and Sasaki (2016) for more details.

Science and technology can and should support the developments in both approaches by creating evidence, with which decision-makers at different levels such as individuals, governments, nongovernmental agencies, supranational donor agencies, etc., can make rational decisions on the investment into disaster risk reductions. Having said that, it should be emphasised that very little evidence to support such decisions have been produced so far, compared with the amount of evidence to support decisions on disaster risk reductions for engineered construction in the developed countries.

Motivated by these, this chapter introduces a procedure for creating such evidence by a probabilistic risk analysis along with an example of the assessment of wind-resistant performance of non-engineered house.[1] The example aims not only at showing the numerical result of the wind-resistant performance of the target house, but also at serving as a guide for future works on the assessment of wind-resistant performance of non-engineered construction. As background information, examples of non-engineered construction in the developing countries and typical damages to structures are first explained.

10.2 Housing in Developing Countries

10.2.1 Non-Engineered Construction

Developing countries located in the areas prone to tropical cyclones suffer from wind damage to infrastructure in general and houses in particular. Part of the reasons why they suffer can be accounted for by the fact that many of the constructions are non-engineered. Here, non-engineered construction refers to construction "*built with little or no intervention by engineers*" (UNESCO, 2016) and therefore, their structural performances are not checked in similar manners as for constructions designed and supervised by engineers. As a consequence, the structural performance of non-engineered constructions is estimated to be relatively low and generally not known quantitatively.

[1] For attempts on seismic risk reduction of non-engineered constructions, see, for example, Arya et al. (2013).

Non-engineered construction may be divided into two groups: traditional constructions using materials easily accessible such as trees, leaves, bamboos, canes, etc. as main construction materials and constructions using industrial materials such as GI sheet, concrete blocks, and lumber. There can also be seen the mixture of these two. Fig. 10.2 shows examples of non-engineered houses. These pictures were taken in suburban areas in each country. In these areas, typical house is one storey (sometimes elevated), has a gable roof as seen in Fig. 10.2A–E, whereas there also exists traditional

(A) Bangladesh (Daulatpur)
(B) Myanmar (Mandalay)
(C) The Philippines (Leyte Island)
(D) The Philippines (Panay Island)
(E) Vanuatu (Tanna Island)
(F) Vanuatu (Tanna Island)[1]

Fig. 10.2 Examples of non-engineered houses in developing countries.

construction that has unique shapes as seen in Fig. 10.2F. Except for urban areas, people are often living in such non-engineered (or partially non-engineered) houses. Since these houses are more prone to natural hazard events, it is important and urgent to install appropriate countermeasures to mitigate meteorological disasters that can be intensified by climate change.

10.2.2 Typical Wind-Induced Damage to Construction

For the purpose of providing a general overview of wind-induced damages, typical wind-induced damages are introduced not only for non-engineered houses but also for engineered construction.

Strong wind causes serious damages to construction. Main causes of the damages are strong wind pressure and flying debris impact. Insufficient resistance is also a source to increase damage occurrence possibility. Fig. 10.3

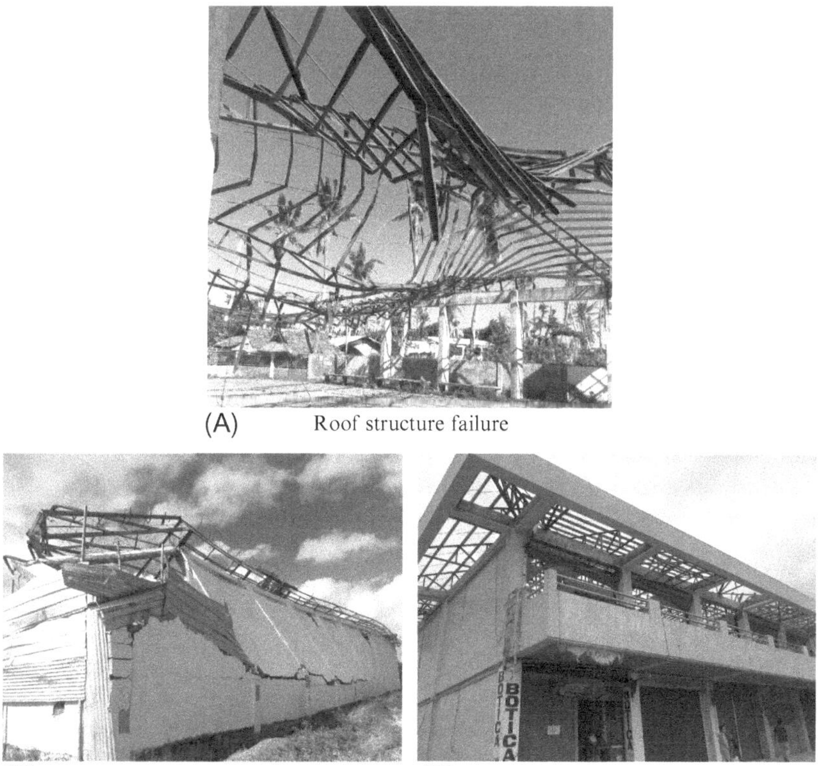

(A) Roof structure failure

(B) Roof and wall structure failure (C) Roofing material failure

Fig. 10.3 Failures due to strong wind during Typhoon Haiyan (2013).

shows examples of failures due to strong wind during Typhoon Haiyan in 2013. Fig. 10.3A shows the failure of a sport facility, which was an open structure, that is, no walls. The roofing materials were lost and the roof structure failed. Fig. 10.3B shows the combination of the failures of roof and wall. Fig. 10.3C shows the failures of roofing material of a commercial building.

Typical damage to smaller buildings and their weak pointes are now explained. The pictures in Fig. 10.4 were taken during a survey after Typhoon Haiyan. As seen in the left picture, ceilings are damaged. The possible failure mechanism is the increase of the internal pressure in the attic space because of openings presented between the wall and roof. These openings are sometimes installed in tropical regions on purpose for natural ventilation but often due to insufficient finishing. In the middle figure, damage to roof along the long eave can be identified. On the one hand, a long eave is preferable since it can create a wider shade area and also it can prevent water from entering into the rooms. On the other hand, a long eave is more vulnerable to strong wind. The right figure shows a piece of failed roof sheet, which is thin and corroded. As these indicate, the problems will not be solved by simply identifying and improving the weak points from a wind engineering viewpoint. There often exist conflicts with the comfortability in normal situations—no ventilation in rooms and no eave on roofs would not be comfortable—and structural performance against natural hazards. Needless to say, the absence or difficulties in accessing quality material and cost pressure are other important issues.

(A) Openings (B) Long eave (C) Thin and corroded roof sheet

Fig. 10.4 Weak points against strong wind.

10.3 Case Study: Assessment of Wind-Resistant Performance of a Non-Engineered House in Leyte Island, The Philippines[2]

10.3.1 Target House

In November 2013, Typhoon Haiyan caused damages to people and assets in southeast Asia, including the Philippines. The typhoon made landfall on Samar Island located in the middle of the Philippines with its peak intensity, and passed through Leyte Island, Cebu Island, and Panay Island. According to the Japan Meteorological Agency (JMA) best track data (as of 18 December 2013), the minimum central pressure is estimated to be 895 hPa and the maximum sustained wind speed is estimated to be 64 m/s (=125 kt). It is reported by the National Disaster Risk Reduction and Management Council (NDRRMC) that 6293 people were killed, 28,689 people were injured, and 1061 people are missing. It is also reported that 550,928 houses were totally destroyed and another 589,404 houses were damaged.

The Disaster Prevention Research Institute, Kyoto University (DPRI-KU) and the University of the Philippines, Diliman (UPD), made joint surveys on houses in Guiuan of Samar Island, the east coast of Leyte Island, eastern part of Panay Island in June 2014. Among other purposes of the survey, information on typical houses and construction materials locally utilised were collected for the purpose to assess the wind-resistant performance of a typical house. Based on the survey, a target house was identified. The target house considered in the case study is shown in Fig. 10.5. In the figure, explanation of this house is added. The following sections introduce to the procedure and results of the assessment of wind-resistant performance of the house.

10.3.2 Assessment Procedure

The assessment of the wind-resistant performance of the target house takes its basis in the standard reliability-based risk assessment procedure, see for example, Zhang et al. (2014). The wind-resistant performance is represented in terms of probability of failure. The steps for the assessment are illustrated in Fig. 10.6. As shown in the figure, the wind vulnerability assessment explicitly considers failure mechanisms by identifying relevant failure modes

[2] The summary on the results shown in this section was included in the Proceedings of the Architectural Institute of Japan Annual Convention in 2015 (in Japanese).

(a) The house is located in a village at the east coast of Leyte island.
(b) The picture was taken approximately 7 months after the impact of typhoon Haiyan.
(c) The house is under reconstruction after the total collapse of the former house.
(d) The house is reconstructed by the inhabitants, who are not experts of construction; a typical non-engineered construction.
(e) Approximate size of the house is 4.0 m width, 4.4 m depth and 3.2 m height.

Fig. 10.5 Target house in case study.

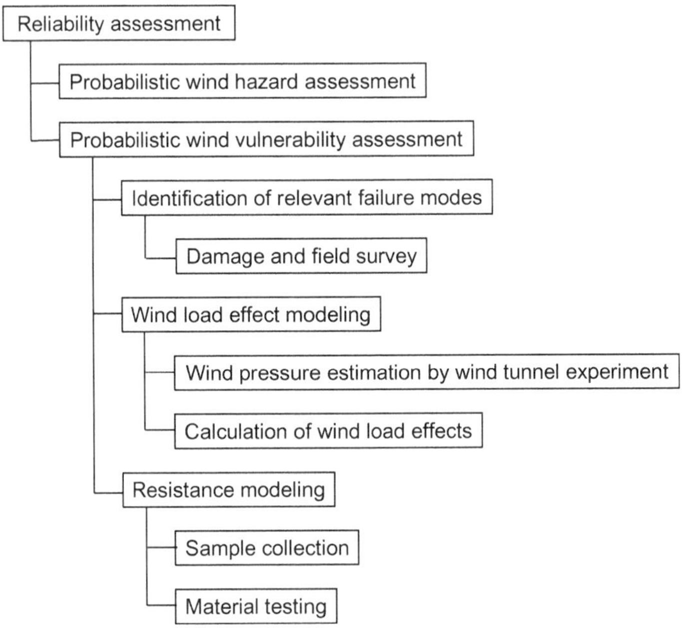

Fig. 10.6 Steps for wind-resistant performance assessment.

and by modelling load effects and resistance that concern the respective failure modes. The advantages of this approach are twofolded: first, the modelling is generic in a sense that parts of the modelling can be replaced so that the modelling can easily adapt to other target structures and second, it enables one to assess the efficiency of the upgrading of certain elements of structures on the increase of the wind-resistant performance. The second advantage is illustrated along with the discussion of the assessment results. In the subsequent subsections, these steps are explained.

10.3.3 Identification of Relevant Failure Modes

Relevant failure modes are identified based on the surveys on typical houses including the target house and their damages in the affected areas conducted by the joint team of the DPRI-KU and UPD after typhoon Haiyan. It was found that most typical failure mode was the failures of fasteners that connect roofing sheet and roof frames. The roofing sheet material is typically made of a rather thin GI sheet, which are often partially or totally corroded. The roof frames are made of coco lumber. The fasteners utilised are nails with so-called "umbrella" washer curved to fit the curvature of corrugated iron sheet. The surface of the nail is smooth or helically carved. The other likely failure modes were found to be failures at connections between structural components such as beams, columns, and foundation. This is because the connections are connected by just a few nails; metal tapes/plates; or bolt joints are not usually utilised. Failure of the construction materials of the structural frame such as beam and columns are not considered, assuming that connections of the structural frame fail before the failure of the structural frames. Fig. 10.7 shows the construction materials utilised in the target house and identified failure modes, for which the probabilities of failures are assessed.

10.3.4 Wind Load Effect Modelling

10.3.4.1 Wind Tunnel Experiment

A wind tunnel experiment is carried out using the Eiffel-type wind tunnel at the DPRI-KU. The cross section of the wind tunnel is 2.5 m width and

Fig. 10.7 Construction materials and identified relevant failure modes.

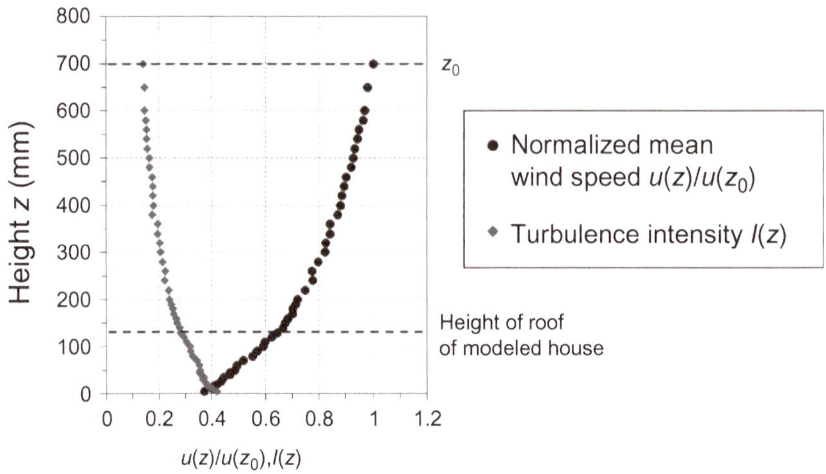

Fig. 10.8 Wind profile in wind tunnel experiment.

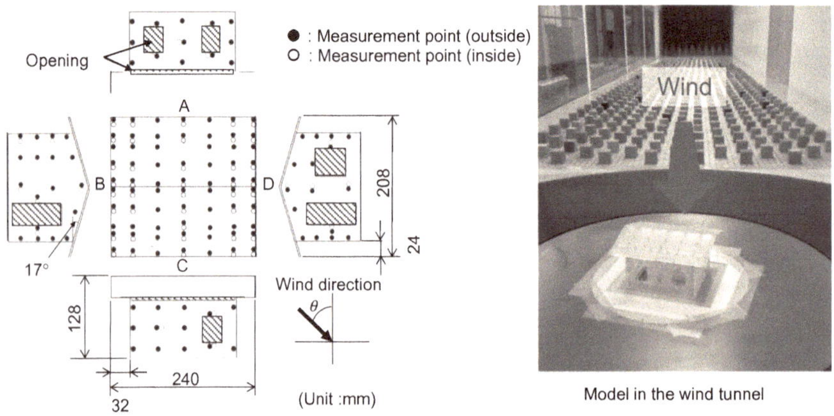

Fig. 10.9 Model detail (left) and model placed in wind tunnel (right).

2.0 m height at the measurement section. A boundary layer flow is developed with the exponent of $\alpha = 0.26$ and the turbulence intensity of $I = 0.3$ at the height of the roof of the wind tunnel model of the target house. The vertical profile of the mean wind horizontal speed and the turbulence intensity are shown in Fig. 10.8.

The model for the wind tunnel experiment is designed and built based on the measurement of the target house during the field survey. The geometrical scale of the model is 1:15. The detail of the model is shown in Fig. 10.9. The model has several openings that represent the windows and doors in the sides A–D, and spaces between the roof and the walls in A and C. The wind tunnel experiment is carried out under the situation where all these openings are open, assuming that the windows and doors could fail under strong wind. In all, 194 measurement points are placed inside and outside of the roof and the walls in order to measure the wind pressures at respective locations.

The model is placed in the middle of the turn table, see Fig. 10.9 (right). The wind direction relative to the model is changed between 0° and 345° at the interval of 15°. For each direction, six time series of the wind pressures at all the measurement points are measured. The duration of the measurement of a single time series corresponds to 10 min in full scale. The effective wind pressures at each location are calculated as the difference between the wind pressure outside and wind pressure inside. Then, the time series of the effective wind pressure coefficients are calculated by normalising the time series of the effective wind pressure with the reference velocity pressure. The parameters of the wind tunnel experiment are summarised in Table 10.1.

An example of the obtained time series of the wind pressure coefficients and the effective wind pressure coefficients are shown in Fig. 10.10. The example is the time series at a measurement point located at the ridge end under the wind direction of 225°. Note that in the right figures, the horizontal axis is the time in the full scale. As seen in the figure, the time series of the wind pressure coefficient inside is close to zero and its fluctuation is minor, whereas the time series of the wind pressure coefficient outside ranges approximately from −1 to −4, showing large negative values and peaks. The calculated time series of the effective wind pressure coefficient is shown in the figure below. The times series of the effective wind pressure coefficients thus obtained at all the measurement points are utilised to model the wind load effects that concern the relevant failure modes.

Table 10.1 Parameters of Wind Tunnel Experiment

Reference Wind Speed at Roof Height	**6.5 m/s**
Exponent of the vertical mean wind speed profile	$\alpha = 0.26$
Wind direction	0° to 345° at 15° interval
Sampling frequency	200 Hz
Geometrical scale	1:25
Time scale	42:200
Assumed wind speed in full scale	34 m/s
Averaging time in full scale	1 s
Evaluation time in full scale	600 s
Number of time series	6

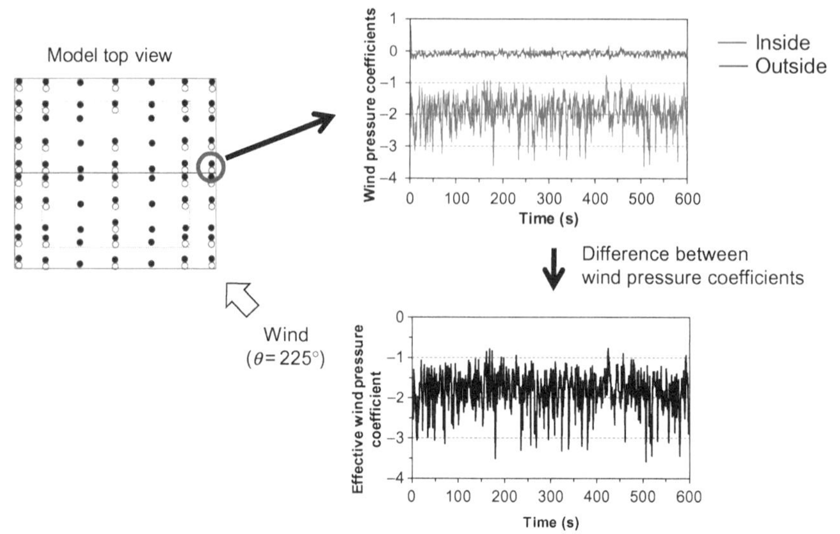

Fig. 10.10 Time series of wind pressure coefficients.

10.3.4.2 Calculation of Wind Load Effects

Wind load effects are calculated for the three failure modes identified in the previous subsection; that is,

- fastener failure,
- roof-column connection failure, and
- column-base connection failure.

The structural frame assumed in the assessment is shown in Fig. 10.11 (left). Since the structural components themselves of the frame are assumed not to fail, fasteners and connections are of most interest. The illustration in the middle of the figure is a model simplified down to the elements related to those failure modes. The simplified model consists of *objects* and *links*. The objects are GI sheet, roof and column, and the links are fasteners,

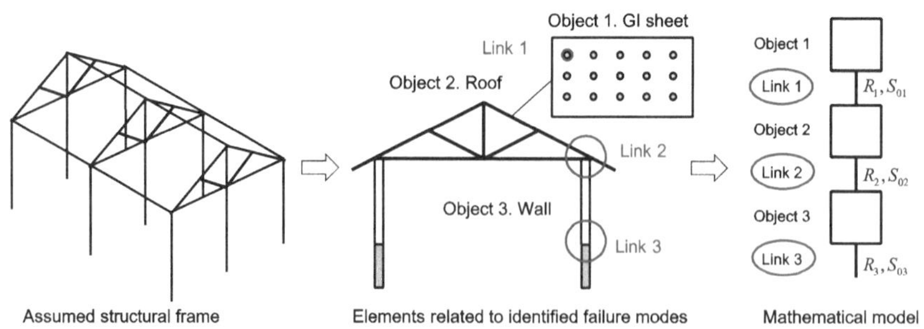

Fig. 10.11 Modelling of structure up to identified failure modes.

the connection between the roof and column, and the connection between the column and foundation. Considering that the objects are assumed not to fail and the links serve to connect the objects, the model is represented as shown in the mathematical model in Fig. 10.11 (right). The fastener failure corresponds to the failure of Link 1, the roof-column connection failure and the column-foundation connection failure corresponds to the failures of Links 2 and 3, respectively. Each of the failures is defined in terms of a limit state function G_{0i} written as:

$$G_{0i} = R_i - S_{0i} \tag{10.1}$$

where R_i and S_{0i} are, respectively, the resistance and the load effects of the link i concerning the failure mode. The subscript "0" signifies the case where all the links are intact. The differentiation by the state (intact or fail) of the links is required because the wind load effects change according to the state of the links. For example, if object 1 (GI sheet) is lost due to the failure of link 1 (fasteners), then wind load acting on the GI sheet is also lost. Therefore, the wind load effects on Links 2 and 3 change. A detailed explanation is provided in Section 10.3.6.1.

The wind load effect for fasteners failure is first considered. The wind load effect on fastener is represented in terms of wind force acting of the tributary area of each fastener. The wind force is obtained by multiplying the effective wind pressure coefficients with the tributary area and reference velocity pressure. The wind force is time varying, since the effective wind pressure is time varying. Thus, the peak wind force is considered as the indicator of the wind load effect on fastener. Note that the peak wind force is a random variable. Here, the evaluation time of the peak value is assumed to be 10 min, which corresponds to the averaging time of the reference wind speed adopted in the probabilistic hazard assessment.

The mean value and the standard deviation of the peak effective wind pressure coefficient are estimated from the time series. Fig. 10.12 shows the estimated mean value and the standard deviation for different wind directions at the point near the verge. As shown, the peak coefficient takes the minimum value at the wind direction of 300°.

The wind load effects for the connections between roof and column and the connections between column and foundation are considered. The wind load effects on these connections considered are the vertical reaction forces, which is the sum of the effects of wind load and dead load. In the following, the modelling of the effects of wind load is explained. In Fig. 10.13, three structural frames at the transverse direction are illustrated. Wind loading acting on the roof is carried by the three structural frames according to their

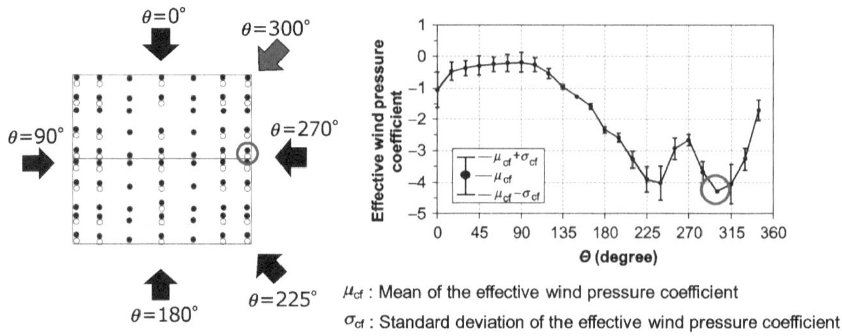

Fig. 10.12 Calculated mean and standard deviation of the wind pressure coefficient.

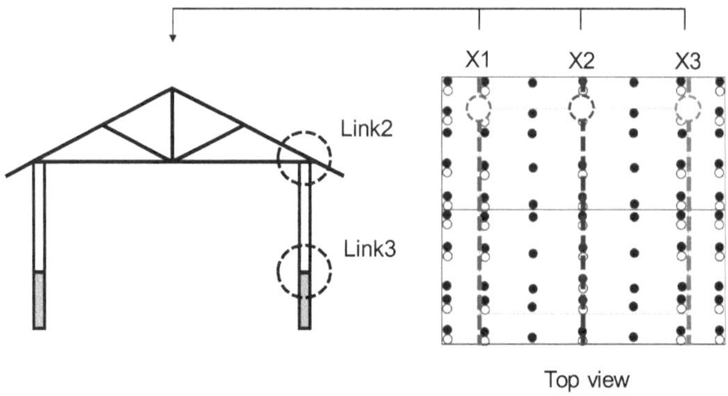

Fig. 10.13 Three frames at transverse direction.

tributary areas. The reaction forces are calculated using the times series of the wind pressure coefficients with the assumed tributary area and reference velocity pressure. The reaction forces are calculated in a quasi-static manner assuming dynamic effects are neglected. The reaction force is time varying, since the wind pressures are time varying. Thus, the peak reaction force is considered as the indicator of the wind load effect on the connections. Note that the peak reaction force is a random variable. Here, the evaluation time of the peak value is assumed to be 10 min, which is the same as the case of wind load effect modelling of fasteners.

The mean values of the normalised reaction forces for the structural frames X1, X2, and X3 at Links 2 and 3 are shown in Fig. 10.14. Note that the normalised reaction forces are wind force coefficients normalised by the product of the reference velocity pressure and the representative area. Negative value indicates that the wind forces act in ways that the structural frame may lift up.

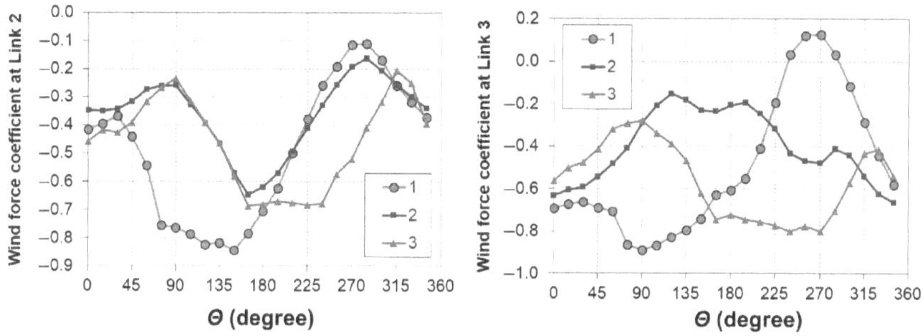

Fig. 10.14 Calculated mean normalised reaction force at Links 2 and 3.

10.3.5 Resistance Modelling

10.3.5.1 Sample Collection

Construction materials utilised for houses in the developing countries are different and often not available in other countries. Further, it is difficult to identify manufacture information of the construction materials. Therefore, for the assessment of the target house, similar construction materials were purchased at a local hardware shop, see Fig. 10.15. The purchased construction materials are corrugated GI sheet, nails with umbrella washer and coco lumber. The thickness of the GI sheet is 0.2 mm (measured at the laboratory), and the height of the GI sheet is 15 mm. The nail has an umbrella washer on the top, the length of the nail is 65 mm, and the diameter of the nail is 3.8 mm. The surface of the nail is either smooth or helical. These samples of the construction materials are employed for the material testing.

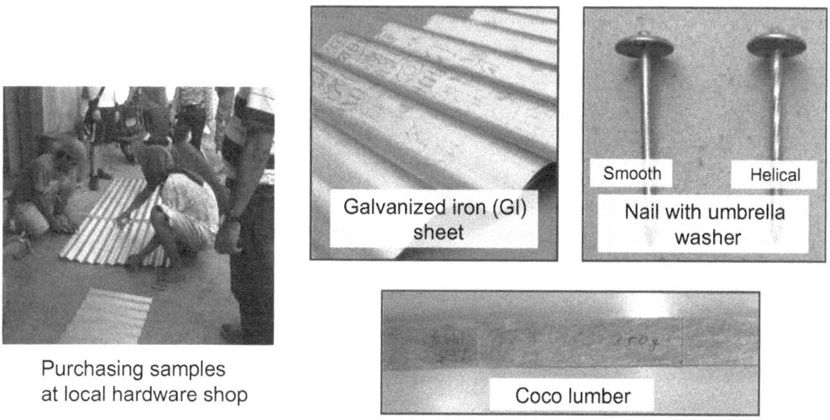

Fig. 10.15 Purchasing samples for material testing at local hardware shop.

10.3.5.2 Material Testing Preparation

The resistances of the materials are estimated for the three failure modes. For the fastener failure mode, three failure mechanisms are considered. For the failures of the connection between roof and column and the connection between column and base, the same single failure mechanism is considered. These failure mechanisms are illustrated in Fig. 10.16.

A testing device was developed in order to measure the resistances for each failure mechanism, see Fig. 10.17 (left). The setup of the specimen for each testing is also illustrated. The testing device consists of an actuator and load cell, to which specimen is attached. These are supported by a frame. The actuator controls the displacement and the load cell measures the force acting on the specimen. The force acting on the specimen is gradually increased by controlling the displacement by the actuator. The resistance of the specimen is estimated as the maximum force acting on the specimen during the testing. The examples of the failures in the testing are shown in Fig. 10.17 (right). The specimens of the testing are prepared as follows:

(1) *Pull-out failure*: a nail is nailed on the flat surface of a piece of coco lumber. The depth of the nail is 50 mm, considering the clearance due to the corrugation of the GI sheet. Two types of nails are utilised (smooth and helically carved shanks, hereafter called types S and H).

(2) *Pull-over failure*: a nail is nailed on the top of a piece of the corrugated GI sheet. The specimen is placed on the testing device upside down as shown in Fig. 10.17. Two types of nails are utilised (smooth and helically carved shanks).

(3) *Shear-tear-out*: a nail is nailed on the top of a piece of the corrugated GI sheet and a piece of a face of the coco lumber, and the force is applied parallel to the face of the coco lumber, see Fig. 10.17.

(4) *Connection failure*: two pieces (A and B) of coco lumber are nailed through with a nail. The depth of the nail to the piece B is 30 mm. Note that for this test, nails obtained in Japan are utilised in this testing.

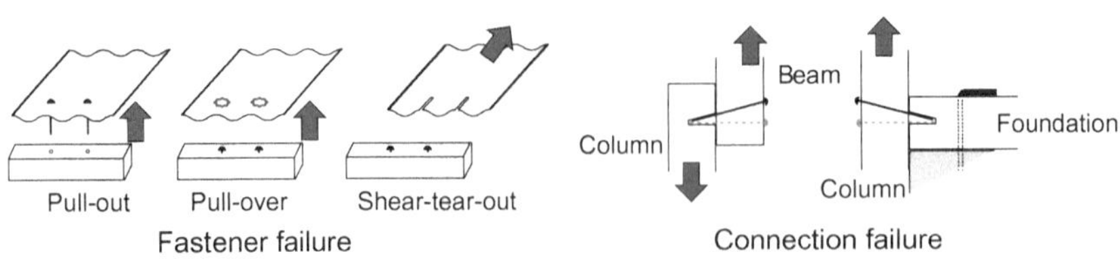

Fig. 10.16 Failure mechanisms considered in material testing.

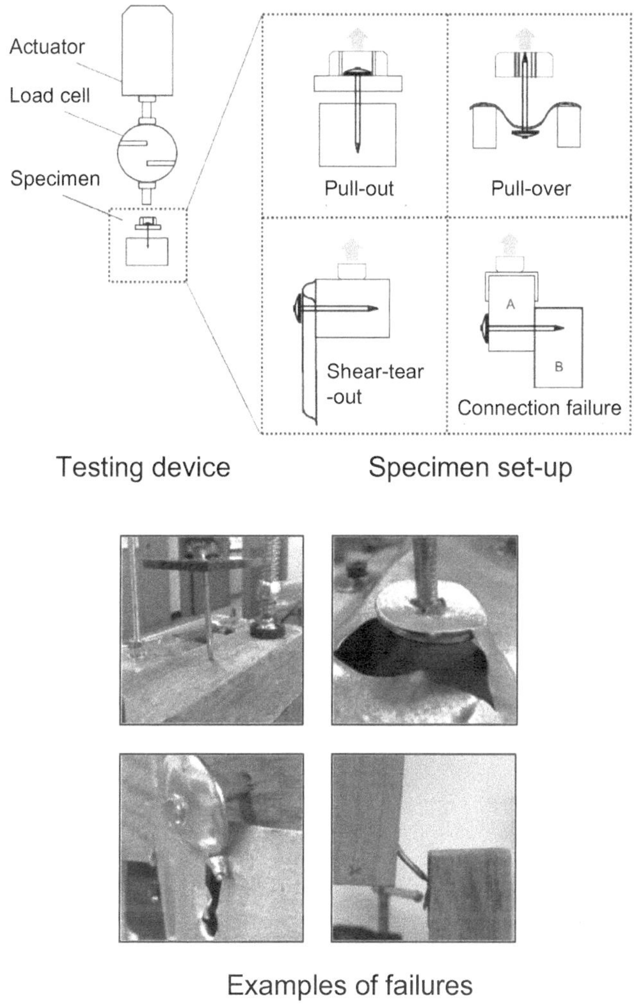

Fig. 10.17 Testing device with specimen setup (left) and examples of failures (right).

10.3.5.3 Material Testing Result

For the testing of pull-out failure, the mean values of the resistance for types S and H are 718 N and 656 N, respectively. In case of type S, the carrying load reduces once it reaches the maximum load, whereas in case of type H, the carrying load fluctuates after it reaches the maximum load, see Fig. 10.18. Large variability of the estimated resistance is observed for both types of nails. Part of the variability may be accounted for by the density of coco lumber, as positive correlation between the resistance and the density is

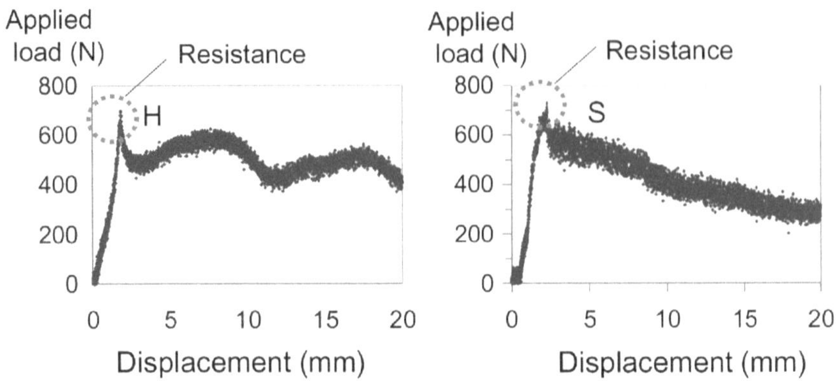

Fig. 10.18 Load–displacement relationships for types H and S in pull-out failure test.

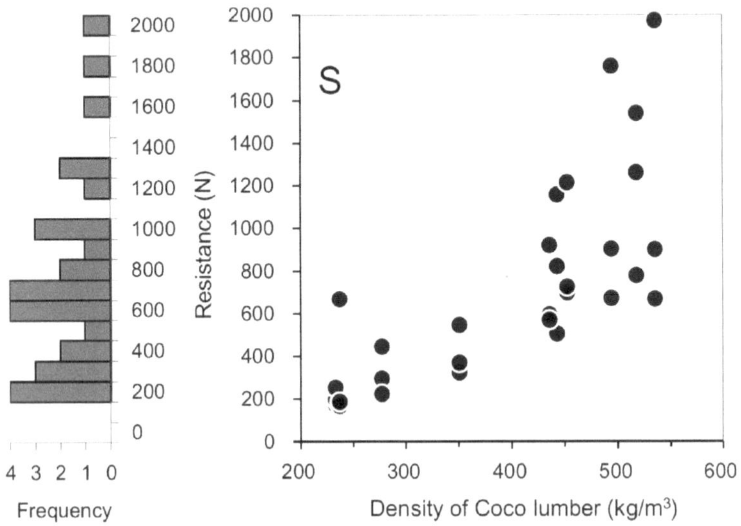

Fig. 10.19 Relationship between resistance and density of coco lumber for pull-out failure in case of type-S nail.

observed. Fig. 10.19 shows the correlation for the case of type-S nail. Note that in the preparation of the specimen, a few pieces of coco lumber are cut from the same lumber. Nevertheless, the large variation in the density is observed. Cross sections of coco lumber with high and low densities are shown in Fig. 10.20.

For the testing of pull-over failure, failure of umbrella washer was also observed, which is not assumed in the design of the testing. Thus, the

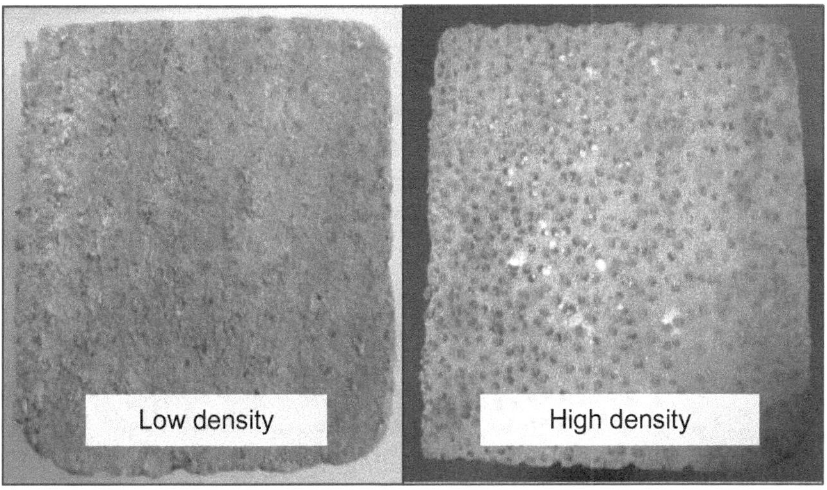

Fig. 10.20 Cross section of coco lumber pieces with low and high densities.

resistance for the pull-over failure is estimated for the case of pull-over failure and the case of umbrella washer failure. The mean value of the resistance for the former failure is 1755 N and the mean value of the resistance for the latter failure is 1038 N.

For the testing of shear-tear-out failure, the mean value of the resistance is estimated to be 514 N. For the testing of connection failure, the mean value of the resistance is estimated to be 1179 N. The resistance shows a positive correlation with the density of coco lumber as shown in Fig. 10.21. The testing results for all the failure modes are summarised in Table 10.2.

By comparing the statistics of the resistances for pull-out failure and pull-over failure, it can be said that the pull-out failure is more likely than the pull-over failure given that the density of coco lumber is unknown. Thus, for the following analysis, the pull-out failure is considered. However, if the density of coco lumber is approximately 500 kg/m^3 or larger, the resistances of the two types of failure modes are comparable.

10.3.6 Reliability Assessment

10.3.6.1 Reliability Assessment Conditions

The wind-resistant performance of the target house is assessed in terms of probabilities of failure for the considered failure modes. The probabilities of failure are calculated based on the mathematical model shown in

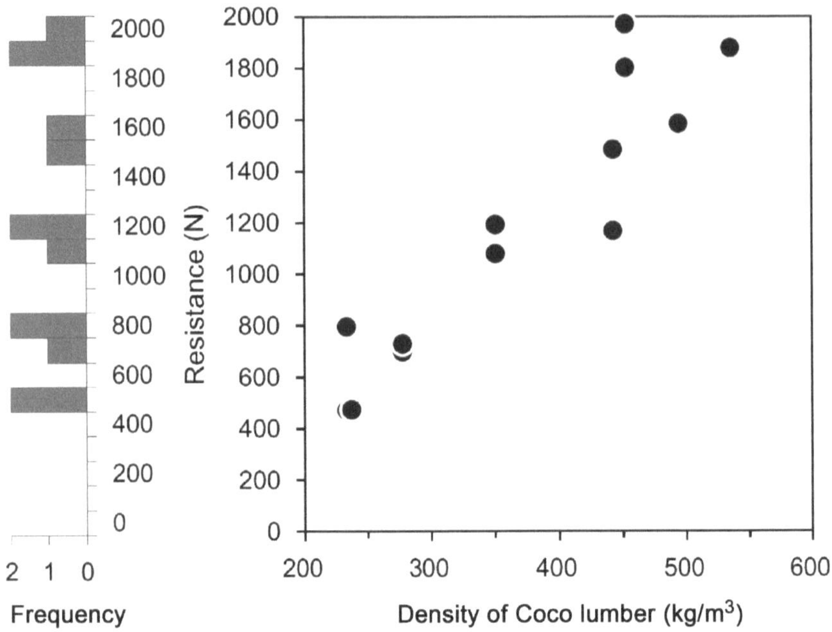

Fig. 10.21 Relationship between resistance and density of coco lumber for connection failure.

Table 10.2 Material Testing Results

	Pull-Out	Pull-Over	Shear-Tear-Out	Connection
Mean value of resistance [N]	718 (type S) 656 (type H)	1,755 (pull-over) 1,038 (umbrella)	514	1,179
Standard deviation of resistance [N]	472 (type S) 458 (type H)	103 (pull-over) 104 (umbrella)	153	529
Number of samples	30 (type S) 30 (type H)	3 (pull-over) 3 (umbrella)	6	13

Fig. 10.6. As previously explained, the loads acting on the links change upon the failure of the other links. In the left figure of Fig. 10.22, the situation where all the links are intact is shown. The first subscript 0 of the variable S_{0i} representing the load acting on Link i ($i=1, 2, 3$) signifies that all the links are intact. In the middle figure, Link 1 failed and Links 2 and 3 are intact. The first subscript 1 of the variable S_{1i} representing the load acting on Link i ($i=2, 3$) signifies that Link 1 failed. In the right figure, Link 2 failed. The first subscript 2 of the variable S_{2i} representing the load acting on Link i ($i=3$)

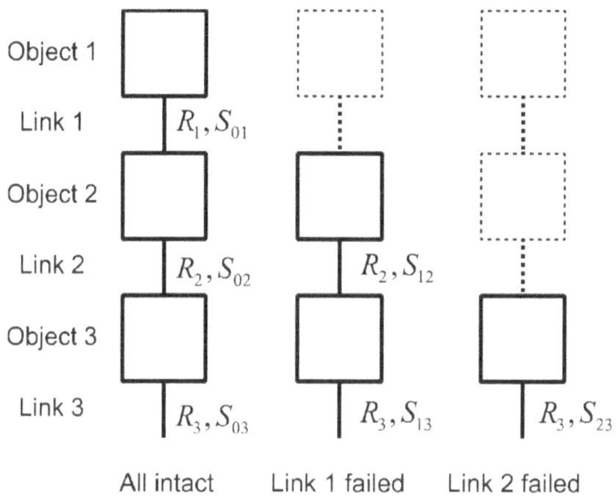

Fig. 10.22 Changes of loads depending on the states of links.

signifies that Link 2 is failed. Note that in general the probability distributions of S_{ji} for each i differ since wind load changes as part of house is lost.

The load acting on each link is assumed to be the sum of the dead load D_{ji} and the wind load W_{ji} (upward positive) as follows:

$$S_{ji} = W_{ji} - D_{ji} \tag{10.2}$$

Note that D_{ji} and W_{ji}, respectively, represent the dead load and wind load acting on Link i upon the failure of Link j. The wind load is expressed as:

$$W_{ji} = C_{ji} Q A_i \tag{10.3}$$

where C_{ji} is the wind pressure/force coefficient, which is obtained from the wind tunnel experiment and the structural analysis as described before, A_i is the reference area, and Q is the reference velocity pressure, which is written as:

$$Q = \frac{1}{2} \rho V^2 \tag{10.4}$$

where ρ is air the density, which is assumed to be $1.2\,\text{kg/m}^3$ and V is the annual maximum 10-min sustained wind speed (m/s) at the reference height, which is assumed to follow the Gumbel distribution with the 50-year return wind speed equal to 36 (m/s) and the coefficient of variation equal to 30%. The dead load is estimated based on the materials used. The limit function is expressed as:

$$G_{ji} = R_i - S_{ji} \tag{10.5}$$

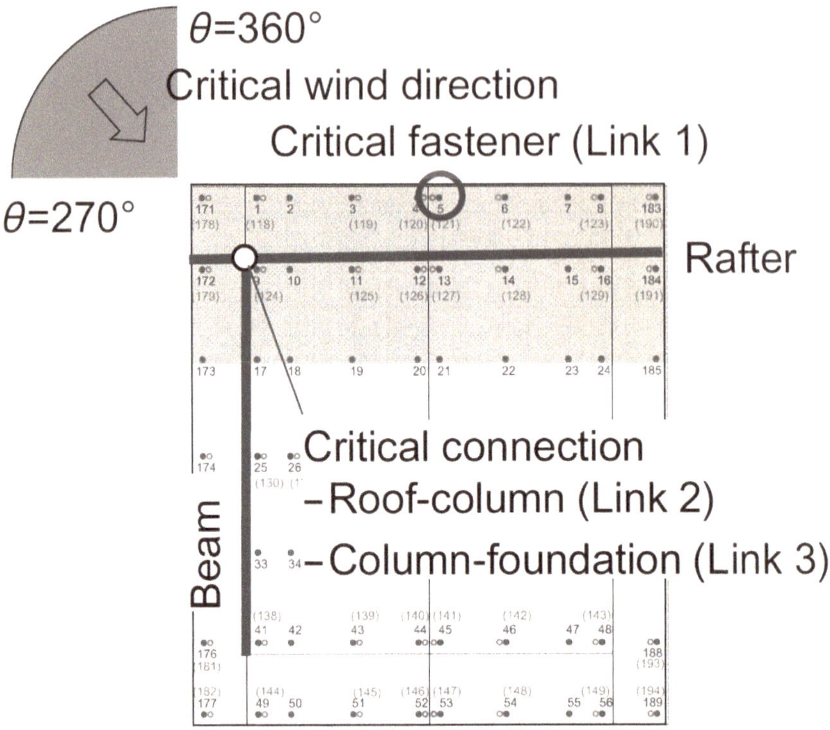

Fig. 10.23 Considered links and assumed wind direction.

where the resistance R_i is modelled based on the material test results. The probabilities of failure for Links 1–3 as indicated in Fig. 10.23 are calculated assuming that the annual maximum wind speed occurs with the most critical wind direction (i.e., between 270° and 360°). Here, it is assumed that all the coloured roof area is lost by a progressive failure if Link 1 is lost. The summary of the random variables and constants assumed in the assessment is shown in Table 10.3.

10.3.6.2 Reliability Assessment Results and Discussion

The calculated annual failure probabilities are shown in Fig. 10.24. Under the given assumption (as indicated in the figure with "nominal"), the probability of failure of Link 1 is approximately 4% and the probability of failure of Link 3 (i.e., the whole main structure is destroyed) is approximately 40%. Considering that the most critical wind direction is assumed in the analysis and other conservative assumptions made herein, the actual probability of failure can be smaller, for example, the order of 10%, which is to say the

Table 10.3 Summary of Random Variables and Constants Assumed in the Assessment
Random Variable and Constants (Unit: [N] except for C_{ji})

		Load		
Link	**Resistance**	**All Intact**	**Link 1 Failed**	**Link 2 Failed**
1	$R_1 \sim LN(6.29, 0.63)$	$C_{01} \sim G(5.4, 0.30)$ $D_{01} = 2$	–	–
2	$R_2 \sim LN(7.62, 0.46)^a$	$C_{02} \sim G(0.44, 0.014)$ $D_{02} = 100$	$C_{12} = 0$ $D_{12} = 40$	–
3	$R_3 \sim LN(7.62, 0.46)^a$	$C_{03} \sim G(0.79, 0.016)$ $D_{03} = 200$	$C_{13} = 0$ $D_{13} = 140$	Fail[b]
$A_1 = 0.105,\ A_2 = 10.60,\ A_3 = 10.69\ [m^2]$				

LN stands for the log normal distribution and *G* stands for the Gumbel distribution.
[a]Links 2 and 3 are assumed to consist of two nails.
[b]Structure is assumed to fail if Link 2 fails.

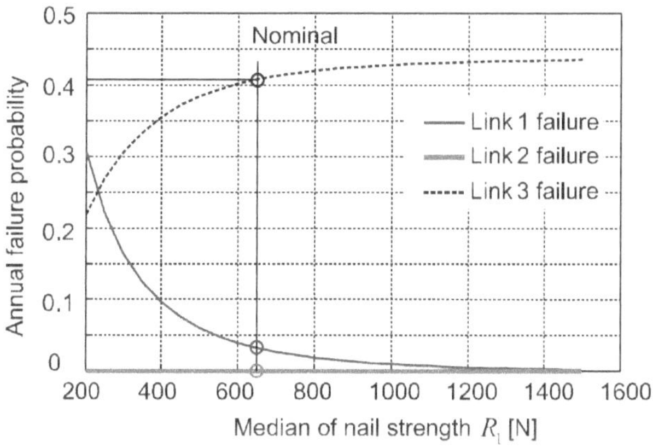

Fig. 10.21 Changes of loads depending on the states of links.

expected return period of the failure is the order of 10 years.[3] Now, consider the impact of the increase of the resistances on the probability of failure by utilising the developed model with changing the parameter values. If the resistances of the three links are increased by a factor of two (more precisely, the median values are increased by a factor of two without changing the coefficient of variation), the probability of failure for Link 3 is decreased by a factor of four; if the resistance of the three links are increased by a factor of three, the probability of failure for Link 3 is decreased by a factor of 10;

[3] Since the failures in different years are not independent due to the dependence of the resistance, this is approximation.

Table 10.4 Changes of Annual Failure Probability for Link 3 as a Function of Resistance

Resistance	Annual Failure Probability	Expected Return Period of Failure (year)
Nominal	0.1	10
Twice the resistance	0.025	40
Triple the resistance	0.01	100

namely, the expected return period becomes the order of 100 years. The increase of the resistance may be achieved by simply increasing the number of nails at the connections or its equivalents. Table 10.4 summarises this.

Fig. 10.24 also shows the sensitivity of the median of the resistance of Link 1 on the probabilities of failures for Links 1–3. As seen in the figure, the annual failure probability for Link 3 increases as the median of the resistance of Link 1 increases. This is because the wind loading effects on Link 3 are more significant when Link 1 does not fail. In other words, the wind loading effects on Link 3 would have been alleviated by having Link 1 failed. The mathematical model presented in the previous section (Fig. 10.22) can capture such characteristics of the system.[4] This poses an important lesson in case where one attempts to increase the wind-resistant performance of houses: Increase of resistance of part of the system to avoid small damages may increase the possibility to cause larger damages.

10.3.6.3 Cost-Effectiveness of Adaptation Measures

The probabilistic risk analysis and benefit–cost analysis can be then conducted by combining the probabilistic modelling as introduced above with cost and consequence evaluation. Here, a simplistic and qualitative analysis is shown: the reduction of the annual failure probability by a factor of 10 can be achieved, for example, by utilising twice the amount of nails according to the results above. The cost of the additional nails is considered to be marginal, compared to the cost for the whole construction. On the other hand, the benefit is quite significant, among others, because the return period of the failure changes from 10 to 100 years, which means house owners can accumulate assets for a substantially longer period until the next failure occurs. Therefore, it is concluded that the benefit over the cost may be significant. Note that the analysis conducted above does not consider the failure

[4] Therefore, it is called "lizard-tail model" named after lizards protect main bodies by abandoning their tails when they face enemies.

of structural components themselves. The underlying assumption is that the connections fail before the structural components fail. However, as the resistances of the connections increase, it may become more likely that structural components fail before the connection fails. Therefore, failures of structural components must be considered in the benefit–cost assessment for more rigorous analysis.

Finally, but not least, the impact of climate change on wind hazard can be integrated in the benefit–cost analysis. Assuming that frequency of the occurrence of strong winds causing damage to construction are increased by climate change and considering that the implied target reliabilities of the current constructions are suboptimal, improvement of wind-resistant performance should be a preferred option. However, it is not straightforward to evaluate the costs and consequences; the costs should include not only the cost for additional construction materials but also social costs such as for increasing public awareness and educating unskilled builders, etc.; and consequences should include not only direct costs such as repair cost but also delayed economic growth when the consequence is not marginal, see Nishijima (2009) for further information.

10.4 Summary

First, the example of housing and their damage in the developing countries is introduced. It is pointed out that (1) many of the houses are built by less-skilled workers or nonexperts including the owners themselves, which often leads to inappropriate choice of construction materials and poor construction quality, and (2) roofing failures are one of the most common wind-induced damage to construction. Second, it is pointed out that there is a lack of close investigation of the failure mechanisms and probabilistic risk analysis of non-engineered construction in similar manners for engineered construction. Motivated by this, a probabilistic risk analysis for the assessment of wind-resistant performance of a non-engineered house is conducted with the target house located at the east coastline of Leyte Island in the Philippines. The analysis takes its basis in the reliability-based wind-risk modelling technique. The information required for the modelling is obtained by wind tunnel experiment and material testing. In the wind tunnel experiment, a model was designed and developed based on the results of the field measurement. For the material testing, most of the samples were collected at local hardware shops. The analysis shows that the annual failure probability of the whole structure is the order of 10%, which, however, can be significantly reduced by adding a few nails at each connection. This appears encouraging for decision-makers to investing into the wind-disaster risk reduction.

10.5 Potential Design and Practice Evolutions

- Non-engineers and non-skilled workers may not know how to design and construct safe housing, even if they wish to do so. Therefore, the development and provision of the set of specifications, which are calibrated for target safety levels, for parts of housing construction may be useful.
- Replacement of local housing construction with foreign construction styles made with industrial materials sometimes does not lead to the improvement of structural performance. An alternative approach may be useful, which takes its basis in the local traditional construction style and to advance them with scientific knowledge and engineering technology.

10.6 Open Research Questions

- It is often the case where it is clear that additional investment into upgrading structural performance will significantly reduce the probability of failure, hence, risk, and such additional investment costs little and that owners can actually afford. Nevertheless, such an investment is often not made. What are the missing elements in persuading owners to invest more into safety? Is there anything engineering researchers can do?
- Two distinct approaches for achieving more resilient construction are presented in this chapter. Which of the two approaches are more reasonable? Are there other approaches?

Acknowledgements

The presented surveys and study were supported by the JST Japan-Philippine Urgent Collaborative Projects regarding "Typhoon Yolanda" within the J-RAPID Program (project title: Investigation of wind damage processes by Yolanda, identification of effective damage reduction measures, and its facilitation to recovery work) and JSPS KAKENHI "16H05905". The author acknowledges Prof. Espina and her colleagues at the College of Architecture, the University of the Philippines, Diliman for the joint survey and valuable inputs for the analysis conducted in the study, and also thanks Mr. Nishimura, Mr. Tomisaka, and Ms. Namigishi at the DPRI, Kyoto University, for assistance to conduct the experiments and tests.

References

Arya, A.S., Boen, T., Ishiyama, Y., 2013. Guidelines for earthquake resistant non-engineered construction. United Nations Educational, Scientific and Cultural Organization.

Fujieda, A., Sasaki, Y., 2016. Japan International Cooperation Agency (JICA) funding: Possible partnerships with Kyoto University, NCURA magazine. NCURA XLVIII (6), 21–22.

Nishijima, K. (2009). Issues of Sustainability in Engineering Decision Analysis, Dissertation, (ETH Zurich).

Nishijima, K., 2016. Learning from disasters—Cyclone pam impacts on the Republic of Vanuatu, March 2015. In: Annual report of Disaster Prevention Research Institute, Kyoto University, vol. 59, A, pp. 91–97.

UNESCO (2016). Toward resilient non-engineered construction, Guide for risk-informed policy making, Editor: Narafu, T., Editorial contribution: Nishijima, K, Sakoda, K.

Zhang, S., Nishijima, K., Maruyama, T., 2014. Reliability-based modeling of typhoon induced wind vulnerability for residential buildings in Japan. J. Wind Eng. Ind. Aerodyn. 124, 68–81.

Further Reading

JMA. RSMC Best Track Data.

NDRRMC, n.d. SitRep No. 108 Effects of Typhoon "YOLANDA" (HAIYAN).

CHAPTER ELEVEN

Balancing Energy Efficiency and Heat Wave Resilience in Building Design

Morshed Alam, Jay Sanjayan, Patrick X.W. Zou
Department of Civil and Construction Engineering, Centre for Sustainable Infrastructure, Swinburne University of Technology, Hawthorn, VIC, Australia

11.1 Introduction

Heat waves, a consequence of climate change, are one of the deadliest natural hazards around the world. The European heat wave of August 2003 has resulted in more than 70,000 additional deaths in 12 countries (Robine et al., 2008). In July and August 2010, even more, intense heat wave scorched vast areas of Eastern Europe and Russia which claimed up to 56,000 lives (Munich, 2011). Two-thirds of continental Australia recorded temperatures above 39°C for over 3 weeks in February 2004, which caused 116 excess deaths (Tong et al., 2010). In 2009, South Australia and Victoria were affected by one of the most extreme heat waves in its history. Over the 5 days, 27–31 January, maximum temperatures were 12–15°C above normal. The Victorian Department of Human Services estimated 374 excess deaths during the first week of the heat wave (Victorian Government Department of Human Services, 2009). The residents of Melbourne experienced another scorching heat wave in 2014 with the temperature above 41°C for four consecutive days which resulted in 167 excess deaths (Victorian Government Department of Human Services, 2014).

The Climate Council of Australia reported that both the frequency and severity of heat waves are increasing significantly in Australia as well as around the world due to climate change which poses a significant risk to human health (Steffen et al., 2014). Over the period 1971–2014, both the duration and frequency of heat waves increased. The average intensity of the heat wave increased by 1–2.5°C in major Australian cities, and the hottest days during heat waves became 1.5–4.3°C hotter. The number of hot days has doubled in Australia in the last 50 years. In Adelaide, Canberra, and Melbourne, the annual average number of days over 35°C has already

Climate Adaptation Engineering
https://doi.org/10.1016/B978-0-12-816782-3.00011-5

exceeded the best estimate projected for 2030 and other capital cities are also heading into that direction (Steffen et al., 2014).

To protect people from the impact of heat wave risk, more attention has been paid recently on overheating issues in the building (Vellei et al., 2017; Baborska-Narożny et al., 2017) and heat wave-resistant building design (Hatvani-Kovacs et al., 2018). The impact of heat waves was reported to be higher in the buildings characterised by low external wall insulation, high infiltration levels, and poor indoor environment quality (Sakka et al., 2012; Vandentorren et al., 2004; Barnett et al., 2013). Energy efficient retrofitting of existing inefficient buildings can reduce the risk of overheating and heat-related health hazards significantly (Alam et al., 2016). However, recent studies also showed that a high-energy efficient house, with high levels of insulation and air tightness, may also have significantly lower resilience to heat stress compared to a relatively inefficient home during a heat wave (Pyrgou et al., 2017; Ren et al., 2014). Therefore, integration of proper climate adaptation features in buildings to reduce both energy consumption and heat wave impact under future climate change scenarios is crucial.

This chapter will describe different adaptation strategies in buildings to reduce the indoor overheating period during a heat wave period. Several available heat stress indicators that can be used to quantify the level of heat stress inside the building are presented. The importance of balancing the energy efficiency and heat wave resilience in building design is then discussed.

11.2 Indoor Heat Stress Criteria

For over a century, attempts have been made to construct an index to define the heat stress and zones of discomfort satisfactorily. Epstein and Moran (2006) reviewed over 40 such indices and divided them into three groups: *rational indices, empirical indices,* and *direct indices*. Rational indices are based on heat balance equation which integrates environmental and behavioural variables. Examples of rational indices include heat stress index (HSI) (Belding and Hatch, 1955), predicted 4-h sweat rate (McArdle et al., 1947), etc. Empirical indices are constructed based on the objective and subjective strain and may incorporate physiological and perceptual responses to increased heat stress (Liang et al., 2011). Examples of empirical indices are physiological strain index (PSI) (Malchaire, 2006), cumulative heat strain index (Frank et al., 1996), etc. The first two groups are sophisticated indices and require regular measurements of environmental and physiological variables which are difficult to calculate and therefore, not feasible for daily use (Epstein and Moran, 2006; Liang et al., 2011).

The third and final group is direct indices which are based on the measurement of basic environmental variables only. Examples of direct indices include discomfort index (DI) (Epstein and Moran, 2006), temperature humidity index (THI) (Unger, 1999), wet bulb globe temperature (WBGT) (Yaglou and Minard, 1957), environmental stress index (ESI) (Moran et al., 2001), equivalent temperature (ET) (Liang et al., 2011), etc. The WBGT index is most widely used around the world and is the basis for ISO 7243 heat stress standard (Parsons, 2006; Holmes et al., 2016; Lemke and Kjellstrom, 2012). The WBGT index includes all six variables of thermal comfort: air dry bulb temperature, humidity, air velocity, radiation (not required for indoor), metabolic rate, and clothing. Lemke and Kjellstrom (2012) reported that Eq. (11.1) is the best for calculating indoor WBGT index:

$$\mathrm{WBGT} = 0.67T_{\mathrm{wb}} + 0.33T_{\mathrm{db}} - 0.048\log v(T_{\mathrm{db}} - T_{\mathrm{wb}}) \tag{11.1}$$

where T_{wb} is the wet bulb temperature in °C, T_{db} is the dry bulb temperature in °C, and v is the indoor air velocity in m/s. The metabolic rate and clothing levels are accounted for in the threshold limit values (TLV) chart (Table 11.1) developed by the American Conference of Government Industrial Hygienists (ACGIH, 2004).

In contrast to WBGT index, the estimation of DI is confined to the dry bulb and wet bulb temperature only and does not take into account other parameters of thermal comfort. However, DI was found to be highly correlated to effective temperature index and to sweat rate both at rest and during exercise which reflects its physiological significance (Epstein and Moran, 2006). A recent Korean study of elderly people in low-income apartments reported that increased body temperatures and reduced blood pressure were significantly associated with increased indoor temperatures, DI, and outdoor temperature in a multivariate model (Kim et al., 2012). Previous researchers used DI to determine the level of heat of stress inside different star-rated houses during heat waves (Ren et al., 2014; Barnett et al., 2013). The DI can be calculated using Eq. (11.2):

$$\mathrm{DI} = 0.5T_{\mathrm{wb}} + 0.5T_{\mathrm{db}} \tag{11.2}$$

Based on a large number of observations on a wide spectrum of population groups and under different climate conditions, the following criteria

Table 11.1 Threshold Values of WBGT Index According to ACGIH

	Acclimated to Environment				Non-Acclimated to Environment			
	Light Work (°C)	Medium Work (°C)	Heavy Work (°C)	Very Heavy Work (°C)	Light Work (°C)	Medium Work (°C)	Heavy Work (°C)	Very Heavy Work (°C)
100% Work	29.5	27.5	26.0		27.5	25.0	22.5	
75% Work and 25% rest	30.5	28.5		27.5	29.0	26.5	24.5	
50% Work and 50% rest	31.5	29.5	28.5	27.5	30.5	28.0	26.5	25.0
25% Work and 75% rest	32.5	31.0	30.0	29.5	31.0	29.0	28.0	26.5

WBGT additions for clothing type are as follows: summer work uniform = 0, woven material overall = +3.5 and double cloth overall = +5.0.

Adapted from ACGIH, 2004. TLVs and BEIs: Threshold Limit Values for Chemical Substances and Physical Agents and Biological Exposure Indicies. ACGIH Signature Publications, Cincinnati.

(Table 11.2) were established to characterise environmental heat stress (Epstein and Moran, 2006).

In addition to abovementioned heat stress criteria, several approaches have been adopted by the researchers to determine the overheating threshold inside the building. The Chartered Institution of Building Services Engineers (CIBSE) recommends that in the UK domestic building thermal comfort threshold temperatures of 28°C for living rooms and 26°C for bedrooms should not be exceeded for >1% of occupied hours (CIBSE, 2006).

Table 11.2 Threshold Values for Discomfort Index (DI)

Discomfort Index (DI)	Classification of Heat Stress
DI < 22	No heat stress is encountered
22 < DI < 24	Mild sensation of heat stress
24 < DI < 28	Moderate heat stress, people feel very hot and physical work may be performed with some difficulties
DI > 28	Heat stress is severe; people engaged in physical work are at increased risk for heat exhaustion and heat stroke.

Adapted from Epstein, Y., Moran, D.S., 2006. Thermal comfort and the heat stress indices. Ind. Health 44, 388–398.

The cumulative number of degree hours over these threshold temperature limit annually has been used to represent the severity of overheating problem inside the building (Porritt et al., 2012). The cumulative number of degree hours arguably gives a better representation of the severity of the overheating problem than simply recording hours over the threshold temperature by quantifying the extent to which the threshold has been exceeded in a period. Recently, Lee and Steemers (2017) proposed to include continuous overheated period as an additional parameter in heat stress calculation because prolonged exposure to even moderately high temperature (>25°C) can have a significant negative impact on health, sleep, and mental and physical well-being. Further study was recommended to identify the appropriate way of including this parameter.

Hatvani-Kovacs et al. (2018) applied an adaptive comfort model to calculate the overheating threshold. In that study, the exponentially weighted running mean of the daily mean temperatures (DMTs) of the last 7 days was adopted, instead of the monthly mean temperatures accepted by the American Society of Heating, Refrigerating, and Air Conditioning (ASHRAE) Standard 55. Their calculation showed maximum allowed indoor temperatures between 30°C and 31°C in Adelaide and between 26.8°C and 28.9°C in Sydney during the studied heat waves.

11.3 Heat Wave Adaptation Measures

The available different building simulation software are useful tools to investigate different heat wave adaptation methods in buildings. The heat stress criteria mentioned in the previous section can be used to determine and rank the effectiveness of each adaptation method for a particular climate zone. A list of different passive adaptation measures and approximate costs of adaptation is presented in Table 11.3. These adaptation methods can be applied independently or in combinations to enhance the heat wave adaptation capacity of the buildings. In this section, the application of these adaptation methods is discussed under the assumptions that the buildings do not have an active air-conditioning system to maintain thermal comfort.

11.3.1 Insulation

The installation of insulation reduces the heat transfer between the indoor and outdoor environment through building fabric and thus helps to maintain stable indoor temperatures during a heat wave. The level of insulation is measured by an *R*-value which is the ratio of the thickness and thermal

Table 11.3 Heat Wave Adaptation Measures in Buildings

Category	Adaptation Methods	Approximate Cost of Adaptations
Insulation	External Wall Insulation	R 4.0 wall insulation: \$11.20 per m^2
		R 2.0 wall insulation: \$5.60 per m^2
	Ceiling Insulation	R 6.0 ceiling insulation: \$16.0 per m^2
		R 4.0 ceiling insulation: \$7.80 per m^2
		R 3.0 ceiling insulation: \$6.75 per m^2
Solar control	External Shutter	600 mm high × 800 mm wide: \$500 to \$660
		800 mm high × 1200 mm wide: \$550 to \$725
		1,200 mm high × 1800 mm wide: \$690 to \$840
		1,800 mm high × 2400 mm wide: \$900 to \$1,000
	Internal Blinds	\$60–\$300 per blind (up to 2.4 m wide)
	Shading	Shade cloth: \$5.25–\$17.98 per m.
		Eaves: \$60 per m^2 (Additional cost in a new build)
	Light coloured roof and walls	\$12–\$60 per m^2 for timber exterior \$15–\$45 per m^2 for brick exterior
	Double- and triple-glazing windows	\$800–\$1,500 per m^2 including window frame (New build)
		\$200–300 per m^2 excluding window frame (Retrofitting)
Airtightness	Seal the gaps	On average \$1,037 per house
Ventilation	Night ventilation Window operation	Not applicable
Thermal energy storage	Phase change materials	BioPCM Mat Q24: \$105 per m^2
Surrounding environment	Vegetation coverage	Not applicable
Building layout	Orientation cool retreat	Not applicable

All the costs are in AUD and are approximated from commercial rates in Australia.

conductivity of the insulation material. Materials with a higher *R*-value provide a higher insulation.

Insulating the external walls of the existing residential building stock in the United Kingdom to comply with current standards (*U*-value 0.35 W/m^2K) can decrease the degree hours over the CIBSE threshold temperature between 20% and 63% during a typical August 2003 heat wave period (Porritt et al.,

2012). The large variation in the percentage reduction was due to the orientation and external wall surface area of the houses. In their other study (Porritt et al., 2011), it was shown that external wall insulation is also one of the most effective adaptation methods for the United Kingdom in future climate change scenarios.

Internal wall insulation is less effective than external wall insulation, and for some building orientations, it increases the number of degree hours. This might be due to the fact that the insulation prevents the internal walls to store thermal energy and provides the radiative cooling effect. Also, it retains the heat gain from the occupants of the room (Porritt et al., 2012).

The increase in ceiling insulation level also showed significant improvement in the thermal performance. Moving from having no insulation to R1.5 insulation level in a typical three bedroom house reduces the total annual hours over severe heat stress (DI > 28) by 24%–78% in Australian cities depending on the local climate (Barnett et al., 2013).

11.3.2 Solar Control

There are several ways of reducing the solar heat gain inside the building during heat wave period which in turn can help to maintain a comfortable indoor environment for extended period.

Coating the external walls and roof with light colour has been found to be one of the most cost-effective heat wave adaptation methods (Saman et al., 2013). The solar reflectance of a typical light colour roof (also known as cool roof) is around 0.3 compared to 0.85 for a dark coloured roof. As a result, the light-coloured roof reflects most of the solar radiation and significantly reduces the heat transmission into the dwellings. On the other hand, roof surfaces which absorb high amounts of solar radiation can readily reach temperatures of 80°C in hot weather (Saman et al., 2013). Changing roof colour from dark to light has a significant impact on reducing the total annual hours of severe heat stress risk to occupants. The modelling results showed that severe heat-related health risk in an Australian household was reduced by a maximum of 95% (in Canberra) and a minimum of 29% (in Mt. Isa) (Barnett et al., 2013).

Windows are a major source of heat gain (up to 87%) and heat loss (up to 40%) which can severely impact the heating and cooling load of a building (Lyons and Hockings, 2013). External shutters are particularly useful in cases where there is a large west-facing window. The scenario analysis showed that during a typical August 2003 heat wave in the United Kingdom, external shutters could reduce the number of degree hours by 39%. Although they are more expensive than the internal blinds, they are more effective than the internal blinds (up to 20% reduction under similar scenarios) (Porritt et al., 2012). This is because the internal blind and curtain cannot prevent some solar radiation being trapped inside the room and converted to longwave radiation. If the blinds and curtains are not in place, the solar radiation will directly be absorbed by internal thermal mass. This heat will then be radiated back to the room air over a period which will create an uncomfortable indoor environment during summer.

Fixed and adjustable shading over the external windows are shown to have similar effects as the internal blinds (Porritt et al., 2012) in reducing indoor overheating hours.

Changing the standard single-glazed windows to double and triple glazing can reduce the heat gain through windows significantly which in turn can minimise the impact of a heat wave (Saman et al., 2013; Porritt et al., 2012). Double- and triple-glazed windows have lower *U*-value, and solar heat gain coefficient (SHGC) value which minimise the solar heat gain through the windows into the room. However, for the climate zone with hot summer and cold winter, the SHGC value should be carefully selected to facilitate heat gain during winter and minimise heat gain during summer (Lyons and Hockings, 2013). Replacing the standard single glazing with double or triple glazing is comparatively expensive than other adaptation options (Porritt et al., 2012).

11.3.3 Airtightness

Air infiltration means unintended airflow in and out of dwellings. It occurs through the cracks in the walls and joints and the gaps surrounding the doors and window frames. Weatherproofing those cracks and joints is a cost-effective way to reduce heating and cooling load and improve thermal comfort. However, the influence of weatherproofing on reducing the impact of the heat wave to occupants is negligible compared with other adaptation strategies (Barnett et al., 2013). Although it is not effective when applied as a lone strategy, it is shown to be highly effective when applied in combination with other strategies, such as, insulation (Alam et al., 2016).

11.3.4 Ventilation

Lack of ventilation is one of the leading contributors to significant indoor overheating issue not only during an extreme heat wave period but also during a typical warm summer day. The number of overheating hours varies significantly with the inhabitant practices regarding window operation. A study in northern England in summer 2013 showed that the occupants who kept their windows closed most of the time and only opened it when hot in the evening experienced severe overheating (>28°C) for longer periods compared to those who kept the windows open all the time (Baborska-Narożny et al., 2017). The study also revealed that only 6% of occupants use the ventilation system (opening windows, operating exhaust fan, etc.) in their houses as intended. Another study in the same area showed that occupants of vulnerable households (i.e., older than 65 years, disabled, long-term illness, etc.) are less likely to open windows and the reduced level of ventilation in these households is a major contributor to overheating issues (Vellei et al., 2017). Appropriate strategies are required to enhance the ventilation rate in both vulnerable and non-vulnerable households. The inhabitants need to be made aware of ventilation practices that can minimise the overheating hours during the hot summer period. The change of occupant behaviour regarding ventilation can significantly reduce the overheating hours with zero capital cost.

In houses with high thermal mass, the strategic use of night ventilation can cool down the thermal mass of the solid walls and floors which can provide radiative cooling during the hotter daytime period. Porrit et al. (2012) showed that night ventilation could reduce the degree hours over the CIBSE comfort threshold temperature by up to 67% in a typical UK dwelling.

11.3.5 Thermal Energy Storage

Integration of thermal energy storage materials (also known as phase change materials (PCM)) in buildings can help to reduce the risk of heat stress in non-air-conditioned buildings during a heat wave period. PCM absorb or release the energy equivalent of their latent heat when the temperature of the material undergoes or overpasses the phase change temperature. When the temperature is lower than phase change temperature at night, the PCM transform from liquid to solid. During the daytime, when the temperature increases over the phase change temperature, the PCM transform from solid to liquid through absorbing the energy equivalent to their latent. In this process, the PCM can provide free cooling to the building during the daytime and reduce the cooling load (Alam et al., 2014).

These passive means of controlling the temperature are particularly useful during heat waves in cities like Melbourne. The integration of PCM in walls and ceilings of a typical Melbourne house (shown in Fig. 11.1) can reduce the severe discomfort hours (DI > 28) during a heat wave period as shown in Fig. 11.2.

In case of 24°C PCM temperature, the severe discomfort hours are reduced by 1 h in living-dining and kids TV zones. No reductions are observed in rumpus and grand master zones. On the other hand, in case of 30°C PCM, severe discomfort hours are reduced by 4 h in living-dining and 3 h in kids TV, grand master, and rumpus zones. Hence, it is evident that 30°C PCM are more efficient than 24°C PCM in reducing severe discomfort hours during a heat wave. This occurs because the outdoor temperature during a heat wave is favourable for regular melting and solidification of 30°C PCM. On the other hand, the 24°C PCM could not solidify at night during a heat wave period because the temperature does not drop below 24°C. Hence, it cannot absorb thermal energy during the day and contribute to the reduction of indoor temperature. To maximise the performance of PCM, it is very important to select the appropriate PCM temperature and PCM layer thickness for the climate where the building is located. Also, regular night ventilation is important to solidify the PCM at night so that it can be ready to absorb the thermal energy during the day and

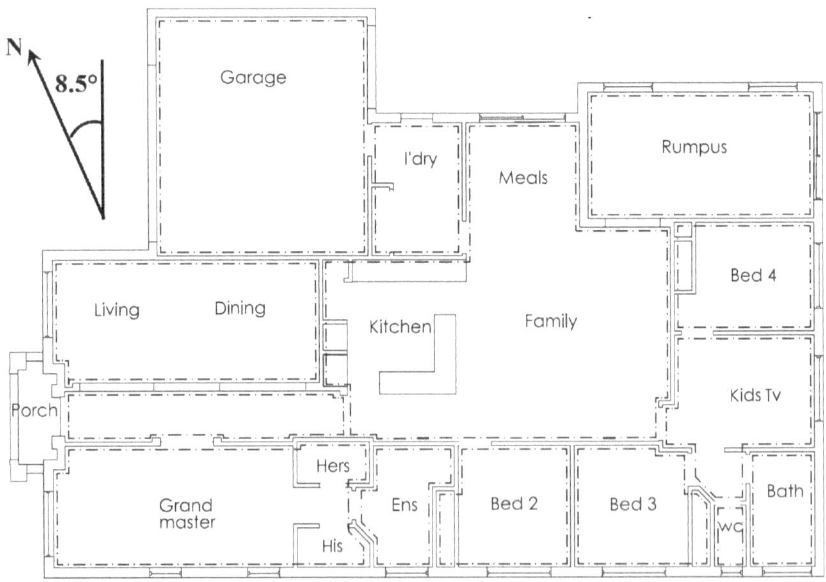

Fig. 11.1 Thermal zones of a typical single-storey house in Melbourne.

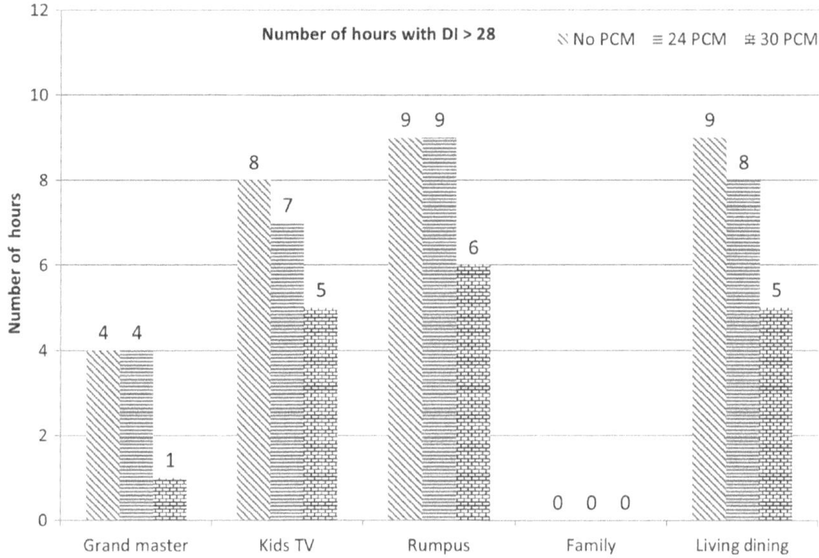

Fig. 11.2 Discomfort index at different zones in Melbourne house during a heat wave.

reduce discomfort hours (Ramakrishnan et al., 2017; Jamil et al., 2016). The effectiveness of PCM in reducing peak zone temperature also depends on the way it is applied in the building. The active application method, where the outdoor air is passed through a PCM heat exchanger before supplying it to the indoor, can significantly reduce the peak indoor zone temperature compared to a passive application case (Alam et al., 2017). However, active application of PCM is more expensive than the passive application method, and therefore, detailed cost-benefit analysis is recommended before choosing a particular method.

11.3.6 Surrounding Environment

Urban vegetation can potentially reduce the heat-related mortality and morbidity during a heat wave period. A case study in the Melbourne Central Business District showed that the average mortality rate in Melbourne increases rapidly when the mean daily indoor temperature exceeds 28.5°C (Chen et al., 2014). An increase in urban vegetation can reduce the average summer outdoor temperature in the range of around 0.5–2°C in Melbourne which in turn contributes to the reduction of peak daily indoor temperature and heat-related mortality. Reduction of average heat-related mortality in the range from 5% to 28% was estimated by increasing the vegetation coverage between 15% and 33% (Chen et al., 2014).

11.3.7 Building Layout

The orientation of a building has a substantial impact on its indoor thermal performance. If the longer external wall of a building faces west and contains large windows, then it will be exposed to solar radiation for most of the day and will have the highest number of overheating hours inside the building (Porritt et al., 2012). The overheating hours vary by almost 100% with the orientation. For example, in Brisbane, Australia, changing from east to north reduces the total annual hours over DI > 28 by 83% (Barnett et al., 2013). However, changing orientation is not a feasible adaptation option for existing buildings and should be considered in the construction of new dwellings.

Saman et al. (2013) proposed the concept of a cool retreat to minimise energy consumption and improve thermal comfort during heat waves for both new and existing dwellings. Cool retreat is a zone inside the dwelling where the occupants can take shelter during an extreme heat wave period. The zone should be designed to minimise the solar heat gain using the approach described in the previous section which can help to maintain a comfortable indoor environment for longer periods. This is particularly useful in the event of power cut during a heat wave period.

11.4 Selection of Optimum Adaptation Measures

The selection of the best single or a combination of the different intervention methods depends on the cost of the intervention, their performance in reducing the discomfort hours, and the practicality of implementing it.

The cost of applying an adaptation measure to minimise the overheating issue depends significantly on the building type, location, surroundings of the building, and the point at which the measure is applied (Zero Carbon Hub, 2016). Some adaptation measures are more cost-effective to include in a new building than in a retrofitting project. For example, selecting best building orientation is cheaper and easy in a new building but more expensive or impractical in an existing building.

Regarding performance, there is currently a lack of experimental evidence showing the effectiveness of the various adaptation measures mentioned in the previous section. In the absence of this knowledge base, the mathematical modelling is largely used to assess the effectiveness of any given combination (Zero Carbon Hub, 2016). The model should capture the dynamics of the thermal processes in the building through using dynamic modelling approach. Attention should be paid to use sufficient details of the house to get an accurate outcome. At the same time, it should not be overcomplicated which might increase the computational time and cost.

Furthermore, one should consider the practicality of implementing adaptation measures. For example, the low-cost solution such as increasing ventilation through opening windows and use of indoor window blind can reduce the degree hours over the CIBSE comfort threshold temperature by 40% during the hot summer period in the United Kingdom (Porritt et al., 2012). However, there are practical issues that must be considered. To operate the window, the occupants need information about when the outside air temperature is higher than the inside air temperature. Some may not want to close the window blind due to the loss of view. Also, for vulnerable occupants, it is not always possible to open the windows for ventilation purposes (Vellei et al., 2017). Hence, the effectiveness is dependent upon the occupants carrying out the procedure.

While selecting the best combination of heat wave adaptation measures, one should also consider their impact on energy consumption. For example, combining window and blind operation with other measures such as painting walls and roof to light colour and application of night ventilation could reduce the degree hours by 94%. Application of external wall and roof insulation along with previously mentioned adaptation measures could result in only further 3% reduction in degree hours, however, the cost of adaptation increases by four times (Porritt et al., 2012). However, the later combination significantly reduces the annual heating energy consumption. Hence, all these parameters should be considered while selecting the optimum adaptation measures for a building.

11.5 Balancing Energy Efficiency and Heat Wave Resilience in Buildings

At present, various building energy rating systems are adopted in different countries to determine the thermal performance of residential buildings. In Australia, the energy star rating of dwellings is determined using the Nationwide House Energy Rating Scheme (NatHERS, 2018). This scheme rates Australian dwellings from 0 to 10 energy stars depending on the total annual heating and cooling energy requirements to maintain comfortable indoor environment. The higher the energy star rating of a building, the lower is the requirement for total annual heating and cooling energy. An increase in building energy rating can also reduce the indoor overheating risk. Alam et al. (2016) showed that during a typical heat wave period in Melbourne like the one in 2009, residents of 0.9 star energy rated house would be exposed to extreme heat stress conditions for almost 25 h compared to only 6 h by the occupants of 5.4 star energy rated house under typical 2009 Melbourne heat wave as shown in Fig. 11.3. Several robust empirical relationships were proposed to predict deaths, ambulance calls, emergency department presentations, and after hour doctor calls during a heat wave (Alam et al., 2016). Using those relationships, impacts of a typical 2009 Melbourne heat wave on the occupants of different energy star houses were predicted and presented in Table 11.4. It was predicted that the mortality rate of 2009-type heat wave in Melbourne, may reduce to 1/10th if the minimum energy star rating of the existing housing stock can be upgraded to a 5.4 energy star (Alam et al., 2016). In other words, risk reduction is a high $\Delta R = 90\%$.

Fig. 11.4 shows that the probabilities of experiencing discomfort hours (according to the WBGT index) depending on mean outdoor temperature which is the average of daily maximum and minimum temperature (Alam et al., 2018). It was chosen here instead of daily maximum temperature only because it better represents the effect of the heat wave on excess mortality (Nicholls et al., 2008). Mean outdoor temperature takes into account the minimum temperature of the day which is usually the early morning temperature of the day.

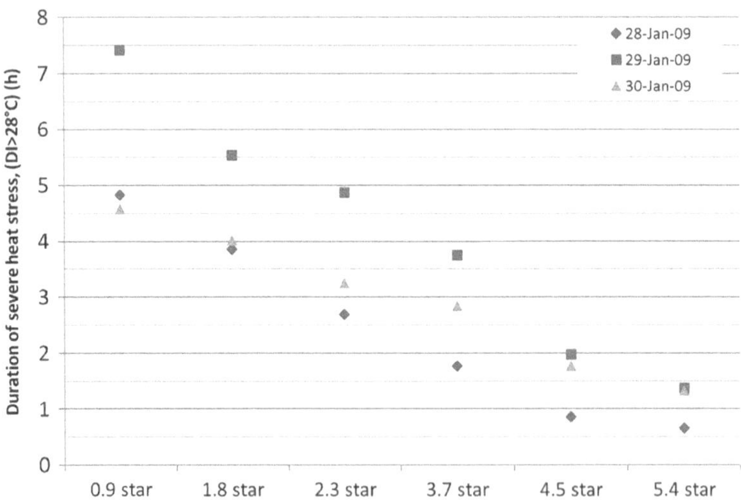

Fig. 11.3 Number of heat stress hours in a typical Melbourne house during a 2009-type heat wave.

Table 11.4 Predicted Health Impacts During a 2009-Type Heat Wave in Different Energy Star Rated Houses in Melbourne

Energy Rating	Deaths	Ambulance Calls	Emergency Department Presentations	After Hour Doctor Consultations
0.9 star	374	514	1,055	71
1.8 star	240	399	864	59
2.3 star	197	347	774	50
3.7 star	96	196	511	28
4.5 star	62	129	394	20
5.4 star	37	63	280	13

The tolerance limit of heat stress hours varies from person to person. Some persons may find it intolerable after 1 h while others may tolerate a longer heat stress period. Hence, it is very difficult to use a constant tolerance limit. Here, three and 6 h were used to demonstrate the variations in probabilities of heat stress exposure for the occupants living in different energy star rated houses. Nicholls et al. (2008) reported that when *Mean T* exceeds 30°C in Melbourne, the mortality rate of the people over 64 years of age increases. Fig. 11.4A and B shows that when the *Mean T* exceeds 30°C, probabilities of exposure to three and six heat stress hours are close to 45% and 33%, respectively, in a 0.9 energy star houses. However, the probabilities are almost zero for 5.4 energy star houses. Fig. 11.4 also shows that probabilities of experiencing three and six heat stress hours in a 0.9 energy star house are 70% and 58%, respectively, when the *Mean T* is 36°C (similar to 2009 heat wave period in Melbourne). On the other hand, in a 5.4 energy star rated house the probabilities are only 13% and 4%, respectively, for the similar outdoor condition. Hence, it can be concluded that risk of heat stress exposure during heat wave period (*Mean T* around 36°C) can be reduced by more than 50% through upgrading the existing 0.9 energy rated houses to 5.4 energy star.

However, several recent studies showed that higher energy rating do not necessarily ensure lower cooling energy consumption and less overheating during heat waves. For example, Australian current residential energy rating system (NatHERS) does not focus on heat stress resistance in new houses while calculating the rating (Hatvani-Kovacs et al., 2018). The new highly insulated and airtight houses, which have been shown to have higher energy efficiency rating according to the NatHERS, may have lower resistance to heat stress during summer as well as during a heat wave period (Hatvani-Kovacs et al., 2018). An increase in fabric insulation and airtightness level lead to more frequent indoor temperature over threshold limit which

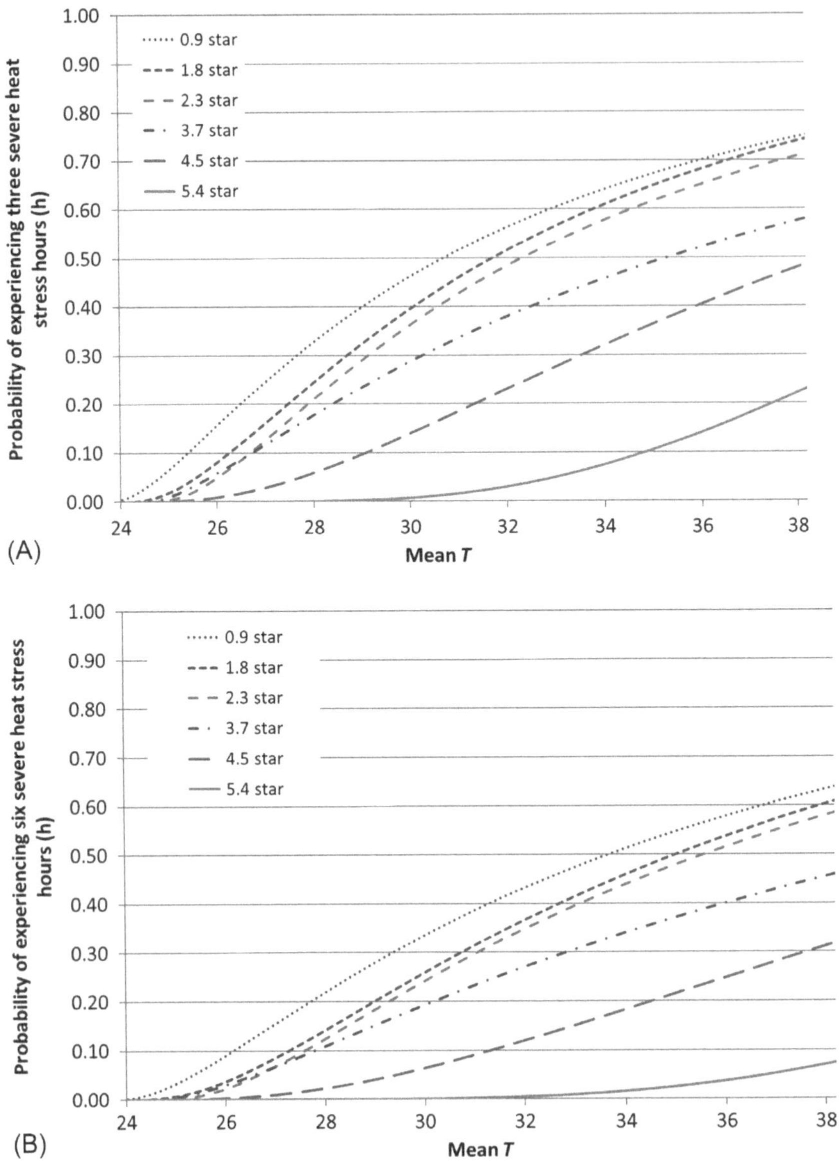

Fig. 11.4 Probability of experiencing (A) three and (B) six heat stress hours in different energy star rated houses in Melbourne during a typical 2009 heat wave period.

increases the risk of cooling energy requirement if the house has an active cooling system (Pyrgou et al., 2017; Willand et al., 2016).

Currently, the building energy rating in Australia is determined based on total heating and cooling energy requirement. In a heating-dominated region like Melbourne, the current rating system encourages building design that minimises the heating load only. The increase in cooling energy load during the hot summer period in a higher energy rated house is much lower than the decrease in heating demand during the winter (Hatvani-Kovacs et al., 2018). As a result, the total energy consumption of those buildings is lower and is rated as higher energy efficient houses although they do not have good heat resistance capacity during a heat wave. This poses an enormous risk to the occupants of the higher energy rated houses if they do not have air conditioners or if there is a power cut during a heat wave. Hence, there is a need to have a building rating system that focuses both on energy efficiency and heat stress resistance.

11.6 Summary

This chapter described the impact of heat waves on human health and outlined different adaptation strategies in buildings to minimise the indoor heat stress which can help to reduce heat-related health hazard during a heat wave period. To quantify the levels of heat stress inside the building, several heat stress indicators have been proposed and used by the previous researchers. Those heat stress indicators are used in conjunction with the building simulation tool to quantify the performance of different heat wave adaptation measures. The heat wave adaptation measures for buildings can be categorised into the following categories: insulation, solar control, air-tightness, ventilation, thermal energy storage, surrounding environment, and building layout. Integration of these adaptation measures, either independently or in combination, has been shown to have the potential to reduce the heat stress condition significantly inside the building, which in turn can reduce the heat-related mortality and morbidity. The optimum combination of the adaptive measures can be calculated considering the cost of the intervention, their performance in reducing the discomfort hours, and the practicality of implementing it. The low-cost solution such as opening windows to allow ventilation, using blinds to minimise heat gain, night ventilation to cool down thermal mass, etc. have a significant impact in reducing the discomfort hours. However, their effectiveness is highly dependent on the behaviour of the occupants living in the house. Other low-cost adaptive measures such as painting walls and roofs with a light colour to reflect solar radiation is also highly effective in reducing the heat wave impact. Furthermore, insulation of walls and ceilings of an energy inefficient house can also effectively reduce the overheating hours during a heat wave. Apart from traditional adaptation measures, new technologies like PCM can also help to reduce the overheating hours significantly if the phase change temperatures and other related parameters are designed appropriately to facilitate the melting and solidification of PCM during a heat wave period. An increase in housing energy rating through incorporating energy efficiency measures can decrease the indoor overheating hours during a heat wave. Although, higher fabric insulation and airtightness level increase the building energy efficiency, it may also lead to more frequent indoor temperature over threshold limit which increases the heat stress risk. Hence, further research is required to optimise the building design for energy efficiency and heat wave resilience.

11.7 Potential Design and Practice Evolutions

- There is a need to develop a guideline to educate the household on the efficient operation (ventilation practices, operation of air conditioners, use of shades, and night ventilation) of their building so that they can minimise the risk of heat stress during a heat wave.
- The building energy rating system should include both the heat stress resilience and energy efficiency criteria. One way might be considering the cooling and heating energy requirement separately rather than considering the total to determine the rating. The rating should also consider the peak cooling load and encourage design that minimised the peak cooling load.

11.8 Open Research Questions

- There is no standard heat stress indicator to determine the level of heat stress inside the building. Different researchers have used different indicators. How can we develop one universal indoor heat stress indicator?
- The higher insulation level in the building helps to minimise the heating load during winter. However, it also increases the risk of heat stress and cooling energy requirement during the hot summer period. The question is how to determine the optimum insulation level in the building?
- How to quickly release the trapped heat in a highly insulated and airtight building? Ventilation through opening the windows is not a practical solution as it depends on user behaviour. Is there any passive way?

- How to efficiently cool the PCM at night during a heat wave period so that it is available at full capacity on the next day to absorb the heat and minimise the indoor temperature?

Acknowledgements

Authors would like to acknowledge the Commonwealth Scientific and Industrial Research Organisation (CSIRO) (grand no. G1101211) for funding the project under climate adaptation engineering for extreme events flagship and PhD student Hasnat Jamil for his help during the experimental study.

References

ACGIH, 2004. TLVs and BEIs: Threshold Limit Values for Chemical Substances and Physical Agents and Biological Exposure Indices. ACGIH Signature Publications, Cincinnati.

Alam, M., Jamil, H., Sanjayan, J., Wilson, J., 2014. Energy saving potential of phase change materials in major Australian cities. Energ. Buildings 78, 192–201.

Alam, M., Sanjayan, J., Zou, P.X.W., Stewart, M.G., Wilson, J., 2016. Modelling the correlation between building energy ratings and heat-related mortality and morbidity. Sustain. Cities Soc. 22, 29–39.

Alam, M., Sanjayan, J., Zou, P.X.W., Ramakrishnan, S., Wilson, J., 2017. Evaluating the passive and free cooling application methods of phase change materials in residential buildings: a comparative study. Energ. Buildings 148, 238–256.

Alam, M., Rajeev, P., Sanjayan, J., Zou, P.X.W., Wilson, J., 2018. Mitigation of heat stress risks through building energy efficiency upgrade: a case study of Melbourne, Australia. Aust. J. Civil Eng. 16, 64–78.

Baborska-Narożny, M., Stevenson, F., Grudzińska, M., 2017. Overheating in retrofitted flats: occupant practices, learning and interventions. Build. Res. Inf. 45, 40–59.

Barnett, G., Beaty, R., Chen, D., McFallan, S., Meyers, J., Nguyen, M., Ren, Z., Spinks, A., Wang, X., 2013. Pathways to Climate Adapted and Healthy Low Income Housing. National Climate Adaptation Research Facility, Gold Coast.

Belding, H.S., Hatch, T.F., 1955. Index for evaluating heat stress in terms of resulting physiological strains. Heat. Piping Air Cond. 27, 129–136.

Chen, D., Wang, X., Thatcher, M., Barnett, G., Kachenko, A., Prince, R., 2014. Urban vegetation for reducing heat related mortality. Environ. Pollut. 192, 275–284.

CIBSE, 2006. CIBSE Guide A: Environmental Design. Chartered Institution of Building Services Engineers, London.

Epstein, Y., Moran, D.S., 2006. Thermal comfort and the heat stress indices. Ind. Health 44, 388–398.

Frank, A., Moran, D.S., Epstein, Y., Belokopytov, M., Shapiro, Y., 1996. The estimation of heat tolerance by a new cumulative heat strain index. In: Environmental ergonomics: recent progress and new frontiers. Freund, London.

Hatvani-Kovacs, G., Belusko, M., Pockett, J., Boland, J., 2018. Heat stress-resistant building design in the Australian context. Energ. Buildings 158, 290–299.

Holmes, S.H., Phillips, T., Wilson, A., 2016. Overheating and passive habitability: indoor health and heat indices. Build. Res. Inf. 44, 1–19.

Jamil, H., Alam, M., Sanjayan, J., Wilson, J., 2016. Investigation of PCM as retrofitting option to enhance occupant thermal comfort in a modern residential building. Energ. Buildings 133, 217–229.

Kim, Y.M., Kim, S., Cheong, H.K., Ahn, B., Choi, K., 2012. Effects of heat wave on body temperature and blood pressure in the poor and elderly. Environ. Health Toxicol. 27e2012013.

Lee, W.V., Steemers, K., 2017. Exposure duration in overheating assessments: a retrofit modelling study. Build. Res. Inf. 45, 60–82.

Lemke, B., Kjellstrom, T., 2012. Calculating workplace WBGT from meteorological data: a tool for climate change assessment. Ind. Health 50, 267–278.

Liang, C., Zheng, G., Zhu, N., Tian, Z., Lu, S., Chen, Y., 2011. A new environmental heat stress index for indoor hot and humid environments based on cox regression. Build. Environ. 46, 2472–2479.

Lyons, P., Hockings, B., 2013. Your Home—Australia's Guide to Environmentally Sustainable Home. Austrlaian Government Department of Industry, Canberra.

Malchaire, J.B., 2006. Occupational heat stress assessment by the predicted heat strain model. Ind. Health 44, 380–387.

McArdle, B., Dunham, W., Holling, H., Ladell, W., Scott, J., Thompson, M., 1947. The Prediction of the Physiological Effects of Warm and Hot Environments. RNP Report 47/391, Medical Research Council, London.

Moran, D.S., Pandolf, K.B., Shapiro, Y., Heled, Y., Shani, Y., Mathew, W.T., Gonzalez, R.R., 2001. An environmental stress index (ESI) as a substitute for the wet bulb globe temperature (WBGT). J. Therm. Biol. 26, 427–431.

Munich, R.E., 2011. Heat wave, drought, wildfires in Russia (Summer 2010) MR Touch Nat. Hazards—Event report. [Online]. Available: http://www.munichre.com/site/touch-naturalhazards/get/documents_E-132629599/mr/assetpool.shared/Documents/5_Touch/_NatCatService/Catastrophe_portraits/event-report-hw-dr-wf-russia-touch-en-update.pdf [Accessed].

NatHERS, 2018. Available from:http://www.nathers.gov.au. [Acessed March 2018].

Nicholls, N., Skinner, C., Loughnan, M., Tapper, N., 2008. A simple heat alert system for Melbourne, Australia. Int. J. Biometeorol. 52, 375–384.

Parsons, K., 2006. Heat stress standard ISO 7243 and its global application. Ind. Health 44, 368–379.

Porritt, S., Shao, L., Cropper, P., Goodier, C., 2011. Adapting dwellings for heat waves. Sustain. Cities Soc. 1, 81–90.

Porritt, S.M., Cropper, P.C., Shao, L., Goodier, C.I., 2012. Ranking of interventions to reduce dwelling overheating during heat waves. Energ. Buildings 55, 16–27.

Pyrgou, A., Castaldo, V.L., Pisello, A.L., Cotana, F., Santamouris, M., 2017. On the effect of summer heatwaves and urban overheating on building thermal-energy performance in Central Italy. Sustain. Cities Soc. 28, 187–200.

Ramakrishnan, S., Wang, X., Sanjayan, J., Wilson, J., 2017. Thermal performance of buildings integrated with phase change materials to reduce heat stress risks during extreme heatwave events. Appl. Energy 194, 410–421.

Ren, Z., Wang, X., Chen, D., 2014. Heat stress within energy efficient dwellings in Australia. Arch. Sci. Rev. 57, 227–236.

Robine, J.-M., Cheung, S.L.K., Le Roy, S., Van Oyen, H., Griffiths, C., Michel, J.-P., Herrmann, F.R., 2008. Death toll exceeded 70,000 in Europe during the summer of 2003. C. R. Biol. 331, 171–178.

Sakka, A., Santamouris, M., Livada, I., Nicol, F., Wilson, M., 2012. On the thermal performance of low income housing during heat waves. Energ. Buildings 49, 69–77.

Saman, W., Boland, J., Pullen, S., de Dear, R., Soebarto, V., Miller, W., Pocock, B., Belusko, M., Bruno, F., Whaley, D., Pockett, J., Bennetts, H., Ridley, B., Palmer, J., Zuo, J., Ma, T., Chileshe, N., Skinner, N., Chapman, J., Vujinovic, N., Walsh, M., Candido, C., Deuble, M., 2013. A Framework for Adaptation of Australian Households to Heat Waves. National Climate Change Adaptation Research Facility, Gold Coast.

Steffen, W., Hughes, L., Perkins, S., 2014. Heatwaves: Hotter, Longer, More Often. (Climate Council of Australia Limited).

Tong, S., Ren, C., Becker, N., 2010. Excess deaths during the 2004 heatwave in Brisbane, Australia. Int. J. Biometeorol. 54, 393–400.

Unger, J., 1999. Comparisons of urban and rural bioclimatological conditions in the case of a central-European city. Int. J. Biometeorol. 43, 139–144.

Vandentorren, S., Suzan, F., Medina, S., Pascal, M., Maulpoix, A., Cohen, J.-C., Ledrans, M., 2004. Mortality in 13 French cities during the august 2003 heat wave. Am. J. Public Health 94, 1518–1520.

Vellei, M., Ramallo-González, A.P., Coley, D., Lee, J., Gabe-Thomas, E., Lovett, T., Natarajan, S., 2017. Overheating in vulnerable and non-vulnerable households. Build. Res. Inf. 45, 102–118.

Victorian Government Department of Human Services, 2009. January 2009 Heatwave in Victoria: An Assessment of Health Impacts. Melbourne, Victoria.

Victorian Government Department of Human Services, 2014. The Health Impacts of the January 2014 Heatwave in Victoria. Melbourne, Victoria.

Willand, N., Ridley, I., Pears, A., 2016. Relationship of thermal performance rating, summer indoor temperatures and cooling energy use in 107 homes in Melbourne, Australia. Energ. Buildings 113, 159–168.

Yaglou, P., Minard, D., 1957. Control of heat casualties at military training centers. AMA Arch. Ind. Health 16, 302–316.

Zero Carbon Hub, 2016. Solutions to Overheating in Homes. London.

PART IV

Conclusions and Recommendations

CHAPTER TWELVE

Recommendations for Infrastructure Decision-Makers Under a Changing Climate

Emilio Bastidas-Arteaga*, Mark G. Stewart†
*Research Institute in Civil and Mechanical Engineering, UMR CNRS 6183, Université de Nantes, Nantes Cedex, France
†Centre for Infrastructure Performance and Reliability, School of Engineering, The University of Newcastle, Callaghan, NSW, Australia

12.1 Introduction

Climate adaptation engineering aims at providing a comprehensive framework to help decision-making under a changing climate. The general framework was described in Chapter 1 and several study cases have been presented and discussed for infrastructure (Chapters 2–7) and buildings (Chapters 8–11).

Decision-making under a changing climate is not an easy task because hard policy choices are needed in the present to deal with consequences in the mid- (20–40 years) or long-term (more than 70 years). The choices should take into account current needs and uncertain and complex future risks under a changing climate. In addition, climate change effects depend on the location. For instance, exposure and related climate change events are different even inside a country (De Larrard et al., 2014) and vulnerability depends on various socioeconomic factors that differ for developed (Chapter 9) or developing countries (Chapter 10). All these aspects should be considered to provide effective decision-making. Fig. 12.1 provides the five key elements needed for decision-making under a changing climate (World Resources Report, 2011). These elements will be discussed in the following sections focusing on decision-making for civil engineering applications.

12.2 Engagement

Early and ongoing public and/or private engagement is essential to effective adaptation. The first step to create this engagement will be to

Climate Adaptation Engineering
https://doi.org/10.1016/B978-0-12-816782-3.00012-7

Fig. 12.1 Key elements for effective decision-making under a changing climate.

identify the different actors involved on the problem. In the case of adaptation of infrastructure and buildings, these actors (or stakeholders) are: governments, academia, owners, users, insurers, communities, designers, and constructors. Engaging these actors will be crucial to prioritise needs, provide and recover information, determine acceptable risk levels, and choosing among the possible adaptation solutions. Discussions with these actors will also give them the possibility to express their views to increase the acceptability and applicability of adaptation measures.

Some challenges for improving public/private engagement in civil engineering are related with the type of climate hazard and affected infrastructure or buildings. Concerning the type of hazard, it is possible to differentiate between extreme (floods, hurricanes, heat waves, etc.) and progressive (corrosion, sea-level rise, etc.) events. For example, Chapter 8 showed that there was a high level of governmental engagement to implement initiatives to reduce vulnerability after an extreme flood event. However, the lack of exchanges with the different actors involved in the problem such as the population (i.e., community), led to the implementation of expensive, unpopular, and unsuccessful adaptation measures after storm Xynthia in France (Mercier and Chadenas, 2012; Pitié and Puech, 2010; Pitié et al., 2011).

For progressive climate hazards (acceleration of corrosion due to climate change), it is very difficult to create strong engagement with major decision-makers (government and owners) because climate change effects are expected in the long-term and decision-makers mainly focus on short-term or mid-term problems. Nevertheless, the framework proposed in Chapter 1 and the case studies in Chapter 4 indicated that a scenario-based approach is very useful to estimate the cost-effectiveness of climate adaptation strategies. It was found that under certain environmental conditions, even if climate change is not expected, some adaptation measures could become cost-effective. Moreover, modest changes in design or construction standards now will allow vulnerability to be gradually reduced overtime as new infrastructure is built or when existing infrastructure is replaced or renovated.

Many structures will be subjected simultaneously to extreme and progressive climate hazards. For example, a bridge could be subjected to deterioration processes and extreme avalanches. In such cases, early engagement of decision-makers that should deal in the short-term with avalanche risks should also consider what might happen if other mid- or long-term hazards will occur to implement more comprehensive adaptation solutions.

Previous discussions focused on given infrastructure subjected to multiple climate change hazards; however, the discussion could be extended by considering simultaneously various infrastructure systems. For example, Val et al. (2014) probabilistically evaluated the performance of two interdependent infrastructure systems (electrical power and water, damaged by flooding). They found that it is possible to optimise the way to operate the systems to minimise the impact on the population.

Communication and education are also useful tools to increase the awareness of these actors concerning climate change risks. The use of innovative methods to disseminate information (videos, articles, talks, incentives, etc.) will be very useful to explain potential climate change risks as well as to propose and discuss different adaptation measures. Future evolutions of civil engineering and public administration education programs should also consider that infrastructure and buildings would be subjected to a changing climate that will modify current practices of design, construction, and management. Towards this objective, the promotion of the methodology and overview of different study cases presented in this book could be used to create the above-mentioned communication tools, provide a basis to revise education programs, and in the mid-term increase the engagement of these actors.

12.3 Decision-Relevant Information

Relevant information is required for different stages of decision-making. It is needed to assess the impacts of climate change up to the implementation of adaptation strategies. Chapter 1 and Bastidas-Arteaga and

Stewart (2015) provided a discussion about climate data and its related uncertainties. The book also presented various case studies that use weather-related information to assess climate change impacts. This information includes temperature, relative humidity, rainfall, wind speed, sea water level, etc. that is estimated from several global circulation models for different climate change scenarios. For many case studies presented in this book, the information was provided from specialised websites or scientific partners and requires a given level of expertise to define if the databases are well suited for the applications. For example, downscaled data are necessary to obtain representative results in some cases. Under these conditions, it would be difficult for decision-makers to use it directly for assessing climate change effects and cost-effectiveness of adaptation strategies. Nevertheless, some national initiatives such as DRIAS (the future of climate) in France (http://www.drias-climat.fr/) provide free downscaled information that could be easily downloaded and used.

In addition to the above-mentioned climate hazards and specific technical information for infrastructure and buildings, other types of data such as demography, economy, social, and environmental information are also needed. This information is crucial for vulnerability assessment as well as for understating if the adaptation measures could be locally implemented and accepted. For example, the level of detail of the diagnosis of non-engineered houses in the Philippines provided in Chapter 10 as well as the knowledge of future wind-related events, served to propose adaptation solutions that will be easily implemented by local communities by using existing construction material and practices.

There is a lack of detailed vulnerability maps concerning the effects of climate change on infrastructure and buildings. It would be interesting to work locally in the development of such maps in a similar format that seismic maps have been development in the past. One could image in the future that some of these maps could be integrated into revised design standards. These maps would be also useful to increase the engagement of all actors involved in the problem.

Vulnerability maps should be managed and used with care. For example, the methodology proposed in Chapter 9 could be used to determine a vulnerability map for houses subjected to extreme winds in a given town. This would need detailed technical information about each house to provide a very accurate assessment. Nevertheless, this information is hard to obtain and many assumptions should be done in the exercise. Under these assumptions, the global assessment would be useful for a decision-maker to know which measures are well adapted to reduce vulnerability in the town. However, communicating the results to the whole community could generate fear, anxiety, and reduced property prices if an owner discovers that his/her house is very vulnerable. Therefore, it is important to communicate which are the assumptions and limits of the study and to take care how the information that will be disseminated to the public.

12.4 Institutional Design

The implementation of cost-effective adaptation strategies should be administratively supported at different institutional levels (ministries, public or private institutions, regions, states, and municipalities). This will require evolutions on current practices to deal in a broad way with the problem. For instance, there is a need of harmonising national policies into the legislative framework and being able to downscale it at province, local, communities, households, and individual levels. One of the major challenges is to maintain its consistency, completeness, and integrity along with the national policies in terms of visions and goals. Towards this aim, the role of centralised agencies is crucial to coordinate these efforts in a comprehensive way.

It is expected that the work of centralised agencies will produce standardised procedures for decision-making under a changing climate sometime in the future. For the moment, some discussions for normalisation have started at national, European, and international levels. In Europe CEN (European Committee for Standardisation) and CENELEC (European Committee for Electrotechnical Standardisation) launched the Adaptation to Climate Change Coordination Group (ACC-CG) in 2014 to coordinate standardisation activities in the field. An effort focused on construction, transport, and energy sectors has been initiated and a first guide for addressing climate change adaptation in standards has been produced in 2016 (CEN-CENELEC, 2016).

12.5 Tools for Planning and Policymaking

Advanced models able to simulate the behaviour of infrastructure and buildings under realistic environmental exposure have been employed in most chapters of this book to assess climate change effects and cost-effectiveness of adaptation strategies. In the current form, these models are useful to provide the information required for the adaptation planning and policymaking in the studied applications. However, their use requires a certain level of expertise that is not currently easily accessible for non-academic users. In the future, it is expected that engineers, software, and standards will provide the capability to directly integrate climate change issues into infrastructure planning. This will require implementing training programs as well as software development.

In the meantime, collaborations between decision-makers, industry, and academia will allow useful outputs to be produced to benefit decision-making. Some examples of these outputs were presented in the book and encompass: climate change impacts, modelling of adaptation measures, cost assessments with and without adaptation measures, evaluation of the risk reduction when adaptation measures are implemented, etc.

Advanced models could be used to generate large databases to develop meta-models for planning and policymaking in a simplified way. These models could be based on different principles (Bayesian belief networks, stochastic processes, fuzzy logic, etc.) depending on the complexity of the problem and the availability of data. It is important that these meta-models could be updated with real data to take into account information on the performance of the structure/building under a realistic exposure. This information could be obtained from embedded or environmental (smart) sensors, and/or nondestructive and destructive measurements. Another advantage of these meta-models is their ability to account for: (i) multiple hazards with limited computational resources (Chapter 2, Shabarchin and Tesfamariam, 2017) and (ii) the interdependency of infrastructure systems highlighted in Chapter 7.

For extreme events, it is important to identify system conditions where the system performance could change dramatically. For example, when sea-level rise is high enough to suggest retreat of housing. Tools for identifying these thresholds are crucial for improving adaptation decisions and outcomes. The above-mentioned models enhanced with monitoring will also be essential to evaluate when these thresholds could be reached and therefore, to implement more efficient adaptation actions.

12.6 Resources

In the aftermath of extreme events that directly affect communities such as floods, heat waves, and extreme winds, decision-makers should urgently provide the necessary financial resources to restore the situation as before as these events. Or preferably, to reinstate infrastructure to a higher (less vulnerable) standard–that is, to "build back better" (NCCARF, 2013). Additional policies may also be adopted to reduce the vulnerability to future extreme events in the mid-term (see Chapter 8 for further details). Nevertheless, increasing the resilience of infrastructure and buildings to face climate change hazards would require targeted and sustained long-term funding. Public and/or private engagement motivated by financial incentives would help to promote the culture of long-term planning.

Prioritising spending is a complex task for decision-makers that should optimise their budget to deal with unexpected, short-, mid-, and long-term expenditures. To help them in this task, Chapter 1 provided a risk-based decision support that is useful for the evaluation of adaptation costs. It was applied in this book to many study cases to evaluate the cost-efficiency of adaptation measures including even the computation of the risk of the investment. For example, in Chapter 4, this framework allowed estimating two measures of cost-efficiency: (i) the mean benefit-to-cost ratio and (ii) the probability that the benefit-to-cost ratio be larger than one for corroding infrastructure. The results of the study were useful to determine under what conditions the investment on an adaptation measure is cost-effective with the largest probability to attain a benefit. This risk-based support is also helpful to evaluate the co-benefits and/or to consider the 'no regrets' or 'win-win' policies that would encourage the implementation of adaptation measures. The above-mentioned risk-based decision support is paramount for effective decision-making under a changing climate. Therefore, promoting and funding technical training on this subject among decision-makers will allow them to integrate climate risks into existing decision-making processes.

12.7 Summary

This chapter discussed the five key elements for effective decision-making: engagement; decision-relevant information; institutional design; tools for planning and policymaking; and resources. The implementation of each of these elements is not without challenges. The task of policymaking for an uncertain climate future is a risky business. The chapter has discussed some recommended practices to allow for more effective climate adaptation engineering, both now and into the future. This included lessons learnt from the case studies presented in the book, and examined how the risk informed decision-making process may be improved.

The underlying principles exposed in this chapter, and throughout the book, will help decision-makers make more robust decisions, and climate proof new and existing infrastructure through to the next century and beyond.

References

Bastidas-Arteaga, E., Stewart, M.G., 2015. Damage risks and economic assessment of climate adaptation strategies for design of new concrete structures subject to chloride-induced corrosion. Struct. Saf. 52, 40–53.

CEN-CENELEC, 2016. Guide for addressing climate change adaptation in standards. European Committee for Standardization, European Committee for Electrotechnical Standardization, ftp://ftp.cencenelec.eu/EN/EuropeanStandardization/Guides/32_CENCLCGuide32.pdf.

De Larrard, T., Bastidas-Arteaga, E., Duprat, F., Schoefs, F., 2014. Effects of climate variations and global warming on the durability of RC structures subjected to carbonation. Civil Eng. Environ. Syst. 31 (2), 153–164. https://doi.org/10.1080/10286608.2014.913033.

Mercier, D., Chadenas, C., 2012. La tempête Xynthia et la cartographie des «zones noires» sur le littoral français: analyse critique à partir de l'exemple de La Faute-sur-Mer (Vendée). Norois 222 (1), 45–60.

NCCARF, 2013. Climate proofing Australia's infrastructure. Policy Brief 7, National Climate Change Adaptation Research Facility, Griffith University, Australia.

Pitié, C., Puech, P., 2010. Expertise complémentaire des zones de solidarité délimitées en Vendée suite à la tempête Xynthia survenue dans la nuit du 27 au 28 février 2010. CGeDD/MEEDDM, 80.

Pitié, C., Bellec, P., Maillot, H., Nadeau, J., Puech, P., 2011. Expertise des zones de solidarité Xynthia en Charente-Maritime. Rapport principal, CGEDD/MEEDDM, 186.

Shabarchin, O., Tesfamariam, S., 2017. Risk assessment of oil and gas pipelines with consideration of induced seismicity and internal corrosion. J. Loss Prev. Process Ind. 47, 85–94.

Val, D.V., Holden, R., Nodwell, S., 2014. Probabilistic analysis of interdependent infrastructures subjected to weather-related hazards. Civ. Eng. Environ. Syst. 31 (2), 140–152.

World Resources Report, 2011. Decision Making under a Changing Climate: Adaptation Challenges and Choices. United Nations, http://pdf.wri.org/world_resources_report_2010-2011.pdf.

Further Reading

Department of Climate Change, 2009. Climate Change Risks to Australia's Coast: A First Pass National Assessment. Commonwealth of Australia, Canberra.

Index

Note: Page numbers followed by *f* indicate figures, and *t* indicate tables.

F

G

H

J

L

M

N

O

T

FSC
www.fsc.org
MIX
Papier | Fördert gute Waldnutzung
FSC® C083411

Kolibri 360 GmbH
Ferdinand-Jühlke-Straße 7
99095 Erfurt, Deutschland
produktsicherheit@kolibri360.de